Mass Transfer
in Engineering Practice

Mass Transfer
in Engineering Practice

AKSEL L. LYDERSEN

Professor of Chemical Engineering,
The University of Trondheim, Norway

A Wiley–Interscience Publication

JOHN WILEY & SONS

Chichester · New York · Brisbane · Toronto · Singapore

Library of Congress Cataloging in Publication Data:

Lydersen, Aksel.
   Mass transfer in engineering practice.

   'A Wiley–Interscience publication'.
   Includes bibliographical references and index.
   1. Mass transfer. I. Title.
TP156.M3L9   6602'8423   82-7086
                               AACR2
ISBN 0 471 10437 X (cloth)
ISBN 0 471 10462 O (paper)

British Library Cataloguing in Publication Data:

Lydersen, Aksel L.
   Mass transfer in engineering practice.
   1. Mass transfer
   I. Title
531'.1137   QC175.2
ISBN 0 471 10437 X (cloth)
ISBN 0 471 10462 O (paper)

Photosetting by Thomson Press (India) Limited, New Delhi
and printed at Page Bros (Norwich) Ltd.

Contents

Preface

In the first half of the twentieth century the chemical process industry had already grown to an extent where it was impossible for an individual to have detailed knowledge of it in its entirety. Industrial processes, however, are built up of unit operations that can be studied and treated separately, irrespective of the process in question. Thus the principles of absorption are the same whether or not this unit operation is used to remove soluble components from a flue gas, or as a step in the production of a fertilizer.

This realization that unit operations are the building-blocks of industrial processes fundamentally changed the study of industrial chemistry and led to the development of chemical engineering as a discipline in its own right, first in the universities of America and then of Europe.

In the second half of the twentieth century the advent of high-speed digital computers, the accumulation of more empirical data, and the increase in the number of unit operations have made it impossible for an individual to be a specialist in all aspects of unit operations and difficult to keep abreast with developments in even a few.

This text seeks therefore to provide a short refresher course for practising engineers in each of the more common unit operations. It highlights the basic equations, presents empirical data for practical cases, and gives references to suitable literature where more detailed treatment can be found. It also contains worked examples from engineering practice which stress important aspects that may be easily overlooked by engineers who have not experienced similar problems.

Being in a condensed form, giving important fundamentals and examples, it is hoped that the text should appeal both to practising engineers and to the student of chemical engineering. Used as a textbook, it is recommended to start with Chapter 2 and take Chapter 1 in connection with Chapter 3.

The book *Fluid Flow and Heat Transfer*[†], by the same author, attempts to satisfy the needs of both the practising engineer and the student in the field of mechanical unit operations and heat transfer in chemical engineering. This book seeks to do likewise in the field of unit operations with mass transfer. The text has deliberately been kept short and a high proportion of the examples stem directly from industrial applications. Numerous other books are available to the reader who seeks detailed description of equipment and specialist calculations. In particular *Chemical Engineers' Handbook* by R. H. Perry and C. H. Chilton should never be far from hand, with *Handbook of Separation Techniques for Chemical Engineers* by P. A. Schweitzer as a more recent supplement.

[†]A. L. Lydersen, *Fluid Flow and Heat Transfer*, John Wiley & Sons, Chichester, 1979.

Several unit operations are carried out as stage processes. The calculations involved are in principle the same whether in connection with absorption, stripping, distillation, or liquid extraction. This favours a separate chapter on stage processes. However, for two reasons, such a chapter is not included here. Firstly, the basic principles of stage processes are so fundamentally important to several unit operations that it may be an advantage for students to meet them repeatedly. Secondly, the practising engineer may prefer a refresher course where attention can be concentrated on the chapter of particular interest without having to read previous chapters first.

Textbooks often include some methods and types of equipment which are obsolete or infrequently used. Examples of this are the Ponchon–Savarit method for distillation calculations and equipment as sugar-beet diffusion batteries, the last two of which in the U.S.A. were shut down in 1971. In this text, efforts are made to try to avoid this kind of pitfall.

The goal of an industrial chemical engineer is profitable production. This depends both on final product quality and on price. Hence the choice of a particular unit operation must be based on knowledge of efficiency and price. This requires, in turn, information on equipment size which in some cases may only be obtained from pilot-plant studies. This book concentrates on the calculation of the major dimensions of equipment and, to a certain extent, on the consumption of energy. Stress calculations and the specification of design details are, in general, the role of the mechanical engineer.

Engineers must also economize in the use of their own time. Techniques should be used which give the necessary accuracy in the minimum time. For example, in designing a packed absorption tower, uncertainty with respect to liquid distribution and possibly also equilibrium and enthalpy data may make it desirable to add 30–35 per cent to the calculated packing height. A practising engineer could well feel that the time spent in refining the calculations by a few per cent was an avoidable luxury.

In the design of multistage distillation units for multicomponent mixtures, digital computers make possible more accurate rigorous solutions where equilibrium and enthalpy data are sufficiently accurate to warrant the expense. For distillation of multicomponent mixtures only short-cut calculations are included in this text. In extensive optimization studies such techniques are useful to define quickly and cheaply an indication of optimum specifications. These can then be investigated exactly by a rigorous method with a considerable saving of computer time.

It is also wise to use common sense, which includes the habit of checking results. A good guess may be all that is needed to reveal an error in the calculations carried out.

The author is grateful for information and suggestions from firms and individuals, and in particular from the publisher's reviewer, Dr. R. Krishna. The author will also be grateful to readers helping to eliminate errors and suggesting improvements in possible future revisions of the book.

Trondheim, October 1982

AKSEL L. LYDERSEN

Introduction

Industrial *mass transfer operations* utilize the transfer of one or more components of mixtures from one phase to another. The different processes may be classified according to whether the phases are in gaseous, liquid, or solid state, as indicated in Table 1.

Table 1

Phases	Processes
Gas–liquid	Absorption–stripping Distillation Humidification–drying
Liquid–liquid	Extraction
Gas–solid	Adsorption–desorption Drying
Liquid–solid	Leaching Adsorption–desorption Crystallization–dissolving

Diffusion is also important in other processes, such as heterogeneous reactions, and in diffusion between miscible phases separated by porous or semipermeable membranes.

The *driving force* for mass transfer is a concentration difference or a concentration gradient. To evaluate this difference or gradient, *equilibrium data* are required.

The *phase rule* is useful in order to classify equilibria. This rule gives the degrees of freedom, i.e. the number of independent variables, as

$$F = C - P + 2 \tag{0.1}$$

where $F =$ number of degrees of freedom, $C =$ number of component, and $P =$ number of phases.

Taking distillation of a binary mixture of A and B as an example, we have two components ($C = 2$) and a liquid and a vapour phase ($P = 2$), giving $F = 2$. There are four variables—pressure, temperature, and fraction of A in the liquid and in the vapour phase. If one is fixed, for instance the pressure, only one variable can be changed independently, e.g. the fraction of A in the liquid, and the temperature and fraction of A in the vapour phase follow. (The fraction of B is one minus the fraction of A.)

CHAPTER 1
Principles of Diffusion

Figure 1.1 shows a box with a mixture of a component A and a component B at pressure P and temperature T. At time $t = 0$, the highest mole fraction of molecules A is in the left-hand half of the box.

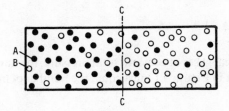

Figure 1.1   Molecules A and B confined in a box at constant pressure and temperature

The molecules move around in all directions, randomly. But since most of the A-molecules are to the left of the centre line C–C, more of the molecules A will pass the line C–C going from left to right, than those passing the line in the opposite direction. This is the *molecular diffusion* of A in the direction of decreasing concentration. At the same time, there is also a net diffusion of B in the opposite direction. This diffusion continues until the concentration is uniform throughout.

Fick's law

The molecular diffusion flux of component A in the z-direction in a mixture of A and B, is given by Fick's law:

$$J_A = -D_{AB}\frac{dc_A}{dz} \tag{1.1}$$

where J_A = rate of molecular diffusion, kmol/(m²s), D_{AB} = diffusion coefficient (or diffusivity) of A in a mixture of A and B, m²/s, and dc_A/dz = concentration gradient of A, $\dfrac{\text{kmol/m}^3}{\text{m}}$

The concentration gradient is the driving force, as the temperature gradient in Fourier's law for heat flux by conduction.

1

For gases it may be convenient to have Fick's law expressed in terms of partial pressure of component A, p_A. The ideal gas law gives

$$p_A v = n_A RT$$

corresponding to a concentration of A,

$$c_A = n_A/v = p_A/RT$$

with the derivative

$$\frac{dc_A}{dz} = \frac{1}{RT}\frac{dp_A}{dz} \tag{1.2}$$

Equations (1.1) and (1.2) give the molar flux of component A at isothermal conditions in the direction of flow,

$$J_A = -\frac{D_{AB}}{RT}\frac{dp_A}{dz} \tag{1.3}$$

In a mixture of ideal gases at constant total pressure, the number of moles of A diffusing to the right in Figure 1.1 must equal the number of moles of B diffusing to the left,

$$J_A = -J_B = -\frac{D_{AB}}{RT}\frac{dp_A}{dz} = \frac{D_{BA}}{RT}\frac{dp_B}{dz} \tag{1.4}$$

With total pressure $P = p_A + p_B = \text{constant}$,

$$dp_A/dz = -dp_B/dz$$

and

$$D_{AB} = D_{BA} \tag{1.5}$$

This shows that for a binary mixture of ideal gases the diffusivity D_{AB} for A diffusing in B is the same as the diffusivity D_{BA} for B diffusing in A.

Diffusion with bulk of mass in motion

Figure 1.2 shows a pipe with a binary mixture of A and B.

With uniform concentration of A, the molar flux of A past a stationary observer is $c_A V$. A concentration gradient dc_A/dz gives an additional flux due to molecular

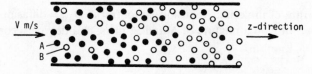

Figure 1.2   Mixture of A and B with bulk velocity V m/s and a concentration gradient of A in the direction of flow

diffusion, J_A, or total molar flux of A,

$$N_A = J_A + c_A V \tag{1.6}$$

The total molar flux is

$$N = N_A + N_B = cV$$

or

$$V = \frac{N_A + N_B}{c} \tag{1.7}$$

where c = total molar concentration, kmol/m³.

Equations (1.1), (1.6), and (1.7) give

$$N_A = -D_{AB}\frac{dc_A}{dz} + \frac{c_A}{c}(N_A + N_B) \tag{1.8}$$

The same reasoning also gives the total molar flux of B,

$$N_B = -D_{BA}\frac{dc_B}{dz} + \frac{c_B}{c}(N_A + N_B) \tag{1.9}$$

The last two equations are valid for gases as well as for liquids and solids. The fluxes N_A and N_B refer to a stationary coordinate reference frame. The concentration ratios c_A/c and c_B/c may be substituted by the mole fractions x_A and x_B, and the concentrations c_A and c_B by the products $x_A c$ and $x_B c$.

Diffusion of A through a stagnant B (Stefan diffusion)

Figure 1.3 shows an arrangement for making a mixture of air with a low content of an organic solvent—to be used in testing of gas masks. At point 1 the partial

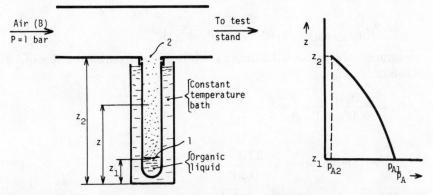

Figure 1.3 Arrangement for constant feed of a small amount of solvent vapour into an air stream. The diagram on the right-hand side shows the partial pressure of solvent A, $p_A = f(z)$

pressure of A, p_{A1}, is the vapour pressure of A at the temperature of the bath, and at point 2, p_{A2}, which may be close to zero. At steady state, the air in the tube is stagnant, and $N_B = 0$. Hence, equation (1.8) becomes

$$N_A = -D_{AB}\frac{dc_A}{dz} + \frac{c_A}{c}N_A \tag{1.10}$$

Substituting dc_A/dz from equation (1.2) and $c_A/c = p_A/P$ for an ideal gas, gives

$$N_A = -\frac{D_{AB}}{RT}\frac{dp_A}{dz} + \frac{p_A}{P}N_A \tag{1.11}$$

or

$$N_A \int_{z=z_1}^{z=z_2} dz = \frac{-D_{AB}P}{RT}\int_{p_{A1}}^{p_{A2}}\frac{dp_A}{P-p_A} \tag{1.12}$$

Integration gives

$$N_A = \frac{D_{AB}P}{RT(z_2-z_1)}\ln\frac{P-p_{A2}}{P-p_{A1}} \tag{1.13}$$

This is the final equation to be used to calculate the net flux of A through a stagnant B. It is sometimes written

$$N_A = \frac{D_{AB}P}{RT(z_2-z_1)}\ln\frac{p_{B2}}{p_{B1}} \tag{1.14}$$

where p_B is the partial pressure of the stagnant inert. Another form of the same equation is

$$N_A = \frac{D_{AB}}{RT(z_2-z_1)}\frac{P}{p_{B1m}}(p_{A1}-p_{A2}) \tag{1.15}$$

where

$$p_{B1m} = \frac{p_{B2}-p_{B1}}{\ln(p_{B2}/p_{B1})} = \frac{p_{A1}-p_{A2}}{\ln[(P-p_{A2})/(P-p_{A1})]} \tag{1.16}$$

Integration of equation (1.12) from $z = z_1$ to z gives the partial pressure of A as a function of z:

$$N_A = \frac{D_{AB}P}{RT(z-z_1)}\ln\frac{P-p_A}{P-p_{A1}}$$

or

$$p_A = P - (P-p_{A1})\exp\frac{N_A RT(z-z_1)}{D_{AB}P} \tag{1.17}$$

Thus, the partial pressure of A diffusing through a stagnant B falls along an exponential curve, as indicated in the diagram on the right-hand side in Figure 1.3.

Multicomponent mixtures

The net flux of component A in a multicomponent mixture is given by equation (1.8) with all net fluxes added,

$$N_A = -D_A \frac{dc_A}{dz} + \frac{c_A}{c}(N_A + N_B + \ldots)$$

where D_A is the effective diffusivity of component A in the multicomponent mixture. This effective diffusivity is generally a function of the mixture composition.

In terms of partial pressure,

$$N_A = -\frac{D_A}{RT}\frac{dp_A}{dz} + \delta N_A \frac{p_A}{P} \qquad (1.18)$$

or integrated with D_A assumed to be a constant,

$$N_A = \frac{D_A P}{RT(z_2 - z_1)\delta} \ln \frac{P - \delta p_{A2}}{P - \delta p_{A1}} \qquad (1.19)$$

where the product $\delta N_A = (N_A + N_B + \ldots)$, and δ is the net number of moles diffusing in the same direction as A per mole A transferred. The equations (1.13)–(1.17) represent the special case of a binary mixture with stagnant B, i.e. $\delta = 1.0$. In steady state and equimolar counterdiffusion, as in distillation of components with the same molar heats of vaporization, $\delta = 0$ and integration of equation (1.18) from $z = z_1$ to $z = z_2$ gives

$$N_A = \frac{D_A}{RT(z_2 - z_1)}(p_{A1} - p_{A2}) \qquad (1.20)$$

Varying cross-section

Figure 1.4 is a spherical drop of A evaporating from its surface into a stagnant layer of the gas B.

With varying area, the net flux N_A must be referred to a given area, for instance the surface area of the sphere, $4\pi r_1^2$. The net flux through the area $4\pi r^2$ is

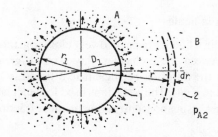

Figure 1.4 Evaporation from a drop
A into a stagnant gas B

$N_A(r_1/r)^2$, and equation (1.11) for the sphere becomes

$$N_A\left(\frac{r_1}{r}\right)^2 = -\frac{D_{AB}}{RT}\frac{dp_A}{dr} + \frac{p_A}{P}N_A\left(\frac{r_1}{r}\right)^2$$

Separation of the variables give

$$N_A r_1^2 \int_{r_1}^{r_2} r^{-2}dr = -\frac{D_{AB}P}{RT}\int_{p_{A1}}^{p_{A2}}\frac{dp_A}{P-p_A}$$

$$N_A = \frac{D_{AB}P}{RTr_1(1-r_1/r_2)}\ln\frac{P-p_{A2}}{P-p_{A1}} \tag{1.21}$$

Eddy diffusion

Turbulent flow is characterized by the random motion of the particles constituting the fluid stream. Short-lived eddies break up in fragments which then form new eddies. Mixing and diffusion within an eddy may be of minor importance, but material may diffuse rapidly by the process of eddy transfer and disintegration. This transfer can be expressed by analogy with Fick's law of diffusion,

$$J_A = -E\frac{dc_A}{dz} \tag{1.22}$$

where E = eddy diffusivity, m^2/s.

Prandtl and Taylor[1] worked out an expression for eddy diffusivity in terms of a characteristic mixing length. Based on experimental data with water vapour, carbon dioxide, and helium in air, Sherwood and Woertz[2] obtained values corresponding to the equation

$$E = (2.37 \times 10^{-8} Re + 2.7 \times 10^{-4})/\rho \tag{1.23}$$

in the region $20\,000 < Re < 120\,000$. Here the Reynolds number $Re = \rho V_{rel}d/\mu$, where ρ and μ are the density and dynamic viscosity of the gas, and V_{rel} is the relative velocity between the gas and the liquid flowing down at the surface of a duct with constant hydraulic diameter d.

Molecular diffusion in liquids

Since the molecules in a liquid are packed very close together, the density and attractive forces between the molecules play a much greater role than in the gaseous phase, and the diffusion coefficient in a liquid is 10^{-4}–10^{-5} times the diffusion coefficient in a gas. The equations used for calculating molar fluxes are the same as for gases, using concentrations or mole fractions instead of partial pressure,

$$J_A = -D_A\frac{dc_A}{dz} \tag{1.1}$$

and

$$N_A = -D_A \frac{dc_A}{dz} + \frac{c_A}{c}(N_A + N_B + \ldots) \tag{1.24}$$

In operations such as distillation of components with the same molar latent heat of vaporization, we have equimolar counterdiffusion, and the sum in parentheses in equation (1.24) is zero. Assuming constant c, integration gives

$$N_A = \frac{D_A(c_{A1} - c_{A2})}{z_2 - z_1} = \frac{D_A c(x_{A1} - x_{A2})}{z_2 - z_1} \tag{1.25}$$

where c is the average value over the distance $z_2 - z_1$,

$$c = \sum_{i=1}^{n} \frac{x_i \rho_i}{M_i} \tag{1.26}$$

where x_i = mole fraction of component i, ρ_i = density of liquid component i, kg/m^3, and M_i = molecular weight of component i, kg/kmol.

In operations such as liquid–liquid extraction, component A may diffuse through a stagnant B, and the parentheses in equation (1.24) becomes N_A. Integration gives

$$N_A = \frac{D_A c}{z_1 - z_2} \ln \frac{c - c_{A1}}{c - c_{A2}} = \frac{D_A c}{z_1 - z_2} \ln \frac{1 - x_{A1}}{1 - x_{A2}} \tag{1.27}$$

The integrated equations are based on both an average molecular weight of the liquid, and an average diffusivity, which may vary with the concentration.

For diffusion in laminar flow in pipes, see the article by Hunt.[3]

Molecular diffusion in solids

Diffusion in solids is much slower than diffusion in gases and liquids, but it does govern processes such as leaching of foods and metal ores, fluid transfer through polymer films, and case-hardening of steel. It may be diffusion through pores or capillaries, sometimes filled with liquid, and it may be through a more or less homogeneous solution, such as diffusion of zinc through copper.

For calculations it is common to use simple equations such as Fick's law, and the last term in equation (1.24) is usually negligible. This gives

$$N_A \approx -D_A \frac{dc_A}{dz} \tag{1.28}$$

Integration for a solid slab at steady state gives equation (1.25).

In bodies such as cylinders and spheres the cross-sections for diffusion vary with the radii, and the flux must be referred to a fixed area. The differential equation for diffusion through a spherical shell with flux N_{A1} referred to radius r_1, is

$$N_{A1}\left(\frac{r_1}{r}\right)^2 = -D_A \frac{dc_A}{dr}$$

8

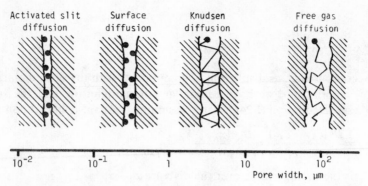

Figure 1.5 Dominating transport mechanisms as a function of the pore width[4]

or

$$N_{A1} \int_{r_1}^{r_2} r^{-2} dr = -\frac{D_A}{r_1^2} \int_{c_{A1}}^{c_{A2}} dc_A$$

giving

$$N_{A1} = D_A \frac{c_{A2} - c_{A1}}{r_1(1 - r_1/r_2)} \tag{1.29}$$

where r_1 and r_2 are the inner and outer radii of the shell and N_{A1} the flux at radius r_1.

Diffusion in pores or capillaries may be governed by different mechanisms, depending on shape, radii, branching of, and connections between the pores. Figure 1.5 is a schematic picture of four different mechanisms.

The *Knudsen diffusion*[5] can be treated as molecular diffusion in a fluid, but with a different diffusivity, and with no convective contribution to the flux N_A. Knudsen diffusion occurs when the molecular mean free path λ_m is large compared to the diameter of the capillary, i.e. only molecule–wall collisions are important. The Knudsen diffusivity is

$$D_{AK} = 97.0r(T/M_A)^{1/2} \ \text{m}^2/\text{s}. \tag{1.30}$$

where r = capillary radius, m, and M_A = molecular weight of A, kg/kmol.

Knudsen diffusion is dominating for mean free paths

$$\lambda_m = \frac{3.2\mu}{P} \sqrt{\frac{RT}{2\pi M}} \tag{1.31}$$

more than one-tenth the diameter of the capillary, while it accounts for less than 10 per cent if the capillary diameter is more than 100 times the free path.[6]

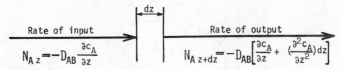

Figure 1.6   Unsteady-state diffusion of component A in a stagnant slab with thickness dz

Unsteady-state diffusion

All mass transfer processes will have an initial period of time with unsteady-state conditions, i.e. the concentration at a certain point varies with time, until a steady state is reached.

Figure 1.6 shows a slab of solid or stagnant fluid with diffusion of A in the z-direction.

The difference $N_{Az} - N_{Az+dz} = D_{AB}(\partial^2 c_A/\partial z^2)dz$ in Figure 1.6 is in kmoles of A accumulated in the slab per unit area and per unit time. It equals $(1 \times dz)\partial c_A/\partial t$, where $1 \times dz$ is the volume of 1 m^2 of the slab,

$$\frac{\partial c_A}{\partial t} = D_{AB}\frac{\partial^2 c_A}{\partial z^2} \tag{1.32}$$

This is the fundamental equation of unsteady-state molecular diffusion in one direction through a fluid at rest. It relates the concentration c_A to position z and time t. The same equation for an ideal gas is

$$\frac{\partial p_A}{\partial t} = D_{AB}\frac{\partial^2 p_A}{\partial z^2} \tag{1.33}$$

and for diffusion in all three directions,

$$\frac{\partial p_A}{\partial t} = D_{AB}\left(\frac{\partial^2 p_A}{\partial x^2} + \frac{\partial^2 p_A}{\partial y^2} + \frac{\partial^2 p_A}{\partial z^2}\right) \tag{1.34}$$

These equations have the same form as the partial differential equations for unsteady state heat transfer. Hence, solutions worked out for heat transfer[7-9] can be applied directly when the thermal diffusivity α is substituted by the diffusivity D_{AB} and the temperature T or θ is substituted by the partial pressure p_A.

The film concept

Figure 1.7 is a schematic drawing of a liquid film that flows downwards along a solid wall, countercurrent to a gas containing a component A that is absorbed by the liquid.

Without any sharp boundaries, three flow regimes can be visualized in the gas. It is a fully developed turbulent region where most of the mass transfer takes place by eddy diffusion, a transition zone still with some turbulence, and a laminar film

10

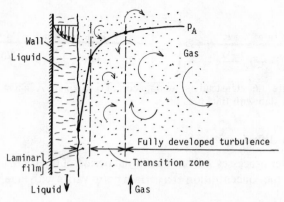

Figure 1.7   Liquid in countercurrent flow with a gas
containing a component A being absorbed in a liquid

with molecular diffusion. A rigorous theory for mass transfer has to take into account the different mechanisms of transfer in all three flow regimes, and also factors such as the ripples or small waves that usually develop at the surface of the liquid. But a simplified model has nevertheless proven useful as a basis for correlation of mass transfer data. It is the so-called two-film theory, which relates the rate of mass transfer to molecular diffusion in a stagnant film with a sharp boundary to a well-mixed fluid where concentration gradients are negligible. Such a 'film concept', assuming an equivalent film thickness, is also applied to the liquid side, even if it is difficult to visualize a liquid film at the boundary between liquid and gas.

In addition, it is also assumed that equilibrium exists at the interface, i.e. that resistance to mass transfer across the interface is negligible. In practice this is probably true in distillation and absorption, if there is no additional resistance caused by concentration of a surfactant at the interface. In extraction, crystallization, and leaching, care should be taken in the use of the concept of equilibrium at the interface.

Mass transfer coefficients

The rate of mass transfer from the bulk of the gas in Figure 1.7 to the surface of the liquid obviously increases with increased turbulence that promotes eddy diffusion and reduces the thickness of the laminar film. It also increases with increasing difference in driving force, expressed as the difference of partial pressure of component A in the bulk of the gas and at the liquid surface, or as the difference in the mole fraction of A. Also, with all other conditions the same, the rate of mass transfer is highest if there is no counterdiffusion from the liquid surface into the gas stream. A fundamental approach is given by Bird et al.,[10] and an interpretative review by Krishna and Standart.[11]

In engineering practice it is common to calculate the rate of mass transfer as the product of a mass transfer coefficient and a driving force. With the driving force expressed by the partial pressure, the mass transfer per unit area and per unit time is

$$N_A = k_g(p_A - p_{Ai}) \tag{1.35}$$

where k_g = gas phase mass transfer coefficient by diffusion through a 'stagnant' inert, $kmol/(m^2 s\ Pa)$, p_A = partial pressure of A in bulk of gas, $Pa = N/m^2$, and p_{Ai} = partial pressure of A at the interface, $Pa = N/m^2$. With the driving force expressed as a mole fraction,

$$N_A = k_{ye}(y - y_i) \tag{1.36}$$

where k_{ye} = gas phase mass transfer coefficient by equimolar counterdiffusion, $kmol/(m^2 s\ mol\ fraction)$, y = mole fraction of A in bulk of gas, and y_i = mole fraction of A in gas at the interface.

The gas phase mass transfer coefficient is often given as k_y or k_G without clarification, whether it refers to mass transfer with or without counterdiffusion.

Integration of equation (1.4) for equimolecular counterdiffusion over film thickness $z_2 - z_1$ and with p_A substituted by yP, and the molecular diffusion J_A substituted by the total molar flux N_A, gives

$$N_A = \frac{D_{AB}P}{RT(z_1 - z_2)}(y_2 - y_1) \tag{1.37}$$

Comparison with equation (1.36) and with $y_1 = y_i$ and $y_2 = y$ in the bulk of the liquid, gives

$$k_{ye} = \frac{D_{AB}P}{RT(z_1 - z_2)} \tag{1.38}$$

Equation (1.13) for diffusion through a stagnant film (Stefan diffusion) can be written as

$$N_A = \frac{D_{AB}P}{RT(z_2 - z_1)}\ln\frac{1 - y_2}{1 - y_1} \tag{1.39}$$

The logarithmic mean mole fraction of components other than A in the film is defined by the equation

$$(1 - y)_{lm} = \frac{(1 - y_2) - (1 - y_1)}{\ln\dfrac{1 - y_2}{1 - y_1}} \tag{1.40}$$

Equations (1.39) and (1.40) give

$$N_A = \frac{D_{AB}P}{RT(z_1 - z_2)}\frac{y_2 - y_1}{(1 - y)_{lm}} \tag{1.41}$$

or with $D_{AB}/[RT(z_1 - z_2)]$ substituted by the mass transfer coefficient in Stefan diffusion, k_{ys} kmol/(m^{2}s mol fraction), and $y_1 = y_i$ and $y_2 = y$,

$$N_A = k_{ys}\frac{y - y_i}{(1 - y)_{lm}} \tag{1.42}$$

Experimental mass transfer coefficients are often reported as k_y without reference to whether it is determined by Stefan or by equimolecular counterdiffusion. Except for dilute solutions where $(1 - y)_{lm} \approx 1.0$, k_y-values determined by equation (1.36) and by Stefan diffusion should be multiplied by $(1 - y)_{lm}$ if applied to equimolar counterdiffusion, i.e.

$$k_{ye} = k_y(1 - y)_{lm} \tag{1.43}$$

or

$$k_y = k_{ye}/(1 - y)_{lm} \tag{1.44}$$

where $1/(1 - y)_{lm}$ is called the *drift factor*. It allows for the enhancement of the rate of transfer arising from the drift of the main body of gas towards the gas/liquid interface.

The same procedure applied to the liquid with mass transfer coefficient k_x and calculated by the equation

$$N_A = k_{xe}(x_i - x) \tag{1.45}$$

gives

$$k_{xs} = k_x/(1 - x)_{lm} \tag{1.46}$$

where

$$(1 - x)_{lm} = \frac{(1 - x_i) - (1 - x)}{\ln\dfrac{1 - x_i}{1 - x}} \tag{1.47}$$

and $k_{xs} =$ liquid phase mass transfer coefficient by Stefan diffusion, kmol/(m^{2}s mol fraction), $x_i =$ mole fraction of A in the liquid at the interface, $x =$ mole fraction of A in the bulk of liquid.

Overall mass transfer coefficients are referred either to the gas or to the liquid phase. These overall coefficients are defined by the equations

$$N_A \equiv K_y(y - y^*) \tag{1.48}$$

and

$$N_A \equiv K_x(x^* - x) \tag{1.49}$$

where $K_y =$ overall gas phase mass transfer coefficient, kmol/(m^{2}s mol fraction), $y^* =$ mole fraction of A in a gas in equilibrium with a liquid with mole fraction x of A, $K_x =$ overall liquid phase mass transfer coefficient, kmol/(m^{2}s mol fraction), and $x^* =$ mole fraction of A in a liquid in equilibrium with a gas with mole fraction y of A.

With steady-state mass transfer through layers in series, it is convenient to

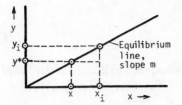

Figure 1.8 Mole fraction y of A in the gas as a function of mole fraction x in the liquid

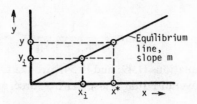

Figure 1.9 Mole fraction y of A in the gas as a function of mole fraction x in the liquid

write the equation for each layer, solve for the driving force, and add the driving forces. This makes the usually unknown intermediate potentials cancel out. To obtain this, however, all driving forces must be expressed in the same units.

Assuming equilibrium at the interface, it is seen from Figures 1.8 and 1.9 that

$$\frac{y_i - y^*}{x_i - x} = m \tag{1.50}$$

and

$$\frac{y - y_i}{x^* - x_i} = m \tag{1.51}$$

where m = the slope of the equilibrium line.

Equation (1.36) solved for the driving force, and equation (1.45) with $(x_i - x)$ substituted by $(y_i - y^*)/m$ (equation (1.50)) and solved for the driving force, give

$$y - y_i = N_A/k_{ye}$$
$$y_i - y^* = mN_A/k_{xe}$$
$$\text{Sum } y - y^* = N_A(1/k_{ye} + m/k_{xe})$$

or

$$N_A = \frac{1}{1/k_{ye} + m/k_{xe}}(y - y^*) \tag{1.52}$$

Equations (1.48) and (1.52) give the important relationship

$$\frac{1}{K_{ye}} = \frac{1}{k_{ye}} + \frac{m}{k_{xe}} \tag{1.53}$$

The term $1/K_{ye}$ is the total resistance to mass transfer based on the overall gas phase driving force. The equation gives the total resistance as the sum of the gas side resistance $1/k_{ye}$ and the liquid side resistance m/k_{xe}.

Equation (1.36) with $y - y_i$ substituted by $m(x^* - x_i)$ (equation (1.51)) and solved for the driving force, and equation (1.45) solved for the driving force, give

$$x^* - x = N_A/mk_{ye}$$
$$x_i - x = N_A/k_{xe}$$
$$\text{Sum } x^* - x = N_A\left(\frac{1}{mk_{ye}} + \frac{1}{k_{xe}}\right)$$

or

$$N_A = \frac{1}{1/mk_{ye} + 1/k_{xe}}(x^* - x) \tag{1.54}$$

Equations (1.49) and (1.54) give the total resistance to mass transfer based on the overall liquid phase driving force, as

$$\frac{1}{K_{xe}} = \frac{1}{mk_{ye}} + \frac{1}{k_{xe}} \tag{1.55}$$

By Stefan diffusion in both phases and the definitions for overall mass transfer coefficients given by equations (1.48) and (1.49), the individual mass transfer coefficients must be multiplied by the drift factors, giving

$$\frac{1}{K_{ys}} = \frac{1}{k_{ys}/(1 - y)_{lm}} + \frac{m}{k_{xs}/(1 - x)_{lm}} \tag{1.56}$$

and

$$\frac{1}{K_{xs}} = \frac{1}{mk_{ys}/(1 - y)_{lm}} + \frac{1}{k_{xs}/(1 - x)_{lm}} \tag{1.57}$$

The equations were derived for films at rest, and they may be modified when applied to fluids in motion. As an example, equation (1.38) gives a first power dependence of k_y on D_{AB}. But for fluids in turbulent flow, experimental data indicate an approximate square-root relationship, which is explained by the penetration theory.

The *penetration theory* was first suggested by Higbie.[12] He proposed as a major mechanism of mass transfer the motion of turbulent eddies from the core of the fluid towards the interface with a short interval of unsteady-state molecular diffusion, until the fluid was displaced by subsequent eddies. This gives an average rate of mass transfer that depends on the exposure time of each eddy. Higbie assumed that all eddies that reach the interface have the same exposure time. Danckwerts[13] modified the theory. He proposed a distribution function for the exposure time of the eddies. Both models give the mass transfer coefficient proportional to the square root of the diffusion coefficient D_{AB}.

Further modifications have been proposed by other investigators, such as that by Harriott,[14] which takes into account both the random distribution of eddy lifetimes or exposure times, and the random distances from the surface. It has also been shown[15] that the mass transfer in the liquid can be influenced by the Marangoni effect, i.e. surface movements caused by local differences in the surface tension.

Owing to the complexity of mass transfer, calculations of actual mass transfer equipment are often based on empirical methods and equations derived by dimensional analysis and semi-theoretical analogies. Data reported for turbulent flow by numerous investigators are well reproduced by the dimensionless equation

$$Sh = CRe^m Sc^n \tag{1.58}$$

where Sh = Sherwood number, $k_c d/D_A$, Re = Reynolds number, $\rho V d/\mu$, Sc = Schmidt number, $\mu/(\rho D_A)$, k_c = mass transfer coefficient for equimolar counterdiffusion with concentration as the driving force. kmol/(m²s kmol/m³), d = hydraulic diameter, m, m = exponent reported in the range from 0.75 to 0.83,[16,17] n = exponent reported in the range from 0.33 to 0.50,[16,17] and C = constant depending on the geometry, for gas in a smooth tube C = 0.023, m = 0.80, and n = 0.33.[16]

The Sherwood number for ideal gases can be expressed in terms of different mass transfer coefficients

$$Sh = \frac{k_c d}{D_A} = \frac{k_g RTd}{D_A} = \frac{k_{ye} dRT}{D_A P} = \frac{k_{ys} RTd p_{B1m}}{D_A P} \tag{1.59}$$

where p_{B1m} = logarithmic mean partial pressure of non-diffusing components, Pa.

Rate-controlling resistance

In equations (1.53) and (1.55) the values of k_x and k_y depend on the physical data of the system and the gas and liquid velocities. If $1/k_y > 20\,m/k_x$, equation (1.53) may be approximated by

$$K_y \approx k_y \tag{1.60}$$

and if $1/k_x > 20/mk_y$, equation (1.55) may be approximated by

$$K_x \approx k_x \tag{1.61}$$

Equation (1.60) reflects a high solubility of the gas in the liquid, i.e. a low value of the slope $m = (y_i - y^*)/(x_i - x)$ of the equilibrium line in Figure 1.8.

Diffusion coefficients

Binary gas mixtures

Experimental values of D_{AB} should be used whenever available. The most comprehensive review and evaluation of empirical data from 1860 to 1970 is given by Marrero and Mason.[18] Table 3-299 in Perry[8] is another source. Some of the data are reproduced in Appendix 1.

Estimation methods are compiled by Reid et al.[19] The empirical correlation by Fuller et al.[20] seems to be among the better ones,

$$D_{AB} = \frac{0.0101 T^{1.75}(1/M_A + 1/M_B)^{1/2}}{P[(\Sigma\Delta_v)_A^{1/3} + (\Sigma\Delta_v)_B^{1/3}]^2} \tag{1.62}$$

where D_{AB} = low-pressure diffusivity, m^2/s, T = absolute temperature, K, P = total pressure, Pa = N/m², M_A and M_B = molecular weights of the two components, and $\Sigma\Delta_v$ = sum of atomic and structural diffusion volume increments for each of the components (Table 1.1).

Table 1.1 Atomic and molecular diffusion volumes in equation (1.62). Parentheses indicate value based on only a few data points

A. *Diffusion volumes for simple molecules,* $\Sigma\Delta_v$

Molecule	$\Sigma\Delta_v$	Molecule	$\Sigma\Delta_v$	Molecule	$\Sigma\Delta_v$	Molecule	$\Sigma\Delta_v$
H_2	7.07	Air	20.1	CO_2	26.9	SF_6	(69.7)
D_2	6.70	A	16.1	N_2O	35.9	Cl_2	(37.7)
He	2.88	Kr	22.8	NH_3	14.9	Br_2	(67.2)
N_2	19.9	Xe	(37.9)	H_2O	12.7	SO_2	(41.1)
O_2	16.6	CO	18.5	CCl_2F_2	(114.8)		

B. *Atomic and structural volumes for other molecules,* Δ_v

Atom	Δ_v	Atom	Δ_v	Atom	Δ_v	Structural increment	
C	16.5	O	5.48	Cl	(19.5)	Aromatic ring	-20.2
H	1.98	N	(5.69)	S	(17.0)	Heterocyclic ring	-20.2

For simple, non-polar systems and moderate temperatures, equation (1.62) probably gives errors less than 12 per cent for pressures up to the critical pressure. The equation should be used with caution as one departs from room temperature.

For diffusion in multicomponent gaseous mixtures, refer to the literature.[11,19,21]

Liquids

The diffusion coefficient D_{AB} in non-viscous liquids is in the order of magnitude 10^{-5} times the diffusion coefficient of common gases at atmospheric pressure. In

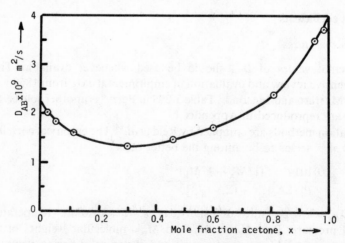

Figure 1.10 Experimentally determined diffusion coefficients in the acetone–cyclohexane binary liquid mixture at 298 K

contrast to gases, the diffusion coefficient often varies substantially with concentration, as seen in Figure 1.10 for the acetone–cyclohexane system.[22]

An extensive tabulation of diffusion coefficients in non-electrolytes is given by Ghai et al.[23] Diffusion coefficients for a large number of amino acids, peptides, and sugars at low concentrations in water are given by Tuwiner,[24] and for dilute organic solutes in aliphatic alcohols by St. Dennis and Fell.[25] Table 3-300 in Perry[8] is another source. Appendices 2 and 3 give a limited number of the experimental data for dilute solutions in water and in some organic solvents.

Estimation methods are compiled by Reid et al.[19] For infinite dilution of non-electrolytes in water, Hayduk and Laudie[26] gave the correlation

$$D_{AW}^0 = 1.42 \times 10^{-10} \mu_w^{-1.4} V_A^{-0.59} \tag{1.63}$$

where D_{AW}^0 = diffusion coefficient with solute A in infinite dilution in water, m^2/s, μ_w = dynamic viscosity of water, $N s/m^2$, and V_A = molar volume of solute A at the normal boiling-point, $m^3/kmol$ (Table 1.2).

Wilke and Chang[27] recommended the equation

$$D_{AB}^0 = 1.17 \times 10^{-16} (\varphi M_B)^{1/2} \frac{T}{\mu_B V_A^{0.6}} \tag{1.64}$$

where D_{AB}^0 = mutual diffusion coefficient of solute A at very low concentrations in

Table 1.2 Molecular volumes at the normal boiling-point for use in equations (1.63) and (1.64) ($m^3/kmol$)

A. Approximate molar volumes of simple molecules, V_A ($m^3/kmol$)

Molecule	V_A	Molecule	V_A	Molecule	V_A	Molecule	V_A
H_2	0.0143	CO	0.0307	NO_2	0.0364	COS	0.0515
O_2	0.0256	CO_2	0.0340	NH_3	0.0258	Cl_2	0.0484
N_2	0.0312	SO_2	0.0448	H_2O	0.0189	Br_2	0.0532
Air	0.0299	NO	0.0236	H_2S	0.0329	I_2	0.0715

B. Atomic and structural additive volume increments for other molecules, ΔV_A ($m^3/kmol$)

Atom	ΔV_A	Atom	ΔV_A
C	0.0148	Br	0.027
H	0.0037	Cl	0.0246
O (except as below)	0.0074	F	0.0087
In methyl esters and ethers	0.0091	I	0.037
In ethyl esters and ethers	0.0099	S	0.0256
In higher esters and ethers	0.011		
In acids	0.012	Ring, 3-membered	− 0.006
Joined to S, P, N	0.0083	4-membered	− 0.0085
N		5-membered	− 0.0115
Doubly bonded	0.0156	6-membered	− 0.015
In primary amines	0.0105	Naphthalene	− 0.030
In secondary amines	0.012	Anthracene	− 0.0475

solvent B, m^2/s, φ = 'association parameter' for solvent B: 2.6 for water, 1.9 for methanol, 1.5 for ethanol, and 1.0 for benzene, ether, heptane, and other unassociated solvents, M_B = molecular weight of B, T = absolute temperature, K, μ_B = dynamic viscosity of B, N s/m^2, and V_A = molar volume of solute A at its normal boiling-point (Table 1.2), $m^3/kmol$.

Most experimental data are obtained in the neighbourhood of the ambient temperature. For extrapolation to higher temperatures, Tyn[28] recommends the following equation for the diffusion coefficient of A in dilute solution with solvent B,

$$\frac{(D_{AB}^0)_1}{(D_{AB})_2} = \left[\frac{T_c - T_2}{T_c - T_1}\right]_B^n \tag{1.65}$$

where the indices 1 and 2 refer to the two temperatures T_1 and T_2, K, T_c = critical temperature of solvent B, K, and n = exponent given in Table 1.3.

Table 1.3 Exponent n in equation (1.65) as a function of the latent heat of vaporization of solvent B at the normal boiling-point, $\Delta H_{vb}(kJ/kmol)$

ΔH_{vb}	7900–30000	30000–39700	39700–46000	46000–50000	> 50000
n	3	4	6	8	10

The effect of concentration on liquid diffusion coefficients in ideal or nearly ideal mixtures is well reproduced by the correlation suggested by Vignes,[29,30]

$$D_{AB} = (D_{AB}^0)^{x_B}(D_{BA}^0)^{x_A} \tag{1.66}$$

Solids and gels

Diffusion coefficients and solubilities for gases in polymers are collected in the book by Rogers[31] and by Crank and Park.[32] Diffusivities of solutes in dilute biological gels in aqueous solution are given by Friedman and Kramer,[33−35] and by Spalding.[36]

Example 1.1 DIFFUSION THROUGH A VENT

A condenser for the overhead methanol vapour from a distillation column has a small cooler in the vent, as shown in Figure 1.11.

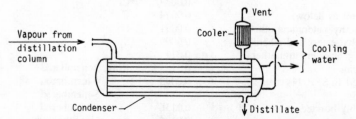

Figure 1.11 Methanol condenser with vent to the atmosphere. The outlet of the vent is covered with a screen for fire protection

The temperature in the cooler and the vent is $21\,^{\circ}\text{C} = 294\,\text{K}$ and the partial pressure of methanol in the cooler is $100\,\text{mm Hg} = 13.3\,\text{kPa}$. The inner diameter of the vent is 38 mm and the length 0.4 m. The vapour volume in the cooler is $0.005\,\text{m}^3$. The atmospheric pressure is $1\,\text{bar} = 10^5\,\text{Pa}$.

(a) Estimate the monthly loss of methanol due to molecular diffusion through the vent at steady state.

(b) Estimate the additional loss due to 'breathing' caused by variation in the heat input to the column, assuming that the 'breathing' corresponds to replacement of the vapour-side volume of the cooler twice every hour.

Solution

(a) Appendix 1 gives the product of pressure and diffusion coefficient for air–methanol mixtures at 273 K, $PD_{AB} = 1.34\,\text{Pa m}^2/\text{s}$. Equation (1.62) indicates that D_{AB} is proportional to $T^{1.75}$, i.e.

$$PD_{AB} = 1.34(294/273)^{1.75} = 1.53\,\text{Pa m}^2/\text{s at 294 K}$$

Assuming stagnant air and vapour pressure at the outlet of the vent $P_{A2} = 0$, equation (1.13) gives the flux

$$N_A = \frac{D_{AB}P}{RT(z_2 - z_1)}\ln\frac{P - p_{A2}}{P - p_{A1}} = \frac{1.53}{8314 \times 294 \times 0.4}\ln\frac{10^5}{10^5 - 1.33 \times 10^4}$$

$$= 2.2 \times 10^{-7}\,\text{kmol/(m}^2\text{s)}$$

Loss, $30 \times 24 \times 3600\,\dfrac{\pi}{4}0.038^2 \times 2.2 \times 10^{-7} = 6.5 \times 10^{-4}\,\text{kmol/month}$

$$\text{or } 32 \times 6.5 \times 10^{-4} = 0.02\,\text{kg/month.}$$

(b) The ideal gas law gives the number of kmoles of methanol vapour in the cooler,

$$n = pv/(RT) = 1.33 \times 10^4 \times 0.005/(8314 \times 294) = 2.7 \times 10^{-5}\,\text{kmol}$$

Loss due to 'breathing', $30 \times 24 \times 2 \times 2.7 \times 10^{-5} \times 32 = 1.25\,\text{kg/month.}$

Example 1.2   INERT IN A CONDENSER

The presence of small amounts of inert may cause a drastic reduction of the capacity of a condenser. Figure 1.12 shows a section of a water-cooled steam condenser with clean surfaces and pure steam. Figure 1.13 shows the same condenser with some air that reduces the surface temperature of the condensate film from 28 °C to 24 °C. By this, the

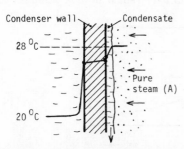

Figure 1.12   Section of a clean vacuum condenser for pure steam (no fouling)

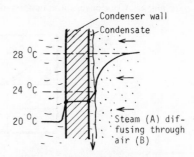

Figure 1.13   The same condenser as in Figure 1.12, but with some air present at the vapour side

temperature difference between the cooling water and the condensing steam is halved. Also, the rate of condensation will be approximately half the rate without inert.

The vapour pressure of water at 28 °C is 3.779 kPa and at 24 °C 2.983 kPa. Without air the rate of condensation is approximately 3×10^{-4} kmol/(m²s).

(a) Plot the partial pressure of the air as a function of the distance z from the condensate surface in the case where the temperature at the condensate surface is 24 °C, assuming stagnant air.

(b) Estimate the total amount of air per m² surface.

Solution

(a) Appendix 1 gives $PD_{AB} = 2.92$ Pa m²/s for air–water vapour mixtures at temperature 313 K. According to equation (1.62), D_{AB} is proportional to $T^{1.75}$. Using the average temperature 299 K, $PD_{AB} = 2.92(299/313)^{1.75} = 2.70$ Pa m²/s. This value and $N_A = 1.5 \times 10^{-4}$ kmol/(m²s) in equation (1.14) give

$$1.5 \times 10^{-4} = \frac{2.70}{8314 \times 299 z} \ln \frac{p_{B2}}{p_{Bz}} \qquad\text{(a)}$$

where p_{Bz} = partial pressure of air (B) at distance z, p_{B2} = partial pressure of air at the interface, $3779 - 2983 = 796$ Pa. Equation (a) solved for p_{Bz} gives

$$p_{Bz} = \exp(6.68 - 138z) = 796\, e^{-138z} \qquad\text{(b)}$$

z	m	0	0.001	0.002	0.004	0.008	0.015	0.03	
p_{Bz}	Pa	796	693	604	458	264	100	13	eqn (b)

In Figure 1.14 the partial pressure p_B of air is plotted as a function of the distance z from the surface.

(b) The ideal gas law applied to $1\,\text{m}^2$,

$$p_B(1\,dz) = dn\, RT$$

$$\int_0^n dn = \frac{1}{RT} \int_0^\infty p_B\, dz = \frac{796}{8314 \times 299} \int_0^\infty e^{-138z}\, dz$$

$$n = \frac{796}{8314 \times 299 \times 138} = 2.3 \times 10^{-6}\ \text{kmol air/m}^2$$

Comment: Condensers are usually designed to have vapour flowing along the cold

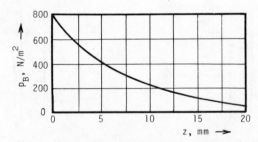

Figure 1.14 Partial pressure of air, p_B, as a function of the distance z from the surface of the condensate film

surface in order to sweep the inert to the end or ends of the condenser where accumulated inert can be removed.

Example 1.3 UNSTEADY-STATE DIFFUSION

An $L = 0.8$ m high vertical gas cylinder with volume $0.014\,m^3$ is filled with $0.058\,x$ kmol krypton and $0.058(1-x)$ kmol neon and stored at temperature 293 K.
(a) Estimate the diffusion coefficient D_{AB}.
(b) Assuming ideal gases with only krypton in the bottom and only neon in the rest of the cylinder at time zero and only molecular diffusion, derive an equation for the partial pressure of krypton at the top of the cylinder as a function of the initial mole fraction x of krypton, and of time.
(c) Insert $x = 0.2$ and calculate the time to obtain 95 per cent of the final pressure of krypton at the top.

Solution

(a) Total pressure,

$$P = \frac{nRT}{v} = \frac{0.058 \times 8314 \times 293}{0.014} = 1.01 \times 10^7 \text{ Pa}$$

Appendix 1 gives the product of pressure and diffusion coefficient, $PD_{AB} = 2.26\ \text{Pa m}^2/\text{s}$ for mixtures of krypton and neon at 273 K. With the influence of the temperature, as given by equation (1.62)

$$D_{AB} = \frac{2.26(293/273)^{1.75}}{1.01 \times 10^7} = 2.53 \times 10^{-7} \text{ m}^2/\text{s}$$

(b) The problem is solved by the partial differential equation (1.33)

$$\frac{\partial p_A}{\partial t} = D \frac{\partial^2 p_A}{\partial z^2} \tag{a}$$

with the initial conditions

$$p_A(z, 0) = f(z) = \begin{cases} P \text{ for } 0 < z \leqslant xL \\ 0 \text{ for } xL < z \leqslant L \end{cases} \tag{b}$$

The function $f(z)$ is extended to be a periodic function as indicated in Figure 1.15, where the heavy solid line is the partial pressure of krypton in the cylinder. The dotted lines may be visualized as the partial pressure in identical cylinders placed after each other, and with no diffusion between them. This is an even function, and its Fourier series is given by (see Kreyzig,[37] p. 482, or any other book on Fourier analysis)

$$f(z) = a_0 + \sum_{n=1}^{\infty} a_n \cos\left(\frac{2n\pi}{2L} z\right) \tag{c}$$

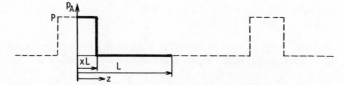

Figure 1.15 Partial pressure of krypton (heavy line) at time $t = 0$, extended to a periodic function (dotted line)

with

$$a_0 = \frac{2}{2L} \int_0^L f(z)dz = xP \tag{d}$$

and

$$a_n = \frac{4}{2L} \int_0^L f(z)\cos\left(\frac{2n\pi}{2L}z\right)dz$$

$$= \frac{2}{L} \int_0^{xL} P\cos\left(\frac{n\pi}{L}z\right)dz = \frac{2P}{n\pi}\sin(n\pi x) \tag{e}$$

Equation (a) is solved by the method of separation of variables, that is, we seek solutions of the type

$$p_A(z, t) = F(z)T(t) \tag{f}$$

Equation (f) inserted in equation (a) gives

$$F(z)T'(t) = DF''(z)T(t) \tag{g}$$

or

$$\frac{T'(t)}{DT(t)} = \frac{F''(z)}{F(z)} \tag{h}$$

The left-hand side is a function of t only, and the right-hand side of z only. This can only be valid if both sides are a constant, say $\pm c$. The equation for $T(t)$ yields

$$T(t) = C\,e^{\pm cDt}$$

Only the minus sign gives a finite value of $T(t)$ when $t \to \infty$. Thus

$$T(t) = e^{-cDt} \tag{i}$$

and

$$F(z) = A\cos(\sqrt{c}z) + B\sin(\sqrt{c}z) \tag{j}$$

where the integration constant C is considered built-in in A and B.
Since

$$p_A(z, 0) = F(z)T(0) = F(z)$$

we want to find values of A, B and c such that

$$f(z) = \sum_c F(z;c) \tag{k}$$

Comparing equations (j) and (c), it follows that $B = 0$, and equation (j) simplifies to

$$F(z) = A\cos(\sqrt{c}z) \tag{l}$$

Equations (l) and (c) give $\sqrt{c} = n\pi/L$, and the constant A is given by equations (d) and (e). Thus, the solution of (a) with the initial conditions equation (b), is

$$p_A(z, t) = P\left\{x + \frac{2}{\pi}\sum_{n=1}^{\infty}\frac{1}{n}\sin(n\pi x)\cos\left(\frac{n\pi}{L}z\right)\exp\left[-D\left(\frac{n\pi}{L}\right)^2 t\right]\right\} \tag{m}$$

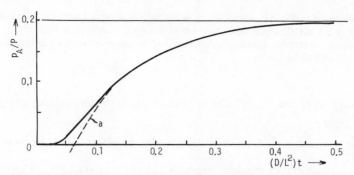

Figure 1.16   Ratio p_A/P as a function of $(D/L^2)t$ calculated by equation (n) for $x = 0.2$. The dotted curve a corresponds to only the first term in the infinite sum in equation (n)

or at the top of the cylinder with $z = L$ and $\cos(n\pi L/L) = (-1)^n$,

$$p_A(L, t) = P\left\{ x + \frac{2}{\pi} \sum_{n=1}^{\infty} \frac{(-1)^n}{n} \sin(n\pi x) \exp\left[-D\left(\frac{n\pi}{L}\right)^2 t \right] \right\}$$ (n)

(c) Figure 1.16 gives the ratio of the partial pressure of krypton to the total pressure, p_A/P, as a function of $(D/L^2)t$ calculated by equation (n) for $x = 0.2$. Calculations for different values of x show that only the first term in equation (n) needs to be included for $(D/L^2)t > 0.15$.

The partial pressure at time $t = \infty$ is $xP = 0.2P$, and 95 per cent of this pressure corresponds to $p_A/P = 0.95 \times 0.2P/P = 0.19$. The first term in equation (n) yields

$$0.19 = 0.20 - \frac{2}{\pi} \sin(0.2\pi) \exp\left[-2.53 \times 10^{-7} \left(\frac{\pi}{0.8}\right)^2 t \right]$$

$t = 930\,000\,\text{s}$ or 11 days

Comment: The calculations show that molecular diffusion in gases under high pressure is extremely slow.

The colour of 'neon lights' depends on the composition of the gas, and a small change in composition may give an appreciable change in colour. A producer of mixtures of noble gases received complaints from a customer claiming that the gas he had started to use from one of the cylinders did not have the specified composition. Subsequently the producer had all cylinders filled with mixtures of noble gases placed in a rocking machine overnight before they were shipped, and the customers seem to have been satisfied ever since.

Example 1.4   MASS TRANSFER COEFFICIENTS

A pilot plant runs with absorption of CO_2 in an aqueous solution containing 4 per cent NaOH in a column with 38 mm 'Hy-Pak' packing rings, air rate $1.22\,\text{kg}/(\text{m}^2\,\text{s})$, and liquid rate L from 2.7 to $40\,\text{kg}/(\text{m}^2\,\text{s})$, gave.[38]

$$K_g a = 0.07 L^{0.25}$$ (a)

where K_g = overall mass transfer coefficient, $\text{kmol}/(\text{m}^2\,\text{s}\,\text{MPa})$, and a = effective contact area, m^2/m^3 column volume.

The tests were carried out with a water temperature of 24 °C and 1 per cent CO_2 in the entering air.

(a) Estimate the product $k_g a$ by absorption of chlorine instead of carbon dioxide in the same column and with all other conditions unchanged.

(b) Calculate the product $k_{ys}a$ with the mole fraction as driving force, a total pressure of 1 bar ($P = 0.1\,\text{MPa}$), and a mole fraction of chlorine, y, less than 0.01.

(c) Assuming the interfacial area a independent of the gas mass velocity, estimate $k_{ys}a$ at a gas rate of $0.82\,\text{kg/(m}^2\,\text{s)}$.

Solution

(a) Except for k_c and D_{AB}, all data in equation (1.58) remain constant, giving

$$k_c \sim D_{AB}^{1-n} \tag{b}$$

where the exponent of the Schmidt number, n, is in the range 0.33–0.5 with the average value 0.42.

For air–carbon dioxide mixtures, Appendix 1 gives the product $PD_{AB} = 1.44\,\text{Pa(m}^2/\text{s)}$ at 276 K and $PD_{AB} = 1.79\,\text{Pa(m}^2/\text{s)}$ at 317 K. According to equation (1.62) $D_{AB} \sim T^{1.75}$, and at the temperature 297 K, the two data from Appendix 1 give

$$PD_{AB} = 1.44(297/276)^{1.75} = 1.64\,\text{Pa(m}^2/\text{s)}$$

and

$$PD_{AB} = 1.79(297/317)^{1.75} = 1.60\,\text{Pa(m}^2\,\text{s)}$$

Arithmetic average, $PD_{AB} = 1.62\,\text{Pa(m}^2/\text{s)}$

For chlorine in air at 297 K, equation (1.62) and Table 1.1 give

$$PD_{AB} = \frac{0.0101 \times 297^{1.75}(1/70.9 + 1/29)}{(37.7^{1/3} + 20.1^{1/3})^2} = 1.08\,\text{Pa(m}^2/\text{s)}$$

Substituting carbon dioxide with chlorine reduces the coefficient 0.07 in equation (a) to $0.07(1.08/1.62)^{1-0.42} = 0.055$

$$k_g a = 0.055L^{0.25} \quad [\text{kmol/(m}^2\,\text{s MPa)}]\text{m}^2/\text{m}^3 \tag{c}$$

(b) With zero pressure of chlorine at the interface, the mass transfer per m^2 interface is given by equations (1.35) and (1.42) with p_{Ai} and y_i equal to zero and $(1-y)_{lm} \approx 1.0$,

$$k_g p_A = k_{ys} y \tag{d}$$

where p_A is in MPa, i.e.

$$p_A = yP = 0.1y \tag{e}$$

Equations (d) and (e) give

$$k_{ys} = 0.1\,k_g$$

which is inserted in equation (c)

$$k_{ys}a = 0.0055L^{0.25}$$

(c) According to equation (1.58), the mass transfer coefficient is proportional to the Reynolds number to the mth power. With $m = 0.8$,

$$k_{ys}a = \left(\frac{0.82}{1.22}\right)^{0.8} 0.0055L^{0.25} = 0.0040L^{0.25}\,[\text{kmol/(m}^2\,\text{s mol fraction)}]\text{m}^2/\text{m}^3$$

Note: The assumption of interfacial area a independent of the gas mass velocity is valid for velocities below the loading point, see Chapters 2 and 3.

Problems

1.1 Figure 1.17 shows a tube with cross-section 6 mm^2 in the lower 40 mm and 2 mm^2 in the upper 20 mm height. In the bottom is 0.003 g water and the tube is filled with air and

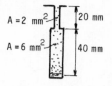

A = 2 mm^2 20 mm
A = 6 mm^2 40 mm

Figure 1.17   Molecular diffusion in a tube with varying cross-section

water vapour with a total pressure of 1.0 bar. The temperature is constant 40 °C and the partial pressure of water vapour at the bottom is 7370 Pa and at the top 2950 Pa. Assuming only molecular diffusion, calculate the time until the water is completely evaporated.

1.2 Using the equation by Fuller et al.,[20] estimate the diffusion coefficient for hydrogen in ammonia at temperatures of 263 K and 473 K at 1 bar total pressure and compare with the data in Appendix 1.

1.3 Using the equation proposed by Wilke and Chang, estimate the diffusion coefficient of acetone in dilute solution of cyclohexane, and of cyclohexane in dilute solution of acetone at temperature 298 K. The dynamic viscosity of cyclohexane at 298 K is 1.06 cP. Using these diffusion coefficients and the correlation by Vignes, estimate the diffusion coefficient in a mixture with mole fraction acetone $x_A = 0.3$ and mole fraction cyclohexane $x_B = 0.7$.
Compare the results with the experimental data given in Figure 1.10.

1.4 Component A passes by molecular diffusion through a film of thickness $z_2 - z_1$. The total pressure is $P = 100 \, \text{kPa}$ and the partial pressure of A is $p_{A1} = 0$ at point 1 and $p_{A2} = 30 \, \text{k Pa}$ at point 2. Assuming all diffusion coefficients to be equal and constant, calculate at distances $z - z_1 = 0.25, 0.50$, and 0.75 times $z_2 - z_1$ the partial pressure p_A of component A:
(a) by Stefan diffusion;
(b) by equimolar counterdiffusion;
(c) by counterdiffusion of two moles for each mole of diffusing A.

Symbols

a	Contact area per unit volume, m^2/m^3
c	Concentration, kmol/m^3
D	Diffusion coefficient, m^2/s
D^0	Diffusion coefficient by infinite dilution, m^2/s
D_{AB}	Diffusion coefficient in a binary mixture of A and B, m^2/s
d	Hydraulic diameter, m
E	Eddy diffusivity, m^2/s
J_A	Molar flux of A relative to bulk of fluid, $\text{kmol}/(\text{m}^2 \, \text{s})$
K_g	Overall mass transfer coefficient, $\text{kmol}/(\text{m}^2 \, \text{s Pa})$ or $\text{kmol}/(\text{m}^2 \, \text{s MPa})$
K	Overall mass transfer coefficient, $\text{kmol}/(\text{m}^2 \, \text{s mol fraction})$
k_c	Mass transfer coefficient for equimolar counterdiffusion, $\text{kmol}/(\text{m}^2 \, \text{s kmol}/\text{m}^3)$
k	Individual transfer coefficient, $\text{kmol}/(\text{m}^2 \, \text{s mol fraction})$
M	Molecular weight, kg/kmol
m	Slope of equilibrium line
N	Molar flux relative to a fixed point, $\text{kmol}/(\text{m}^2 \, \text{s})$
n	Number of kmoles
P	Total pressure, Pa
p	Partial pressure, yP, Pa
R	Gas constant, 8314.3 $\text{J}/(\text{K kmol}) = \text{N s}/(\text{K kmol})$
Re	Reynolds number, $\rho V d / \mu$

r	Radius, m
Sc	Schmidt number, $\mu/(\rho D)$
Sh	Sherwood number, $k_c d/D$
T	Absolute temperature, K
V	Velocity, m/s
v	Volume, m^3
x	Mole fraction
y	Mole fraction
z	Distance, m
λ_m	Mean free path, m
μ	Dynamic viscosity, N s/m^2 = kg/(s m)
ρ	Density, kg/m^3

Subscripts

A	Component A
B	Component B
e	Equimolar counterdiffusion
i	Component i
i	Condition at gas–liquid interface
lm	Logarithmic mean, equations (1.40) and (1.47)
s	Stefan diffusion, i.e. diffusion through stagnant inert
w	Water
x	Liquid phase
y	Gas phase

References

1. Taylor, G. I., *Proc. Roy. Math. Soc.*, **20**, 196 (1921).
2. Sherwood, T. K., and B. B. Woertz, Mass transfer between phases. Role of eddy diffusion, *Ind. Eng. Chem.*, **31**, 1034–1041 (1939).
3. Hunt, B., Diffusion in laminar tube flow, *Int. J. Heat Mass Transfer*, **20**, 393–401 (1977).
4. Richter, E., K. Knoblauch, and H. Jüntgen, Modellierung isothermer Festbettadsorber, *Chem.-Ing.-Techn.*, **50**, 600–611 (1978).
5. Knudsen, M., Die Gesetze der Molekularströmung und der inneren Reibungsströmung der Gase durch Rohren, *Ann. Physik*, **28**, 75–130 (1909).
6. Geankoplis, C. J., *Transport Processes and Unit Operations*, Allyn and Bacon, Boston, 1978.
7. Lydersen, A. L., *Fluid Flow and Heat Transfer*, John Wiley & Sons, Chichester, 1979.
8. Perry, H. R., and C. H. Chilton, *Chemical Engineers' Handbook*, 5th edn, McGraw-Hill, New York, 1973.
9. Carslaw, H., and J. C. Jaeger, *Conduction of Heat in Solids*, 2nd edn, Clarendon Press, Oxford, 1959.
10. Bird, R. B., W. E. Stewart, and E. N. Lightfoot, *Transport Phenomena*, John Wiley & Sons, New York, 1960.
11. Krishna, R., and G. L. Standart, Mass and energy transfer in multicomponent systems, *Chem. Engn. Commn.*, **3**, 201–275 (1979).
12. Higbie, R., The rate of absorption of a pure gas into a still liquid during short periods of exposure, *Trans. Am. Inst. Chem. Eng.*, **31**, 365–388 (1935).
13. Danckwerts, P. V., Significance of liquid–film coefficients in gas absorption, *Ind. Eng. Chem.*, **43**, 1460–1467 (1951).
14. Harriott, P., A random eddy modification of the penetration theory, *Chem. Eng. Sci.*, **1962**, 149–154.

15. Sternling, C. V., and L. E. Scriven, Interfacial turbulence: hydrodynamic instability and the Marangoni effect, *A.I.Ch.E.J.*, **5**, 514–523 (1959).
16. Bennett, C. O., and J. E. Myers, *Momentum, Heat and Mass Transfer*, 2nd edn, McGraw-Hill, New York, 1974.
17. Yilmaz, T., Flüssigkeitseitiger Stoffübergang in berieselten Füllkörperschüttungen, *Chem.-Ing.-Tech.*, **45**, 253–259 (1973).
18. Marrero, T. R. and E. A. Mason, Gaseous diffusion coefficients, *J. Phys. Chem. Ref. Data*, **1**, 3–118 (1972).
19. Reid, R. C., J. M. Prausnitz, and T. K. Sherwood, *The Properties of Gases and Liquids*, 3rd edn, McGraw-Hill, New York, 1977.
20. Fuller, E. N., P. D. Schettler, and J. C. Giddings: A new method for prediction of binary gas-phase diffusion coefficients, *Ind. Eng. Chem.*, **58** (No. 5), 19–27 (1966).
21. Obermeier, E., and A. Schaber, A simple formula for multicomponent gaseous diffusion coefficients derived from mean free path theory, *Int. J. Heat Mass Transfer*, **20**, 1301–1305 (1977).
22. Tasić, Z., B. D. Djordjević, S. P. Šerbanović, and D. K. Grozdanić, Diffusion coefficients for the liquid system acetone–cyclohexane at 298.15 K, *J. Chem. Eng. Data*, **26**, 118–120 (1981).
23. Ghai, R. K., H. Ertl, and F. A. L. Dullien, Liquid diffusion in nonelectrolytes, *A.I.Ch.E.J.*, **19**, 881–900 (1973).
24. Tuwiner, S. B., *Diffusion and Membrane Technology*, A.C.S. Monogr. 156.
25. St. Dennis, C. E., and C. J. D. Fell, Diffusivity of oxygen in water, *Can. J. Chem. Eng.*, **49**, 885 (1971).
26. Hayduk, W., and H. Laudie, Prediction of diffusion coefficients for nonelectrolytes in dilute aqueous solutions, *A.I.Ch.E.J.*, **20**, 611–615 (1974).
27. Wilke, C. R., and P. C. Chang, Correlation of diffusion coefficients in dilute solutions, *A.I.Ch.E.J.*, **1**, 264–270 (1955).
28. Tyn, M. T., Temperature dependence of liquid phase diffusion coefficients, *Trans. I. Chem. E*, **59**, 112–118 (1981).
29. Vignes, A., Diffusion in binary solutions, *Ind. Eng. Chem. Fundam.*, **5**, 189–199 (1966).
30. Dullien, F. A. L., Statistical test of Vignes' correlation of liquid-phase diffusion coefficients, *Ind. Eng. Chem. Fundam.*, **10**, 41–49 (1971).
31. Rogers, C. E., *Engineering Design for Plastics*, Reinhold, New York, 1964.
32. Crank, J., and G. S. Park, *Diffusion in Polymers*, Academic Press, New York, 1968.
33. Friedman, L., and E. O. Kramer, The structure of gelatin gels from studies of diffusion, *J. Am. Chem. Soc.*, **52**, 1295–1304 (1930).
34. Friedman, L., Diffusion of non-electrolytes in gelatin gels, *J. Am. Chem. Soc.*, **52**, 1305–1310 (1930).
35. Friedman, L., Structure of agar gels from studies of diffusion, *J. Am. Chem. Soc.*, **52**, 1311–1314 (1930).
36. Spalding, G. E., A sensitive method for measuring diffusion coefficients in agarose gels of electrolyte solutions, *J. Phys. Chem.*, **73**, 3380–3383 (1969).
37. Kreyzig, E., *Advanced Engineering Mathematics*, 4th edn, John Wiley & Sons, New York, 1979.
38. Norton, *High Performance Metal Tower Packing Hy-Pak*, Bulletin HY-40, Norton Co., Akron, Ohio, 1978.

CHAPTER 2

Distillation

One of the earliest distillations was carried out by Greek sailors when they boiled sea-water and condensed the vapour to obtain fresh water.[1] Today, distillation is one of the most widely used unit operations for separation of mixtures of liquids in the chemical and petroleum industries. The separation obtained is due to the fact that the vapour of a boiling mixture of liquids in most cases will have a composition that differs from the composition of the boiling liquid.

The Greek sailors obtained a pure product as the dissolved salts did not vaporize. But most boiling liquid mixtures will give off vapours containing more than one component. For instance, the vapour from a boiling mixture of toluene and benzene will contain both components, but the fraction of toluene in the vapour will be less than in the boiling liquid. Further enrichment in benzene is obtained if the vapour is condensed and a fraction of the condensate is vaporized again. This effect is utilized in distillation columns, the main subject of this chapter.

Vapour–liquid equilibria

Calculations of distillation columns are based on vapour–liquid equilibrium data. Experimental determination, and in particular estimation of such data, is a field under continuous development. This chapter gives only an introduction and a few data for use in problems for students.

Ideal systems

In a homogeneous solution the number of molecules of any one component per unit area of surface will be less than if that component exposed the same area of surface in the pure liquid state. For this reason the rate of vaporization of a substance per unit area will decrease with dilution. According to *Raoult's law* the resulting partial pressure of component A in the vapour phase is

$$p_A = x_A P_A \tag{2.1}$$

where x_A = the mole fraction of component A in the boiling liquid, and P_A = saturation pressure of pure component A at the temperature of the liquid mixture.

Dalton's law states that the total pressure of an ideal gas mixture equals the sum of the partial pressures

$$P = \sum p_i \tag{2.2}$$

The partial pressure p_i is, by definition, equal to the product of the gas-phase mole fraction and the total pressure

$$p_i \equiv y_i P \tag{2.3}$$

The first and the last equation give the mole fraction of component A in the gas phase as

$$y_A = x_A P_A / P \tag{2.4}$$

Applied to a binary system, Raoult's law and Dalton's law (equations (2.1) and (2.2)) give

$$P = p_A + p_B = x_A P_A + (1 - x_A)P_B = x_A(P_A - P_B) + P_B$$

or

$$x_A = (P - P_B)/(P_A - P_B) \tag{2.5}$$

Equation (2.5) is used to find x_A for ideal binary mixtures at selected temperatures between the boiling temperatures of the two pure components at pressure P, and equation (2.4) gives the corresponding value of y_A (see Example 2.1).

The distribution coefficient, also called the vaporization equilibrium constant or just the K-value for component i, is defined by the equation

$$y_i = K_i x_i \tag{2.6}$$

The *relative volatility* of component A relative to component B is defined as

$$\alpha_{AB} = \frac{K_A}{K_B} = \frac{y_A/x_A}{y_B/x_B} \tag{2.7}$$

or in combination with equation (2.5) for ideal mixtures,

$$\alpha_{AB} = P_A/P_B \tag{2.8}$$

Relative volatility is a useful concept for calculation of multicomponent mixtures.

Raoult's and Dalton's law give reasonable estimates for cognate molecules such as benzene and toluene when distilled at normal or lower pressures.

Non-ideal homogeneous mixtures

In a close to ideal mixture, such as benzene and toluene, the intermolecular forces between two benzene molecules are approximately the same as between two toluene molecules, and also between a benzene and a toluene molecule.

Deviations from Raoult's law are due to changes in the intermolecular forces, of which the hydrogen bond plays an important role.[2] The hydrogen bonds

$$O \longrightarrow HO \qquad N \longrightarrow HO \qquad O \longrightarrow HN$$

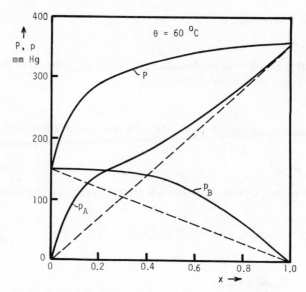

Figure 2.1 Vapour pressure of ethanol (p_A) and water (p_B) and total pressure $P = p_A + p_B$ of ethanol–water mixtures;[3] x = mole fraction of ethanol in the liquid. The dashed lines correspond to Raoult's law

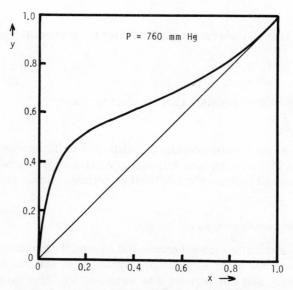

Figure 2.2 Mole fraction y of ethanol in the vapour as a function of mole fraction x of ethanol in the liquid of ethanol–water mixtures boiling at atmospheric pressure[4]

are strong. In water the hydrogen bond is especially strong. A high fraction of a component with less tendency to form hydrogen bonds with water, will reduce the hydrogen bonds in the water and increase its volatility. The increase in vapour pressure corresponds to a positive deviation from Raoult's law, as shown in Figure 2.1 for the ethanol–water mixture at a temperature of 60 °C. The total pressure of this mixture goes through a maximum. A composition giving maximum pressure when boiling at constant temperature, gives a minimum boiling temperature when boiling at constant pressure. At this point the composition of the vapour equals the composition of the liquid, and it is called an *azeotropic mixture*.

Figure 2.2 shows the mole fraction y of ethanol in the vapour as a function of the mole fraction x of ethanol in the liquid when an ethanol–water mixture boils at atmospheric pressure. The mole fraction $x = y = 0.894$ or 96 per cent by volume is the azeotropic mixture of ethanol and water. The boiling-point of the azeotrope is 78.15 °C, which is below the boiling-points of the pure components. The mixture is called a *minimum azeotrope*.

At reduced pressures the azeotrope shifts towards higher alcohol concentrations. It disappears at pressures below 0.08 bar absolute.

A mixture of acetone and chloroform shows a different type of deviation from Raoult's law. The pure components do not have hydrogen bonds. But chloroform has an active hydrogen atom that acts as donor of an electron, thus forming a hydrogen bond between an acetone and a chloroform molecule. This gives a negative deviation from Raoult's law (Figure 2.3). Figure 2.4 shows the mole fraction y of acetone in the vapour as a function of the mole fraction x of acetone in the liquid when the mixture boils at atmospheric pressure. This mixture has a maximum azeotrope, i.e. the boiling temperature of the azeotrope is above the boiling-points of the pure components.

Deviations from ideal behaviour in the liquid phase are taken into account by the liquid phase activity coefficient γ, and equation (2.4) becomes

$$y_A = x_A \gamma_A P_A / P \tag{2.9}$$

Equations (2.6) and (2.9) give

$$K_A = \gamma_A P_A / P \tag{2.10}$$

The activity coefficient is a function of temperature and composition of the liquid phase. Many equations have been suggested for calculation of the activity coefficient. Van Laar's[5] and Margules'[6] are the oldest ones, while Wilson's equation[7] and the UNIQUAC equation[8] are among the better newer ones. Van Laar's third-order equations for binary mixtures of non-electrolytes are

$$\log \gamma_1 = \frac{A_{12} x_2^2}{[x_1(A_{12}/A_{21}) + x_2]^2} \tag{2.11}$$

$$\log \gamma_2 = \frac{A_{21} x_1^2}{[x_2(A_{21}/A_{12}) + x_1]^2} \tag{2.12}$$

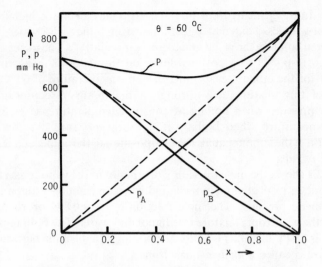

Figure 2.3 Vapour pressure of acetone (p_A) and chloroform (p_B) and total pressure $P = p_A + p_B$ of acetone–chloroform mixtures;[11] x = mole fraction of acetone in the liquid. The dashed lines correspond to Raoult's law

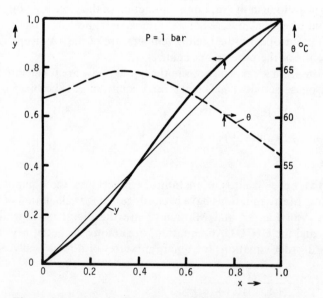

Figure 2.4 Mole fraction y of acetone in the vapour as a function of the mole fraction x of acetone in liquid mixtures of acetone–chloroform boiling at atmospheric pressure. The dotted curve is the boiling-point

Van Laar's equations fit fairly well most binary systems, and the influence of temperature can be neglected for a limited temperature range. The constants A_{12} and A_{21} can be determined from one or a few experimental points (see Example 2.2).

In the estimation of vapour–liquid equilibria at pressures above 7—10 bar, deviations from ideal gas behaviour should also be taken into account. Estimation of K-values for hydrocarbons and a few other compounds at higher pressures are given in the N.G.P.S.A.'s *Engineering Data Book*.[9] References to other sources of vapour–liquid equilibrium data are given in Perry,[10] pp. 13–6 and 13–7.

Heterogeneous mixtures

In a mixture consisting of two liquid phases, condensation of a component can take place only at the restricted areas where the vapour molecules impinge upon its own molecules. Thus, the vapour pressure of one of the liquid phases is unaffected by the presence of the other liquid phase. The total pressure is the sum of the vapour pressures of the two liquid phases.

Equilibrium diagrams

For calculation of distillation columns with binary systems, it is convenient to make use of diagrams, as shown in Figures 2.2 and 2.4. The mole fraction y of the more volatile component in the vapour is plotted as a function of the mole

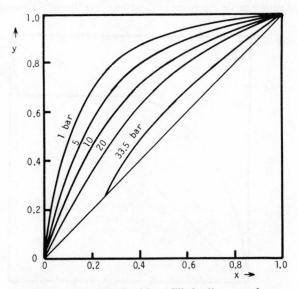

Figure 2.5 Vapour–liquid equilibria diagrams for *n*-butane–*n*-hexane mixtures at different pressures.[12] The coordinates x and y are mole fractions of *n*-butane

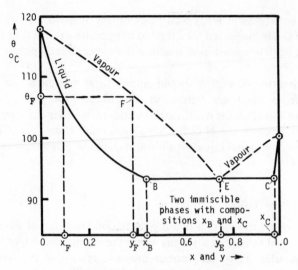

Figure 2.6 Boiling- and dewpoint diagram for water–
n-butanol at 1 atm;[13] *x* and *y* refer to water

fraction *x* of the same component in the liquid. A crossing of the diagonal represents an azeotrope.

In general the relative volatility decreases with increasing pressure. Figure 2.5 shows vapour–liquid equilibrium curves for mixtures of *n*-butane and *n*-hexane. The curves become discontinuous for pressures above 30 bar. Thus, at 33.5 bar, mixtures containing mole fractions of *n*-butane less than 0.24 consist of only one

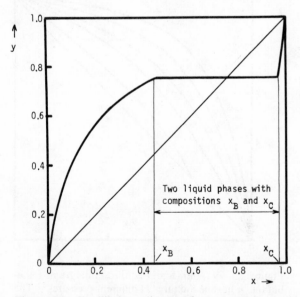

Figure 2.7 Equilibrium diagram for water–*n*-butanol
at 1 atm; *x* and *y* refer to water

phase. Also, the composition of an azeotrope changes with changes in pressure. In the ethanol–water system (Figure 2.2) the azeotrope disappears at absolute pressures below 60 mm mercury.

The behaviour of a non-homogeneous mixture is shown in Figures 2.6 and 2.7. In Figure 2.6 the thick lines represent the boiling temperature of mixtures of water and n-butanol.

When the mole fraction x of water in the liquid is less than x_B or greater than x_C, there is only one liquid phase. The liquid with the mole fraction x_F boils at the temperature θ_F, and the corresponding mole fraction of water in the vapour is y_F. When the mole fraction of water in the liquid, x, is between x_B and x_C, no single liquid phase can exist. The liquid consists of two liquid phases with compositions B and C, and the vapour has the composition E. The composition y_E is said to be a *heterogeneous azeotrope* because of the presence of two liquid phases.

Figure 2.7 is the corresponding equilibrium diagram.

Distillation of binary mixtures

Differential distillation

Figure 2.8 shows the principle of differential distillation. The vapour is removed continuously, while the liquid remaining becomes steadily weaker in the more volatile component.

Let L be the number of moles in the still and x the mole fraction of the more volatile component A. Vaporization of dL moles gives $y\,dL$ moles of component A in the vapour and $d(xL)$ less moles left in the liquid,

$$y\,dL = d(xL) = L\,dx + x\,dL$$

$$\int_{L_0}^{L} \frac{dL}{L} = \int_{x_0}^{x} \frac{dx}{y-x}$$

$$\ln\frac{L}{L_0} = -\int_{x}^{x_0} \frac{dx}{y-x} \qquad (2.13)$$

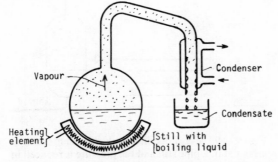

Figure 2.8 Differential distillation of a binary mixture

This integral is solved graphically by plotting $1/(y-x)$ versus x and measuring the area under the curve between the final mole fraction x and the original mole fraction x_0 (see Example 2.3). In some cases the integration can be carried out analytically. With a binary mixture equation (2.7) can be written

$$\alpha = \frac{y/x}{(1-y)/(1-x)} \tag{2.14}$$

or

$$y = \frac{\alpha x}{1+(\alpha-1)x} \tag{2.15}$$

This value inserted in equation (2.13) gives

$$\ln \frac{L}{L_0} = \frac{1}{\alpha-1}\left(\ln \frac{x}{x_0} - \alpha \ln \frac{1-x}{1-x_0}\right) \tag{2.16}$$

where an average value of α has to be used.

Flash or equilibrium distillation

Figure 2.9 shows continuous flash distillation as used on a large scale in petroleum refining. The vaporized fraction of the feed, $f = D/F$, depends on the amount of heat added in the still. A material balance on the more volatile component gives

$$x_F F = y_D f F + x_B (1-f)F$$

or

$$y_D = \frac{x_F}{f} - \frac{1-f}{f}x_B \tag{2.17}$$

With f and x_F known, this is one equation with two unknowns, y_D and x_B. But they are also coordinates of a point on the equilibrium curve. If y_D and x_B are

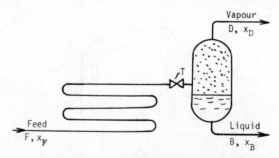

Figure 2.9   Flash distillation. Feed under pressure is heated in a pipe still and the pressure is reduced in the valve T

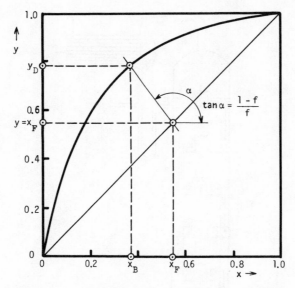

Figure 2.10 Determination of x_B and y_D by a straight line through $y = x_F$ with slope $-(1-f)/f$

replaced by y and x, equation (2.17) becomes

$$y = \frac{x_F}{f} - \frac{1-f}{f}x \tag{2.18}$$

This is the equation of a straight line through the point $y = x_F$, and with the slope $-(1-f)/f$ (Figure 2.10).

Continuous distillation with rectification

Rectification is a distillation process where some condensate from the overhead vapour is returned to the still where it comes into contact with the vapour.

Continuous distillation with rectification in distillation columns of a type used even today, was invented by Cellier Blumenthal in 1808.[1] France was cut off from supplies of cane sugar, and Cellier Blumenthal was attracted by a prize of 1 million francs offered by Napoleon for a good method of obtaining large quantities of white sugar from sugar-beet. The process involved the use of alcohol that had to be recovered.

Figure 2.11 is a schematic drawing of his column. Vapour from the reboiler bubbles through the liquid on the first tray and is partially condensed. The heat liberated produces vapour containing a higher fraction of the more volatile component. This process is repeated on each tray. The overhead vapour is liquefied, and the resultant condensate split in a fraction D withdrawn as distillate and L returned to the column as *reflux*.

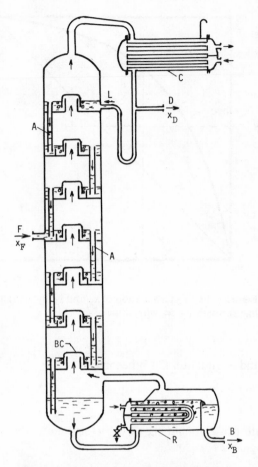

Figure 2.11 Continuous distillation column. Liquid flows down through the downcomers A, while vapour from the reboiler R passes through the risers under the bubble caps BC and bubbles through the liquid on each plate. The vapour from the top plate is liquefied in the condenser C. The condensate is split in two streams, the product (distillate) D and the reflux L. F is the feed and B the bottom product

Number of theoretical stages

McCabe–Thiele method. A *theoretical stage* or a theoretical plate is defined as a plate where the vapour leaving is in equilibrium with the liquid from the same plate. The first step in column design is calculation of the number of theoretical stages in order to give a certain separation. A later step includes an estimate of

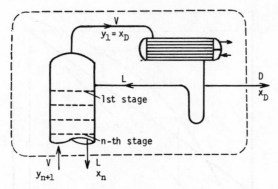

Figure 2.12   Material balance at top of column

plate efficiencies to determine the number of actual plates, or—for a packed column—the height equivalent to a theoretical plate, *HETP*.

The calculations are often based on the concept of *constant molal overflow*, i.e. that the number of moles of liquid entering a plate equals the number of moles of liquid leaving the same plate,—except where changed by additions or withdrawals of material from the column. In general this is almost correct if the molal latent heat of vaporization is approximately the same for all components. Changes in sensible heats of liquid and vapour, in heat of mixing, and the heat loss to the surroundings are usually small compared with the latent heats of vaporization. In addition the temperature of the liquid increases and the temperature of the vapour decreases as they pass a plate, and the two effects of the sensible heats may cancel. A difference in molal heats of vaporization may be taken care of by the use of a fictitious molecular weight or a 'latent heat unit'[14] (see Example 2.5).

Figure 2.12 is the top of a distillation column for a binary mixture. A material balance for the more volatile component gives

$$Vy_{n+1} = Lx_n + Dx_D$$

Thus

$$y_{n+1} = \frac{L}{V}x_n + \frac{D}{V}x_D \qquad (2.19)$$

where

$$V = L + D \qquad (2.20)$$

Equation (2.19) is called the equation for the *operating line* for the rectifying section of the column. It gives the relation between the composition of the vapour entering a plate and the composition of the liquid leaving the same plate. Here, y_{n+1} is a linear function of x_n, and the operating line is a straight line through the two points $(x = x_D, y = x_D)$ and $(x = 0, y = (D/V)x_D)$ (Figure 2.13). The equilibrium curve is plotted in the same diagram.

40

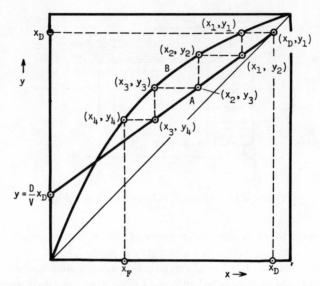

Figure 2.13   McCable–Thiele diagram for rectifying section;
A is the operating line and B the equilibrium curve

The mole fraction y_1 of the more volatile compound in the vapour from the top stage equals the mole fraction of the same compound in the distillate, x_D. Assuming ideal stages, the liquid composition on the top plate is located on the equilibrium curve with coordinate (x_1, y_1). The vapour from the second stage is located on the operating line, point (x_1, y_2). The liquid composition on stage 2 is on the equilibrium curve, point (x_2, y_2). This is continued step by step until the liquid composition on the feed plate is reached. If the feed is at its boiling-point, the composition at the feed plate equals the composition of the feed, x_F.

The number of ideal stages needed between the feed plate and the top plate is thus determined graphically, as shown in Figure 2.13, by counting the number of steps between the operating line A and the equilibrium curve B. The number of ideal stages may also include a fraction of a stage. Figure 2.14 gives a flow chart for computer calculation of the number of ideal stages by use of equations (2.15) and (2.19). The program is valid for intervals of x where α is sufficiently close to a constant.

Figure 2.15 is the bottom part of a distillation column for a binary mixture. A material balance for the more volatile component gives

$$V'y_m = L'x_{m+1} - Bx_B$$

or

$$y_m = \frac{L'}{V'}x_{m+1} - \frac{B}{V'}x_B \tag{2.21}$$

where

$$V' = L' - B \tag{2.22}$$

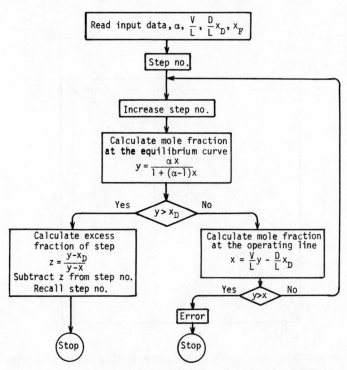

Figure 2.14   Flow chart for a program for calculating the number of theoretical stages in the rectifying section of a column for a binary mixture with constant relative volatility α. The second question ($x < y$?) is introduced to check that the first $x(x_F)$ is not below the point where the operating line crosses the equilibrium curve (not to the left of point x_G in Figure 2.22)

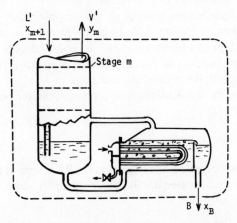

Figure 2.15   Material balance around the bottom of a distillation column cut off between stage $m + 1$ and stage m under the feed plate

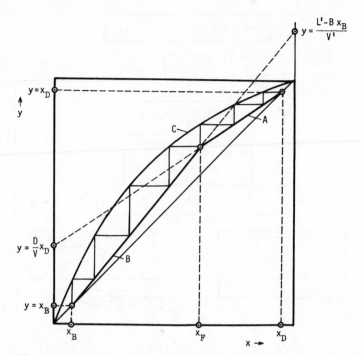

Figure 2.16 McCabe–Thiele diagram for the rectifying and the stripping section; A is the operating line for the rectifying section and B for the stripping section, C is the equilibrium curve. The stripping section has four theoretical stages including the reboiler

Equation (2.21) is the equation for the operating line for the stripping section of the column. It is a straight line through the point ($x = x_B$, $y = x_B$) and ($x = 1.0$, $y = (L' - Bx_B)/V'$), as shown in Figure 2.16.

The number of ideal stages needed in the stripping section is determined graphically, as shown in Figure 2.16. It is the number of steps between the operating line B and the equilibrium curve C, as the liquid mole fraction increases from x_B in the bottom product to x_F at the feed plate. Figure 2.16 shows four ideal stages. Of these the reboiler represents one stage.

A program for the same calculations on a programmable pocket calculator is given by Benenati.[15]

Reflux ratio

The slope L/V of the operating line (equation (2.19)) is termed the *internal reflux ratio*, R_{int}. This ratio is related to the *external reflux ratio*, $R = L/D$, by the equation

$$R = \frac{L}{V - L} = \frac{R_{int}}{1 - R_{int}} \tag{2.23}$$

In this chapter reflux refers to the external reflux ratio R.

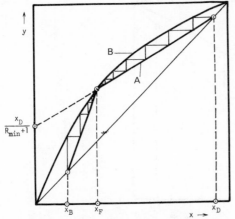

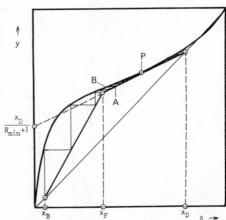

Figure 2.17 Operating line A with minimum reflux ratio and feed at its boiling point; B is the equilibrium curve

Figure 2.18 Operating line A with minimum reflux R_{min} that gives the pinched-in region P. The feed is at its boiling-point; B is the equilibrium curve.

In terms of reflux ratio, the operating line for the rectifying section intersects the ordinate at

$$y = \frac{D}{V} x_D = \frac{x_D}{R+1} \tag{2.24}$$

Minimum reflux ratio

A reduction of the reflux ratio reduces the heat and cooling requirements, but increases the number of theoretical stages needed to obtain a given separation. The minimum reflux ratio, R_{min}, is defined as the ratio that will require an infinite number of stages for the given separation. It is the ratio that gives an operating line that either crosses the equilibrium curve at the feed plate (Figure 2.17), or is a tangent to the equilibrium curve (Figure 2.18).

Total reflux

The minimum number of theoretical stages is obtained with total reflux, i.e. $R = \infty$. In this case, no product or practically no product is withdrawn. All the overhead vapour is condensed and returned to the top of the column as reflux. In equation (2.19) D is zero and L equals V, i.e. the diagonal $y = x$ becomes the operating line as shown in Figure 2.19.

For a binary mixture of components 1 and 2, equation (2.7) can be written as

$$\frac{(y_1)_n}{(y_2)_n} = \alpha \frac{(x_1)_n}{(x_2)_n}$$

where n is any ideal stage or ideal plate in the column. With total reflux ($L = V$, $D = 0$), equation (2.19) simplifies to $y_{n+1} = x_n$ or $y_n = x_{n-1}$. Substituting this in the

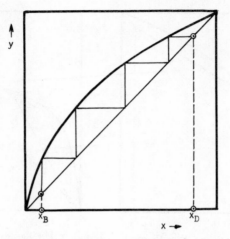

Figure 2.19   McCabe–Thiele   diagram
for total reflux

equation above, gives

$$\frac{(x_1)_{n-1}}{(x_2)_{n-1}} = \alpha \frac{(x_1)_n}{(x_2)_n}$$

Starting at the reboiler where $x_1 = (x_1)_B$ and $x_2 = (x_2)_B$ and going up stage by stage to the condenser where $x_1 = (x_1)_D$ and $x_2 = (x_2)_D$, the general equation then follows as

$$\frac{(x_1)_D}{(x_2)_D} = \alpha^{N_{min}} \frac{(x_1)_B}{(x_2)_B}$$

where N_{min} is the number of theoretical stages including the reboiler. This is the well-known *Fenske equation*,[16]

$$N_{min} = \frac{\log[(x_1/x_2)_D (x_2/x_1)_B]}{\log \alpha} \tag{2.25}$$

This equation is valid for total reflux, constant molal overflow, and constant relative volatility α. Minor variations in α are taken into account by the use of an average value,

$$\alpha_{av} = (\alpha_{top} \alpha_{bottom})^{1/2} \tag{2.26}$$

or

$$\alpha_{av} = (\alpha_{top} \alpha_{middle} \alpha_{bottom})^{1/3} \tag{2.27}$$

Optimum reflux ratio

The reflux ratio R_{opt} is that which gives the lowest sum of capital and operating costs. An increase in the reflux ratio causes increased operating cost. With liquid

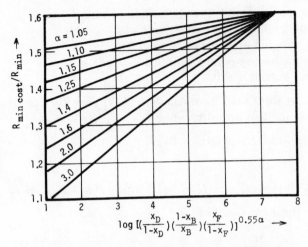

Figure 2.20 The reflux ratio that gives minimum column cost, $R_{\text{min cost}}$, divided by the minimum reflux ratio, R_{min}, as a function of distillate, bottoms and feed composition, and the relative volatility α

feed at its boiling-point, the operating cost is approximately proportional to $R + 1$. With the reflux ratio close to minimum the capital cost decreases with increased reflux, due to a rapid reduction in the number of theoretical stages. At higher reflux ratios, the capital cost increases again, due to increased column diameter and increased size of reboiler and condenser.

According to Van Winkle and Todd,[17] Figure 2.20 gives the reflux ratio $R_{\text{min cost}}$ that gives *minimum capital cost* (column and certain accessory equipment).

The optimum reflux ratio is somewhat lower than the ratio that gives minimum column cost, and usually less than $1.3R_{\text{min}}$. But it is difficult to operate a column with a reflux ratio close to minimum, and most columns are designed for reflux ratios between 1.2 and 1.7 times R_{min}. Even higher values are used in vacuum distillation (see also p. 56).

Feed-point location

The relationship between L and L' and between V and V' depends on the condition of the feed. The molar overflow below the feed point is

$$L' = L + qF \tag{2.28}$$

where

$$q = \frac{\text{energy to convert 1 mol of feed to saturated vapour}}{\text{molar heat of vaporization}} \tag{2.29}$$

The feed can be

—liquid at a temperature below the boiling-point, $q > 1.0$;
—liquid at the boiling-point ('saturated liquid'), $q = 1.0$;
—a mixture of liquid and vapour, $1 > q > 0$;
—vapour at the boiling-point ('saturated vapour'), $q = 0$;
—superheated vapour, $q < 0$.

Equation (2.21) subtracted from equation (2.19) gives the coordinates for the intersection of the two operating lines,

$$(V - V')y = (L - L')x + (Dx_D + Bx_B)$$

where

$$V - V' = (1 - q)F, \quad L - L' = -qF, \quad \text{and} \quad Dx_D + Bx_B = Fx_F$$

This gives the *q-line equation*,

$$y = \frac{q}{q-1}x - \frac{x_F}{q-1} \tag{2.30}$$

Equation (2.30) corresponds to a straight line through the point P at the diagonal in Figure 2.21 ($x = x_F, y = x_F$) with the slope $q/(q-1)$.

Wrong location of the feed increases the number of theoretical stages needed for a certain separation, and the number of stages approach infinity as the liquid

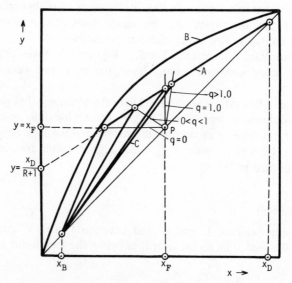

Figure 2.21 Effect of feed condition. The q-lines shown are for cold feed ($q > 1.0$), saturated liquid ($q = 1.0$), mixed feed ($1 > q > 0$), and saturated vapour ($q = 0$). Here A is the operating line of the rectifying section, B the equilibrium curve, and C the operating line of the stripping section with cold feed

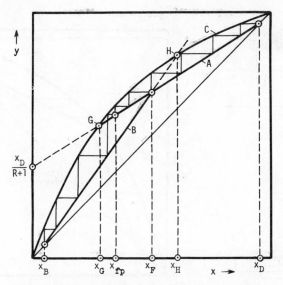

Figure 2.22 Effect of location of feed point. Here A and B are the operating lines and C the equilibrium curve. The diagram is plotted for a binary mixtures with relative volatility $\alpha = 3.0$, the feed at its boiling-point, $x_F = 0.5$, $x_D = 0.95$, $x_B = 0.05$, and $R = 2R_{min}$. The feed point is at $x_{fp} = 0.35$ which gives approximately 1.1 theoretical stage more than the feed point at the intersection of the operating lines where $x = 0.5$. The feed point must be in the interval $x_G < x_{fp} < x_H$ in order to give a finite number of stages

composition at the feed point approaches the intersection between one of the operating lines and the equilibrium curve (Figure 2.22).

Multiple feeds and sidestreams

Multiple feeds and sidestreams change the operating line. Figure 2.23 shows a column with a liquid sidestream. A material balance around the top gives

$$V_s y_{n+1} = L_s x_n + S x_s + D x_D$$

or

$$y_{n+1} = \frac{L_s}{V_s} x_n + \frac{S x_s + D x_D}{V_s} \tag{2.31}$$

This is the equation for the operating line between the feed and the sidestream. Figure 2.24 is the McCabe–Thiele diagram for the column (Figure 2.23) with a liquid sidestream ($V = V_s$), and Figure 2.25 the diagram for the same column with the sidestream withdrawn from the vapour phase ($L_s = L$, $V_s = V - S$). In this

48

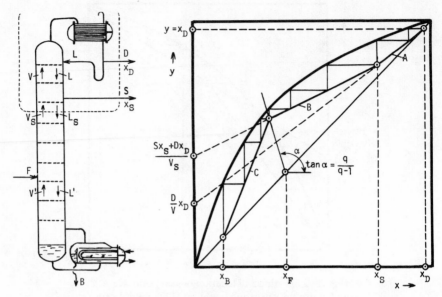

Figure 2.23 Column with liquid sidestream

Figure 2.24 Effect of a liquid sidestream $S \approx D$. The slopes of the operating lines are L/V for A, and L_s/V for B

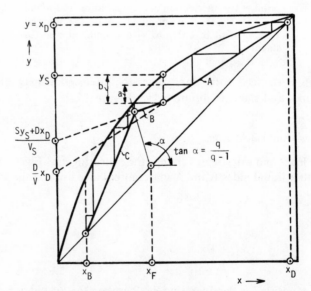

Figure 2.25 Effect of vapour sidestream in a column with the same boundary conditions and the same total number of theoretical stages as in Figure 2.24. It is 3.7 theoretical stages above the sidestream ($a/b \approx 0.7$)

example withdrawal of vapour instead of liquid reduces the reflux ratio from 2.7 to 1.6 without an increase in the number of theoretical stages.

Direct steam distillation

In some cases in which the bottom product is water, and some where the distilled mixture is immiscible with water, steam is introduced directly into the still. As an example Figure 2.26 shows a still for an ethanol–water mixture with direct steam injection instead of a reboiler. A material balance for ethanol around the bottom part gives equation (2.21). But here $V' = S$ and $L' = B$, giving the operating line of the stripping section

$$y_m = \frac{B}{S}x_{m+1} - \frac{B}{S}x_B \tag{2.32}$$

This line has the slope B/S; but at $x = y$, $x = Bx_B/(B - S)$ instead of x_B. Figure 2.27 gives the lower part of the x–y diagram.

Direct steam is also used in vacuum distillation of high-boiling organic materials which would decompose if they were distilled directly at atmospheric pressure. The steam acts as an inert that reduces the partial pressure and thereby the temperature of the organic compounds. It also dilutes oxygen from air leaks, thus reducing oxidation and coke formation in the column. Direct steam is also used to obtain agitation in liquids with poor heat-transfer characteristics. An example is the distillation of fatty acids from tall oil.

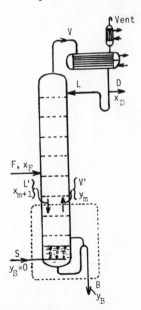

Figure 2.26   Distillation column for ethanol water with direct steam

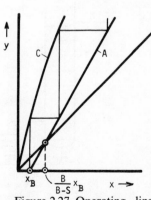

Figure 2.27 Operating line A for steam distillation: C is the equilibrium curve

Partial condensation

In some cases it is an advantage to use a partial condenser. The 'distillate' leaves the condenser as vapour that may be almost in equilibrium with the condensate. The condensate is returned to the column as reflux (see Example 2.5). A partial condenser is also called a dephlegmator.

Azeotropic distillation

Heterogeneous azeotropes deviate considerably from Raoult's law, and partial immiscibility can also occur with two liquid phases at the boiling temperature. At moderate pressures the vapour pressure of each of the liquid phases is independent of the presence of the other liquid phase.

The separation of a heterogeneous azeotrope can be carried out in two distillation columns, as shown in Figure 2.28 for a mixture of n-butanol and water. Feed with water content $x < x_B$ in Figure 2.6 is introduced to the column on the left, feed with $x_B < x < x_C$ is introduced to the decanter, and feed with $x > x_C$ to the column on the right.

Many *homogeneous azeotropes* and mixtures with relative volatility close to 1.0 over a wide range can be separated by ternary distillation where the third component changes the deviations from Raoult's law.

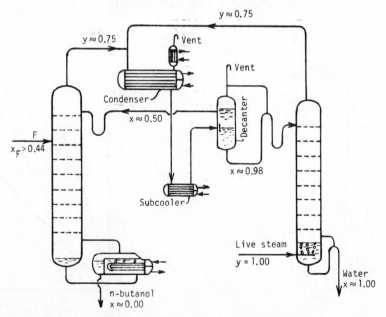

Figure 2.28 Azeotrope distillation of the heterogeneous azeotrope n-butanol–water; x and y are mole fractions of water (see equilibrium diagrams, Figures 2.6 and 2.7)

Summary of calculations

The following is a summary of the calculations of the number of theoretical stages and feed-point location for the column (Figure 2.11) with given composition and enthalpy of the feed F, and given composition of the distillate D and the bottom product B. The letters refer to Figure 2.29:

1. Plot the equilibrium curve E and the diagonal $y = x$.
2. Calculate the slope of the q-line, $\tan \alpha = q/(q - 1)$, and plot the q-line through point P $(x = x_F, y = x_F)$.
3. Determine the minimum reflux ratio R_{min} from the intersection with the ordinate of a straight line through $(x = x_D, y = x_D)$ and the intersection of the q-line with the equilibrium curve. The intersection with the ordinate is $x_D/(R_{min} + 1)$.

    Select a reasonable value of the reflux ratio R, for example 1.3 times R_{min}.
4. Plot the operating line A for the rectifying section as a straight line through the points $(x = x_D, y = x_D)$ and $(x = 0, y = x_D/(R + 1))$.
5. Plot the operating line C for the stripping section as a straight line through $(x = x_B, y = x_B)$ and the intersection of the q-line with the operating line A, point F. This intersection is also the feed point.
6. Starting at the points (x_D, x_D) and (x_B, x_B), plot the step-curves between the operating lines and the equilibrium curve and count the number of steps (= number of theoretical stages) until the feed point F is reached.

The graphical technique with the $x-y$ diagram was developed by McCabe and Thiele in 1925. Despite its age, it is still a useful tool for the calculation of stage-

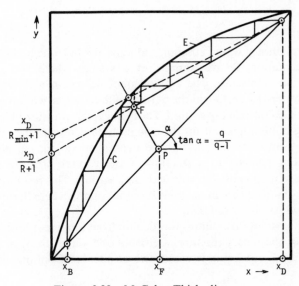

Figure 2.29   McCabe–Thiele diagram

wise processes. In distillation, the influence of reflux ratio, composition of products, and feed condition can readily be seen by adjusting the operating lines and q-lines.

Ponchon[18] and Savarit[19] used a concentration–enthalpy diagram that takes into account deviations from constant molal overflow. Their method is described in most standard texts on distillation. The *Ponchon–Savarit method* is mainly used for binary mixtures with large heats of mixing, such as aqueous solutions of the ionic compounds HCl, HF, and NH_3 for which adequate enthalpy diagrams are available.

Distillation of ternary mixtures

Miscible systems

For component i, material balances, as shown in Figures 2.12 and 2.15, give

$$(y_i)_{n+1} = \frac{L}{V}(x_i)_n + \frac{D}{V}(x_i)_D \tag{2.33}$$

for the rectifying section of the column, and

$$(y_i)_m = \frac{L'}{V'}(x_i)_{m+1} - \frac{B}{V'}(x_i)_B \tag{2.34}$$

for the stripping section of the column.

The calculation of the necessary number of theoretical stages can be carried out as a stage-by-stage calculation, also called the *Lewis–Matheson method*[20] (see Example 2.7, p. 89).

Immiscible systems

The ternary mixture of water, ethanol, and amyl alcohol (one of the 'fusel oils' in alcohol fermentation) is an example of a system where two of the pure components are immiscible. The amyl alcohol will occur as a separate liquid phase in the water-rich section at the bottom of the column, while it is completely soluble in the alcohol-rich top section. The high vapour pressure of the pure amyl alcohol at the bottom forces it up the column until it reaches the point where it dissolves in a homogeneous liquid phase. From there on, the partial pressure of amyl alcohol is reduced to a fraction of the vapour pressure of the pure component, and it flows downwards with the overflow. This results in a build-up of amyl alcohol somewhere in the middle of the column where the amyl-alcohol-enriched liquid can be withdrawn.

In practice, distillation columns for alcohol from fermentation are calculated as columns for the binary mixture of ethanol and water. In operation of such columns 'fusel-oil'-rich liquid is withdrawn at intervals from one of the middle plates.

Calculation procedures for partly miscible liquids and three examples are given by Hegner and Block.[21]

Azeotropic and extractive distillation

Binary, close-boiling mixtures and mixtures forming a homogeneous azeotrope can often be separated in distillation by the addition of a suitable third component.

In *azeotropic distillation* the third component has a volatility comparable to that of the feed mixture. It forms an azeotrope with one or both of the feed components. It has also to be chosen so that the new azeotrope gives two liquid layers after it is cooled. The principle is shown in Figure 2.30 where the feed A + B

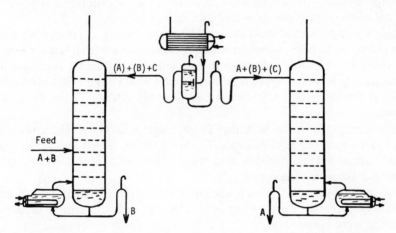

Figure 2.30 Azeotropic distillation. Parentheses around a letter indicate a small amount

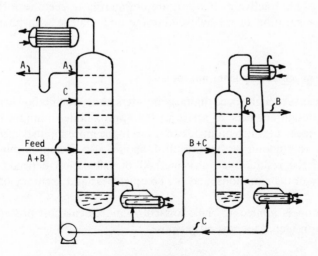

Figure 2.31 Extractive distillation of a close-boiling or azeotropic mixture A + B with solvent C

is an azeotrope, and the third component C, called the *solvent* or the *entrainer*, is added at the top. Parentheses around letters indicate small amounts.

Azeotropic distillation is used for dewatering of azeotropic ethanol–water mixtures with benzene or cyclohexane as entrainer.

Extractive distillation is a process where a high-boiling solvent is added to alter the relative volatilities of the compounds in the feed, as shown schematically in Figure 2.31.

The mixture A + B is the feed to the first column, and the solvent C is added to reduce the volatility of B. The solvent that usually amounts to 65 to 90 mol per cent of the liquid within each stage, is normally introduced a few stages below the top stage in order to reduce the amount of solvent in the distillate. The solvent C and the component B are recovered in a second column.

Stage-by-stage calculations can be carried out, starting at the bottom where the amounts of the heavy solvent and components from the fresh feed are known. The overflow rate of a solvent with low volatility is almost constant. The concentration, however, will change abruptly at the feed point if the fresh feed is a liquid. A vapour feed is sometimes used to avoid dilution of the descending solvent.

With a binary feed, the McCabe–Thiele diagram can be used for short-cut calculations with the equilibrium curve being drawn at the constant solvent concentration. If a liquid feed is used, there will be different equilibrium curves above and below the feed point.

A stage-by-stage calculation is shown in Perry[10], p. 13–45, and a graphical solution is given by Torres-Marchal[22] together with calculation of ternary equilibria from binary data.[23]

If a suitable solvent can be found, extractive distillation is usually to be preferred to azeotropic distillation.

Examples of extractive distillation are dewatering of acetone with ethylene glycol and separation of methylcyclohexane and toluene with phenol as the solvent.

Distillation of multicomponent mixtures

Even with access to high-speed digital computers, short-cut methods are useful to enable the design engineer to arrive at the approximate optimum number of theoretical stages, at which point he can resort to the more sophisticated, iterative calculations to pinpoint his final result. It should also be kept in mind that the accuracy of the results from any method of calculation depends upon the applicability of the equations, and the completeness and accuracy of the input data.

This section is limited to the short-cut, hand calculation procedures that should be considered for several purposes:[24]

1. Scoping studies suitable for preliminary costs.
2. Evaluation of operating variables.

3. Separations having coarse purity requirements (i.e. contaminants > 0.5 per cent by weight).
4. Detailed designs for ideal and close to ideal systems.
5. Designs for systems for which equilibrium data are unavailable.

Rigorous design procedures should be applied if the following are valid:

1. High product purity is required.
2. The system is highly non-ideal and good equilibrium data are available.
3. The relative volatility between key components is less than 1.3.
4. One or more of the components is near the critical pressure.

Key components

Few true binary systems are encountered in industry, and many calculation procedures are based on pseudo-binary mixtures in which two principal components are designated as the light key and the heavy key. In this section, the keys are components boiling adjacent to each other on the temperature scale, excluding components present only in small amounts. Most of the light key is recovered in the distillate, and most of the heavy key in the bottoms.

Short-cut calculations

In some cases, the first rough estimate may be the calculation of a binary system where all low-boiling components are assumed to be the light key and all high-boiling components the heavy key.

A better approach is based on the calculation of a minimum number of theoretical stages and a minimum reflux ratio, using the Fenske and Underwood equations. Both are based on constant molal overflow and constant relative volatilities.

The Fenske equation is the same as for a binary mixture,[16]

$$N_{min} = \frac{\log\left[\left(\frac{x_{LK}}{x_{HK}}\right)_D \left(\frac{x_{HK}}{x_{LK}}\right)_B\right]}{\log \alpha_{LK-HK}} \tag{2.35}$$

where N_{min} = minimum number of theoretical stages (total reflux), x_{LK} and x_{HK} = liquid mole fractions of light key and heavy key, and α_{LK-HK} = average relative volatility, light to heavy key, see equations (2.26) and (2.27). At the total reflux, the distribution of the other components is determined by the same equation,[25]

$$\log(x_D/x_B)_i = N_{min} \log \alpha_i + \log(x_D/x_B)_{HK} \tag{2.36}$$

where α_i is the volatility of the i-th component relative to the heavy key.

The Underwood equation[26,27] gives the minimum reflux ratio,

$$R_{min} + 1 = \sum \frac{(\alpha_i x_i)_D}{\alpha_i - \theta} \tag{2.37}$$

Table 2.1 Range of optimum reflux ratio and optimum number of stages

	R_{opt}/R_{min}	N_{opt}/N_{min}
Low-level refrigeration ($< -100\,°C$)	1.05–1.1	2.5–3.5
High-level refrigeration	1.1–1.2	2.0–3.0
Water- and air-cooled condensers	1.2–1.3	1.8–2.5

where x_i in the distillate can be approximated by the number for total reflux, and θ is the Underwood constant. It is calculated by a straightforward iteration of equation (2.38). The value ranges between α_{LK} and 1.0. Other values of θ are discarded.

$$1 - q = \sum \frac{(\alpha_i x_i)_F}{\alpha_i - \theta} \tag{2.38}$$

where q is given by equation (2.29) and is 1.0 for a liquid feed at its bubble point, and zero for vapour at its dew-point. Subscripts D and F refer to distillate and feed. The difference $(\alpha_i - \theta)$ can be small, and θ should be determined with four decimals (see example 2.7, p. 89).

In Table 2.1 Frank[24] reports reasonably good estimates of optimum reflux ratio, R_{opt} (minimum sum of energy and capital costs), and the corresponding values of the optimum number of theoretical stages, N_{opt}.

The Gilliland empirical correlation for the number of theoretical stages N as a function of R, R_{min}, and N_{min} for short-cut calculations is given in Figure 2.32.

The feed plate location can be estimated by the Kirkbride[29] empirical equation,

$$\frac{N_r}{N_s} = \left[\left(\frac{x_{HK}}{x_{LK}} \right)_F \left(\frac{x_{LK,B}}{x_{HK,D}} \right)^2 \frac{B}{D} \right]^{0.206} \tag{2.39}$$

where N_r is the number of theoretical stages in the rectifying section above the

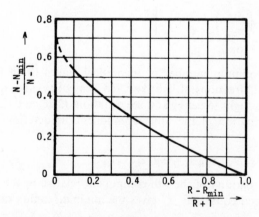

Figure 2.32 Gilliland empirical solution of theoretical stages, based on data for binary and multicomponent hydrocarbon mixtures[28]

feed point, and N_s the number of theoretical stages in the stripping section below the feed point.

Summary of short-cut calculations

1. Select the two key components and calculate the minimum number of theoretical stages, N_{min}, equation (2.35), using average relative volatility, equations (2.26) or (2.27).
2. Calculate distillate and bottoms compositions at total reflux, equation (2.36).
3. Calculate the Underwood constant θ from equation (2.38) and minimum reflux ratio R_{min} from equation (2.37), using the distillate composition at total reflux.
4. Choose a reasonable value of R/R_{min}. Table 2.1, indicates the range of optimum reflux ratio, and Figure 2.20 the reflux ratio that gives minimum column cost, using

$$\log\left[\left(\frac{x_{LK}}{x_{HK}}\right)_D \left(\frac{x_{HK}}{x_{LK}}\right)_B \left(\frac{x_{LK}}{x_{HK}}\right)_F\right]^{0.55\alpha}$$

as the abscissa.

Table 2.2   Some common types of contacting devices

Staged tray columns (separate liquid and vapour flow paths)

Common types	Fig.	Proprietary types
Bubble cap	2.33	Angle
Sieve	2.34	Uniflex
Valve	2.35	Montz
		Linde
		Jet

Differential columns (countercurrent flow on packing or tray surface)

Randomly packed towers	Fig.	Systematically packed towers
Raschig or partition rings	2.36	Flexipac
Saddles	2.36	Goodloe
Slotted rings		Hyperfil
Tellerettes		Sulzer
		Glitsch Grid
		Leva film trays

Pseudo-equilibrium stages (countercurrent flow across discrete trays)

Downcomerless trays	Low-pressure drop trays	Fig.
Perforated	Disc and doughnut (shower deck)	2.37
Turbogrid	Sheds	2.38
Ripple		

58

5. For hydrocarbon mixtures obtain the number of theoretical stages, N, from Figure 2.32.
6. Using equation (2.39), estimate the feed-point location.

Even short-cut calculations are time-consuming if they are carried out by hand, and repeated calculations can preferably be carried out by a calculator program. Kesler[30] gives a 474 steps program for a TI-59 calculator.

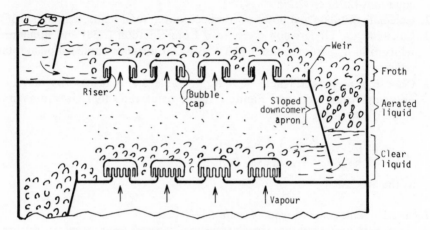

Figure 2.33 Bubble cap column with apron downcomers

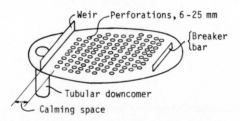

Figure 2.34 Sieve tray. Holes usually placed in triangular pitch. Common hole diameter 4 to 13 mm. Smaller holes lessen weeping and reduce entrainment. Larger holes up to 25 mm are used in fouling service

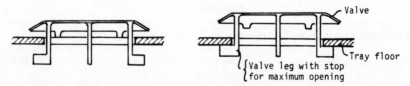

Figure 2.35 Valve, to the left in the lowest (initial) position and to the right in the highest position

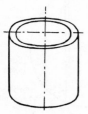

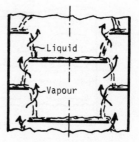

Figure 2.36 Raschig ring and ceramic super Intalox saddle

Figure 2.37 Disc and doughnut shower deck

Table 2.3 Selection guide for distillation column internals. Key: 0—Do not use; 1—evaluate carefully; 2—usually applicable; 3—best selection.[24]

	Staged columns		Differential columns		Pseudo-equilibrium	
	Perforated or valve trays	Bubble cap or tunnel trays	Randomly packed	Systematically packed	Downcomerless	Disc and doughnut
Low pressure (< 100 mm Hg)	2	1	2	3	0	1
Moderate pressure	3	2	2	1	1	1
High pressure (> 50% of critical)	3	2	2	0	2	0
High turndown ratio[†]	2	3	1	2	0	1
Low liquid rates	1	3	1	2	0	0
Foaming systems	2	1	3	0	2	1
Internal tower cooling	2	3	1	0	1	0
Solids present	2	1	1	0	3	1
Multiple feeds and sidestreams	3	3	1	0	2	1
High liquid rates (scrubbing)	2	1	3	0	3	2
Small dia. columns	1	1	3	2	1	1
Columns with dia. 1–3 m	3	2	2	2	2	1
Corrosive fluids	2	1	3	1	2	2
Viscous liquids	2	1	3	0	1	0
Low Δp (efficiency no concern)	1	0	2	2	0	3
Low cost (performance no concern)	2	1	2	1	3	3

[†] The turndown ratio is the ratio between maximum and minimum operating rates.

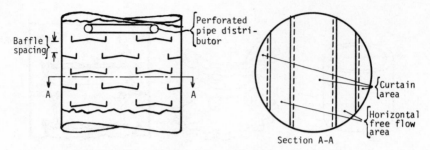

Figure 2.38 Sheds. The top baffles under the distributor have notched weirs. In non-fouling service notched weirs are also on the other baffles

Distillation columns

Over the last half-century, a large number of column internals for vapour or gas and liquid contacting have been developed. Some of the better-known ones are listed in Table 2.2.

Table 2.3 and Figure 2.39 give some general guidelines regarding the suitability of different column internals, depending on the service.

For low and moderate pressures (< 10 bar) the free cross-sectional area of the column (the total column area minus the down-comer area) can be quickly estimated from the simple F-factor,[24]

$$F = V_f \sqrt{\rho_v} \tag{2.40}$$

where V_f = vapour velocity referred to the free cross-sectional area between the trays, and m/s, ρ_v = vapour density, kg/m^3.

Tray columns

Figure 2.40 gives the F-factor in equation (2.40) for non-foaming systems. For foaming systems, this factor should be multiplied by 0.75.[25]

Columns should always be designed for operation at a safe distance from the *flooding* point. Flooding is caused by excessive vapour flow, the *entrainment flooding*, or excessive liquid flow, the *downflow flooding*. Flooding is observed by increased liquid hold-up in the column, and increased pressure drop. Also, control becomes difficult.

Minimum capacity depends on the type of trays. A crossflow sieve plate can be operated at reduced vapour flow down to the point where liquid drains down through the perforations and gas dispersion is inadequate. Valve trays can be operated at lower vapour rates because of valve closing. Bubble cap trays can be operated at even lower vapour rates.

In Perry,[10] p. 18–7, Fair gives the flooding vapour velocity.

$$V_f = C_f \gamma^{0.2} \left(\frac{\rho_1 - \rho_g}{\rho_g} \right)^{0.5} \tag{2.41}$$

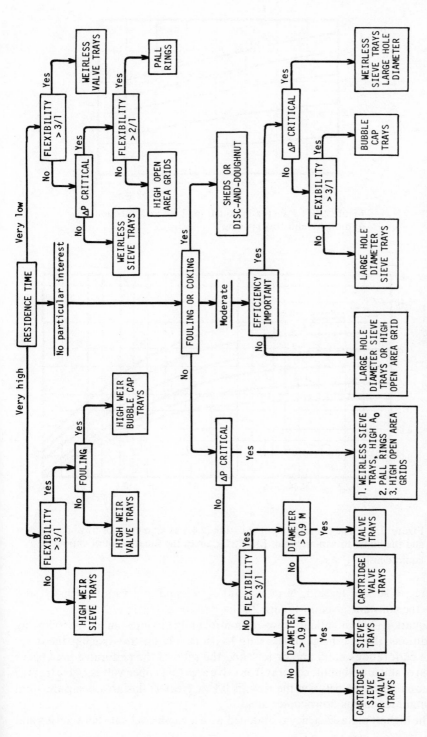

Figure 2.39 Internals selection diagram for fractionation columns. Published by permission from Exxon Research and Engineering Company

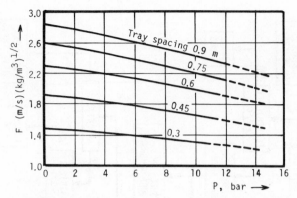

Figure 2.40 F-factor in equation (2.40) as a function of pressure and tray spacing[24]

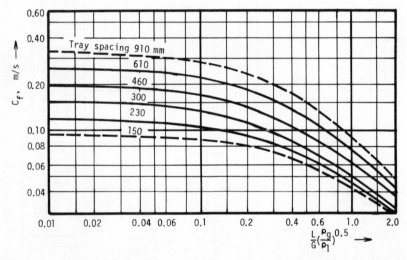

Figure 2.41   The constant C_f in equation (2.41) as a function of tray spacing and the liquid to vapour ratio, L/G kg/kg, times the square root of vapour to liquid density, $\sqrt{\rho_g/\rho_l}$[10]

where γ = surface tension, N/m, ρ_1 and ρ_g = liquid and vapour density, and C_f = the constant given in Figure 2.41.

Equation (2.41) is valid for low- or non-foaming mixtures under the following conditions: the weir height is less than 15 per cent of the tray spacing, the sieve-plate perforations are 6 mm or less, and the ratio of the perforated area (sieve trays) or slot area (bubble cap trays) or valve area (fully open valves, valve tray) to the total flow area between the trays is 0.1 or greater; this area being the total column area minus downcomer area.

The highest plate efficiency is obtained with a vapour velocity between 0.4 and

Table 2.4. Constant a in equation (2.42)[31]

Service	a
Non-foaming, regular systems	1.00
Fluorine systems, e.g. BF$_3$, Freon	0.90
Moderate foaming, e.g. oil absorbers, amine and glycol regenerators	0.85
Heavy foaming, e.g. amine and glycol absorbers	0.75
Severe foaming, e.g. methylethylketone units	0.60
Foam-stable systems, e.g. caustic regenerators	0.3–0.6

0.85 times the flooding velocity. Reasonable design values for low and non-foaming mixtures are in the range 0.6–0.7 times flooding velocity.

The cross-section of the column can be estimated from the equation

$$V = a\varphi V_f \tag{2.42}$$

where V = the superficial vapour velocity, m/s, referred to the total column area minus the downcomer area, φ = fraction of flooding velocity, usually between 0.7 and 0.85, and V_f = flooding velocity from equation (2.41).

The downcomer area may be obtained from the one of the following equations that gives the smallest value of the design liquid velocity in the downcomer,[31]

$$V_d = 0.17a \tag{2.43}$$

$$V_d = 0.007a\sqrt{\rho_1 - \rho_g} \tag{2.44}$$

$$V_d = 0.044a\sqrt{(\rho_1 - \rho_g)t_s} \tag{2.45}$$

where V_d = liquid velocity in downcomer referred to the total downcomer area, m/s, a = system factor (Table 2.4), (with foam-stable systems use the lowest value, $a = 0.3$), $\rho_1 - \rho_g$ = liquid density minus vapour density, kg/m^3, and t_s = tray spacing, m.

The liquid flow over the tray is usually crossflow, as shown in Figure 2.42a. Reverse flow can be used for small liquid rates (Figure 2.42b) and multipass flow for larger columns and large liquid flow rates (Figure 2.42c). Figure 2.43 is a guide to the selection of type of flow in order to obtain a good liquid distribution over the tray. An inlet weir or breaker bar as shown in Figure 2.34 helps to give a more uniform liquid distribution and to collect rust and sediment.

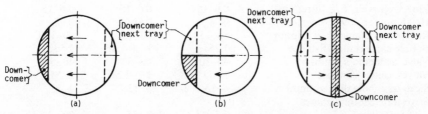

Figure 2.42   (a) Tray with crossflow. (b) Tray with reversed liquid flow. (c) Double-pass tray

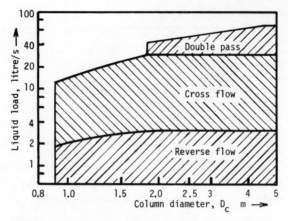

Figure 2.43 Guide to the selection of type of flow as a function of column diameter in metres and liquid flow in litre/s[32]

The height of the outlet weir should not exceed 0.15 times the tray spacing in order to avoid jetting and entrainment. For vacuum and most sieve and valve trays, a 25 mm weir seems to be satisfactory. The downcomer area is usually around 15 per cent of the column area. Sloped downcomers (Figure 2.33) aid froth collapse at the top of the downcomer. It also leaves the maximum active area on the tray below. A reasonable distance between the downcomer apron and the tray deck is half the weir height or 20 mm, whichever is greater. The liquid flow rate through the opening under the downcomer skirt should be less than 0.3 m/s.

A summary of some design data is given in Table 2.5. Further guidelines for the

Table 2.5 Some common design data for tray columns with downcomers.[35-37] D_c = column diameter (m), L = liquid flow rate (litre/s)

	Rectification, vacuum	Rectification, atmospheric pressure	Rectification under pressure, and absorption
Tray spacing, t_s (m)	0.4–0.6	0.4–0.6	0.3–0.4
Weir length, l_w (m)	0.0–0.6D_c	0.6–0.75D_c	0.85D_c
Weir height, h_w (mm)	20–30	25–70	40–100
Flow over weir, L/l_w (litre/(s m))	0.6–15	0.6–15	0.6–15
Height under downcomer skirt	0.7h_w	0.8h_w	0.9h_w
Bubble cap diameter, d_{cap} (mm)	80–150	80–150	80–150
Bubble caps, centre distance	1.25d_{cap}	1.25–1.4d_{cap}	1.5d_{cap}
Valve diameter, d_v (mm)	40–50	40–50	40–50
Valves, centre distance	1.5d_v	1.7–2.2d_v	2–3d_v
Sieve tray, hole dia., d_s (mm)	4–13	4–13	4–13
Sieve tray, centre distance of holes (mm)	2.5–3d_s	3–4d_s	3.5–4.5d_s
Hole area/tray area (m²/m²)	0.1–0.15	0.06–0.10	0.045–0.075

design and layout are given in reference 33, and useful guidelines for successful start-up and eliminating potential trouble spots in reference 34.

Bubble caps

Bubble cap trays were the most common distillation column internals in new installations in the chemical and petroleum industries until the mid 1950s. Today they are installed mainly in columns with extremely low liquid rate (< 0.4 litre/s and per m average flow width), and in columns with extremely high operating range (> 5:1).

The primary reason for this shift is the high cost of bubble cap trays which are two to three times the cost of equivalent sieve trays. However, the cost of a complete column with bubble cap trays is only approximately 15 per cent higher than the cost of a similar column with sieve trays.

Sieve trays

Sieve trays have the following advantages:

—Installed costs are the lowest of all types of trays.
—Fouling tendency is low (with large holes).
—Design procedures are well known.
—With proper design, efficiency is good.

Sieve trays are not suitable where the pressure drop must be less than 2.5 mm Hg/tray. The same is the case when liquid rates are less than 0.4 litre/s and per metre average flow width, or where turndown ratios are higher than 3:1 at high pressure and 2:1 at low pressure.

The holes are usually arranged in a triangular pattern with rows normal to liquid flow and with dimensions as given in Table 2.5. Small holes lessen weeping and reduce entrainment in low-pressure columns. Large holes (20 and 25 mm) are used in fouling service. The first row of perforations should be 50 to 75 mm from both inlet downcomer and outlet weir.

The pressure drop is lowest if the perforations are punched upwards to create a nozzle effect. Despite this advantage, the trays are often installed with the holes punched downwards in order to reduce the risks presented by jagged edges to personnel installing or inspecting the trays.

For low and atmospheric pressure, the weir height is usually 25 to 50 mm, and for high-pressure columns, up to 100 mm. Variations in liquid height over the trays can be neglected for columns less than 2.5 m diameter.

The most critical variable in sieve-tray design is open-hole area. Too large an area causes weeping or even dumping, while too small an area causes high-pressure drop and in extreme cases also jetting. Frank[24] recommends the following procedure: the vapour velocity through the holes V_h, is estimated by equation (2.40) where V_f is substituted by V_h and $F = 13$ for vacuum columns, 16 for atmospheric and moderate-pressure columns, and 18 for high-pressure

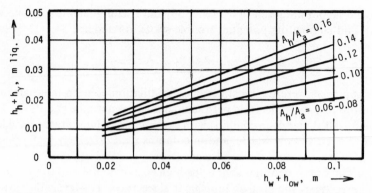

Figure 2.44 Sieve-tray parameters. Here A_h/A_a = ratio of hole area to active column area, h_w = height of weir, m, h_{ow} = liquid height over weir, m, h_h = pressure drop in dry column, m liquid, equation (2.47), h_γ = resistance due to liquid surface tension, m liquid $h_\gamma = 0.4\gamma/(\rho_1 d_h)$ with surface tension γ in N/m, liquid density ρ_1 in kg/m^3, and hole diameter d_h in metres[24]

columns. This gives the hole area,

$$A_h = GV_h/\rho_v \qquad (2.46)$$

where G = vapour rate, kg/s, and ρ_v = density of vapour, kg/m^3.

This result is checked against the curves in Figure 2.44 where A_a is the active area, i.e. the column area minus the area of the inlet downcomer and the area outside the outlet weir. An operating point above the line for A_h/A_a is considered a safe design with less than 25 per cent weeping. A point below the line may be in some doubt, but experience has shown that if the operating point falls above the lowest line ($A_h/A_a = 0.06$–0.08), the column will most probably operate within acceptable efficiency limits.

The pressure drop across the dry holes expressed in metres liquid head is given by the equation

$$\Delta p = C\rho_v \frac{V_h^2}{2}$$

or

$$h_h = C\frac{\rho_v}{\rho_1}\frac{V_h^2}{2g} \quad \text{m liquid head} \qquad (2.47)$$

where V_h = vapour velocity in the holes, m/s, and C = discharge coefficient taking into account losses due to contraction of the streamlines and outlet loss (Figure 2.45).

The head h_{ow} of liquid over the weir is calculated by the equation,[36]

$$h_{ow} = 0.7(Q/l_w)^{2/3} \quad \text{m liquid head} \qquad (2.48)$$

where Q = liquid flow, m^3/s, and l_w = weir length, m.

The sieve-tray pressure drop is the sum of the pressure drop in the

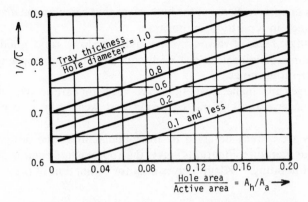

Figure 2.45 The reciprocal of the square root of the orifice coefficient for sieve trays.[36] Normal plate thickness is 2 mm for stainless steel and 2.5–3 mm for carbon steel

perforations and the hydrostatic head of aerated liquid on the tray,

$$h_t = h_h + h_1 \quad \text{m liquid head} \tag{2.49}$$

with

$$h_1 = \beta(h_w + h_{ow}) \quad \text{m liquid head} \tag{2.50}$$

where β is the aeration factor (Figure 2.46).[24]

The normal total pressure drop for a sieve tray in operation is from 40 to 130 mm water gauge or from 370 to 1250 Pa. Outside these limits, trays may weep excessively in vacuum service or jet in pressure columns. In low-pressure columns, total pressure drops 15 to 25 per cent higher than calculated have been observed.

Valve trays

Valve trays have liftable caps or discs (Figure 2.35) that act as variable orifices by adjusting themselves to changes in vapour flow. The movement is limited by

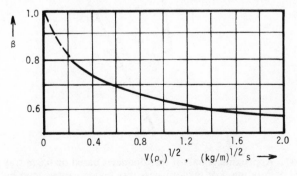

Figure 2.46 Aeration factor in equation (2.50) as a function of $V(\rho_v)^{0.5}$ with V in m/s and ρ_v in kg/m^3

68

retaining lugs or spiders. Some of the first valves that came on the market had a tendency to stick in a partly open position.

The main advantages of valve trays are high turndown ratio, good efficiency over a larger range of flow rates, and a fairly constant pressure drop across a large part of the operating range.

Disadvantages are: higher costs than sieve trays (in carbon steel 40 per cent higher than a sieve tray),[10] and corrosion and wear may give mechanical problems.[25]

Valve trays are usually designed by the manufacturer, and design data are given in vendor literature. Figure 2.47 gives the approximate column diameter at 80 per cent of flooding.[31] Klein[38] gives data for estimation of pressure drop.

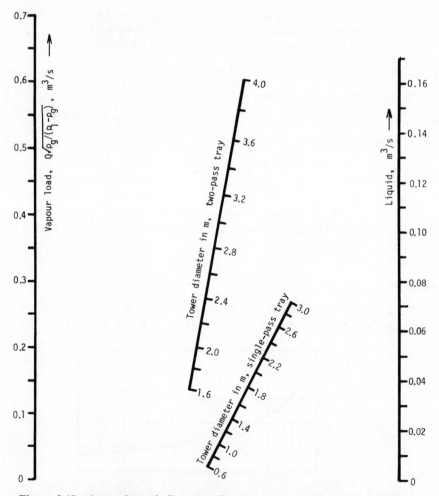

Figure 2.47 Approximate ballast tray diameter based on 0.6 m tray spacing and 80 per cent of flooding. For the four-pass tray divide vapour load by 2 and liquid load by 2 and obtain the diameter from the two-pass tray line. Multiply this diameter by $\sqrt{2}$. Reproduced by permission of Glitsch Inc., Dallas, Texas

Downcomerless trays

In downcomerless trays the liquid and vapour pass through the same openings. The simplest possible design is a sieve tray without downcomers.

Straight, downcomerless trays have an operating range (turndown ratio) which is considerably smaller than that of trays having downcomers. They may be considered when operating conditions are not expected to vary for columns with solid-containing liquids, and when ease of cleaning is important. For design procedures see the literature.[25]

Variations of downcomerless trays are Turbogrid with a flat grating instead of perforations, and Kittle trays whose surface imparts directional flow to the vapour.

Plate efficiencies

In order to transfer the number of theoretical equilibrium stages into the number of actual trays or plates, the plate efficiency has to be known.

The *overall plate efficiency* is defined as

$$\eta_0 = \frac{\text{number of theoretical stages}}{\text{actual number of trays}} \qquad (2.15)$$

With the nomenclature in Figure 2.48, the *Murphree efficiency* is defined as

$$\eta_M = \frac{y_n - y_{n+1}}{y_n^* - y_{n+1}} \qquad (2.52)$$

where y_n = mole fraction in the vapour leaving plate n, y_{n+1} = mole fraction in the vapour entering plate n, and y_n^* = mole fraction in the vapour in equilibrium with the liquid leaving plate n with composition x_n.

The liquid composition varies as the liquid flows over the tray, and the Murphree efficiency may exceed 1.0.

Murphree local or *point efficiency* is defined by the equation

$$\eta_{Mp} = \left(\frac{y_n - y_{n+1}}{y_n^* - y_{n+1}} \right)_{\text{point}} \qquad (2.53)$$

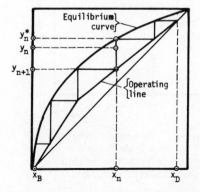
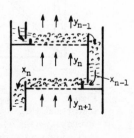

Figure 2.48   Nomenclature in equation (2.52)

where all mole fractions refer to a special location on the tray. The Murphree point efficiency of a binary mixture is the same for both components, and always less than 1.0. The Murphree point efficiency of individual components in a multicomponent mixture, however, may be widely different from one another, and for certain mixtures and compositions, even lie outside the interval 0 to 1.0. Krishna et al.[39] give a method for qualitative prediction of Murphree point efficiencies based on a multicomponent film model and assuming gas phase diffusion control.

The two Murphree efficiencies are the same if there is complete mixing of the liquid on a tray. The Murphree plate efficiency also coincides with the overall plate efficiency for sections where the operating line and the equilibrium line are parallel. In other cases, the overall plate efficiency is less than the Murphree efficiency, as shown in Figure 2.49 where the dotted lines represent 50 per cent Murphree efficiency (steps half-way between the operating line and the equilibrium curve). The stripping section has two theoretical stages after the reboiler and six trays, i.e. over all plate efficiency $\eta_0 = 2/6 = 0.33$. In the rectifying section the operating line and the equilibrium curve are almost parallel, and the three theoretical stages correspond to six actual trays.

At present there is no truly satisfactory method for accurately predicting tray efficiencies, and the estimation of tray efficiency is often the most uncertain part of distillation column calculations. The only rational procedure available in the open literature is the A.I.Ch.E. monograph on bubble cap trays.[40] In lack of

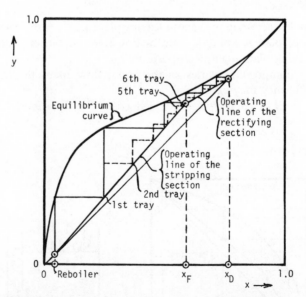

Figure 2.49 McCabe–Thiele diagram showing theoretical stages and actual trays with Murphree efficiency $\eta_M = 0.5$. The stripping section has two theoretical stages and six trays above the reboiler, i.e. $\eta_0 = 2/6 = 0.33$

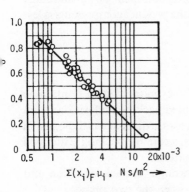

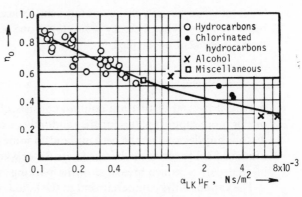

igure 2.50 Overall plate efficiency s a function of feedstock molar ave- ige viscosity at average tower mperature

Figure 2.51 Overall plate efficiency as a function of relative volatility of light key component times viscosity of feed at average column temperature

similar correlations for other designs, Figures 2.50 and 2.51 may serve as a guide. The Drickamer–Bradford correlation[41] in Figure 2.50 is based on overall plate efficiencies on 54 refinery fractionating columns and data on 30 commercial columns obtained from the literature. The correlation by O'Connell[42] in Figure 2.51 is based on data from 32 commercial columns. Figure 2.52 gives overall efficiencies obtained with valve trays and mixtures of isobutane–n-butane and cyclohexane–n-heptane at various pressures.[31]

The plate efficiency is low for the trays near the end if high product purity has to

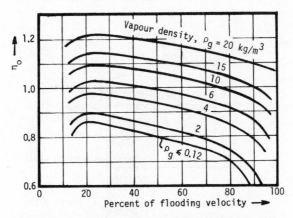

Figure 2.52 Overall plate efficiency obtained in a 1.2 m diameter column with valve trays, weir height $h_w = 50$ mm, and total reflux. Test mixtures isobutane– n-butane and cyclohexane–n-heptane at various pressures. Reproduced by permission of Glitsch Inc. Dallas, Texas

be obtained, especially if the contaminant concentration must be substantially below 100 ppm. The efficiency is also low if one of the product components is less than 10 per cent of the feed.

Packed columns

A typical packed distillation column consists of a cylindrical shell containing one or more support plates for the randomly oriented packing material, a liquid distributing device over the packing, and, in some cases, a demister as indicated in Figure 2.53. The tower can be filled with water when a ceramic packing is dumped in, in order to reduce breakage of the packing material.

The vapour flows countercurrent to the liquid through the packed section, and the packing provides a large area of contacts between the two phases.

The older types of random packings, such as Berl saddles and Rasching rings, have been almost completely superseded by the more efficient Intalox saddles and slotted rings (trade names such as Ballast rings, Flexirings, Hy-Pak, Pall rings, etc.). To keep channelling along the walls at an acceptable level, the ratio of column diameter to packing size should exceed 30 for Rasching rings, 15 for saddles, and 10 for slotted rings. It is claimed[24] that the 50 mm slotted metal ring is generally the best from the standpoint of capacity, and also the most economic packing size and shape for most distillation applications. Ceramic packings should be selected for hot, corrosive services.
Packed columns have the following advantages:

—Greater capacity and with diameter less than 0.6 m usually cheaper than plate columns, except if alloys have to be used.
—Corrosive fluids can be handled with resistant materials such as ceramic or carbon.

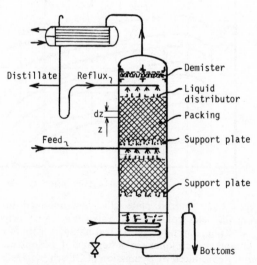

Figure 2.53   Packed distillation column

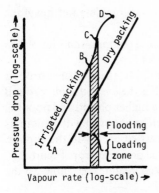

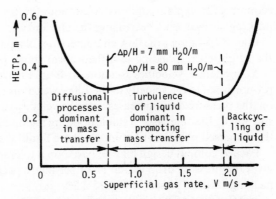

Figure 2.54 Pressure drop in a packed column with constant liquid flow rate, as a function of vapour flow

Figure 2.55 Efficiency of a packed column in terms of *HETP* for a 380 mm diameter column with 25 mm Pall rings[43]

—Packings often exhibit low-pressure drop per theoretical stage, which is important in vacuum distillation.
—Foaming liquids can be handled more readily due to relatively low liquid agitation by the vapour.
—Hold-up of liquid can be low, an advantage if the liquid is heat sensitive.

Packed columns have narrower operating ranges than crossflow plate columns. Hence, both high and low liquid rates can often be handled more

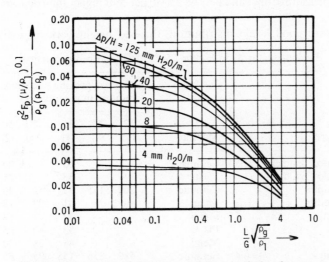

Figure 2.56 Generalized pressure drop correlation for packed columns.[44] Packing factor F_p in Table 2.6. Here, G and L are mass flow rate of vapour and liquid referred to the total cross-section, kg/(m^2 s), ρ_g and ρ_l vapour and liquid density, kg/m^3, and μ the liquid viscosity, N s/m^2

economically in plate columns. A low liquid rate gives incomplete wetting of the packing surface, thus decreasing the efficiency.

Figure 2.54 shows the flow characteristics of a packed column, and Figure 2.55 the height corresponding to a theoretical stage, *HETP*.

At moderate velocities (section A–B in Figure 2.54), the pressure drop is proportional to the velocity to the 1.8 power. Point B is the *loading point*. From B on, the pressure drop increases with the velocity to a higher power, and flooding starts at C. In some cases, a stable flooding region can be obtained with vapour as the dispersed phase, until point D where liquid is carried over with the vapour.

Packed column design may be based on a tolerable pressure drop. Figure 2.56 gives the pressure drop in mm H_2O per metre packed height, and Table 2.6 the packing factor F_p used in Figure 2.56.

The flooding point is not plotted in Figure 2.55 because it varies with different types of packing. Raschig rings give flooding at a pressure drop around 125 mm water/m packed height, compared to 160 to 210 mm water/m for some more efficient packings, such as slotted rings.

For moderate and high pressure distillation, a pressure drop in the range 35–65 mm water/m gives good performance, while values between 8 and 15 are common in vacuum distillation, and 15–30 is a suitable range for absorbers and strippers.

The curves in Figure 2.55 are somewhat uncertain towards the extreme ends of the scale. Passing below 0.02 on the abscissa normally means a vapour velocity that gives excessive entrainment, and above 4.0 vapour back-mixing caused by the high liquid load. In vacuum distillation with extremely low values on the abscissa, pressure drops can easily be 25–30 per cent higher than determined from the graph.

To ensure good liquid distribution, the bed height between redistributors

Table 2.6 Packing factor F_p in Figure 2.56[24,44,45] (dumped packing)

Packing type	Material	Normal packing size (mm, inches or number in parentheses)					
		12 $(\frac{1}{2})$	20 $(\frac{3}{4})$	25 (1)	40 $(1\frac{1}{2})$	50 (2)	75 (3)
Super Intalox Saddles	Ceramic	—	—	60	—	30	—
Super Intalox Saddles	Plastic	—	—	33	—	21	16
Intalox saddles	Ceramic	200	145	92	52	40	22
Intalox saddles	Metal	—	—	41	27	18	—
Hy-Pak rings	Metal	—	—	42	—	18	15
Pall rings	Plastic	—	—	52	40	25	—
Pall rings	Metal	—	—	48	28	20	—
Raschig rings	Ceramic	580	255	155	95	65	37
Raschig rings							
0.8 mm wall	Metal	300	155	115	—	—	—
1.6 mm wall	Metal	410	220	137	83	57	32
Tellerettes	Plastic	—	—	38	—	19	—
Partition rings	Ceramic	—	—	—	—	—	80

should not exceed 2.5–3 times the column diameter for Raschig rings, 5–8 times for saddles, 5–10 times for slotted rings, or 6 m, whichever is the least.

Packing efficiency

In equimolar counterdiffusion equation (1.36) gives the molar flux of component A from the interface to the vapour,

$$N_A = k_{ye}(y - y_i) \tag{1.36}$$

where y is the mole fraction of component A in the vapour and y_i at the interface.

In the packed column (Figure 2.53) with column cross-section S and a as the active interfacial area for mass transfer per unit volume of the column, the area for mass transfer over the height dz is aSdz. The rate of transfer over this height is

$$k_{ye} \, aS(y_i - y)\mathrm{d}z$$

This equals the increase in component A in the vapour phase,

$$GS \, \mathrm{d}y = k_{ye} aS(y_i - y)\mathrm{d}z \tag{2.54}$$

or integrated for the enriching section,

$$\int_{y_F}^{y_D} \frac{\mathrm{d}y}{y_i - y} = \frac{k_{ye} a}{G} z_e \tag{2.55}$$

and for the stripping section,

$$\int_{y_B}^{y_F} \frac{\mathrm{d}y}{y_i - y} = \frac{k_{ye} a}{G'} z_s \tag{2.56}$$

where z_e is the height of packing in the enriching and z_s in the stripping section, y_B is the mole fraction of component A in the vapour entering the stripping section, y_F in the vapour at the feed point, and y_D in the vapour leaving the rectifying section.

The integrals on the left-hand side in the two last equations were defined by Chilton and Colburn[46] as the number of *transfer units* (*NTU*), corresponding to a column efficiency expressed in *height of a transfer unit*, $HTU = G/(k_{ye} a)$. The transfer unit may be viewed as a section in which the change in composition of the vapour stream is numerically equal to the average driving force in that section.

In spite of the fact that Chilton and Colburn pointed out the advantages of this method as early as 1935, manufacturers of packing materials still report packing efficiencies for distillation only in terms of *HETP*. This may be because of reluctance to change the procedure, and because *HETP*-values are remarkably constant for both organic and inorganic liquids, provided low liquid viscosity, liquid flow rates above 1.4 kg/(m² s), and plastics which are difficult to wet, such as fluorocarbons, are avoided. (Easily wetted plastics often swell and lose strength.)

In commercial columns, the *HETP*-values will be somewhat higher than values obtained by manufactures under closely controlled pilot-plant tests. Frank[24] recommends *HETP*-values for commercial distillation columns approximately 1.5 times those obtained in closely controlled pilot-plant columns. For high-efficiency packings such as slotted rings and Intalox saddles in commercial columns, *HETP* run about 0.45 m for 25 mm, 0.65 m for 40 mm, and 0.9 m for 50 mm nominal size. In vacuum distillation with low irrigation efficiency, another 0.15 m to the listed *HETP* may be added as a safety factor.

For a given type of packing, the pressure drop per *HETP* is fairly constant for the different sizes. Hence, in vacuum distillation where pressure drop is important, little is gained by replacement of 50 mm slotted rings with a smaller size.

Figure 2.57 give the *HETP* obtained in closely controlled test with three sizes of Hy-Pak rings.

Absorption columns usually have *HETP* in the range 1.5–1.8 m.

Systematically packed columns have either stacked grids or stacked rings, which is rarely used in distillation service, or they have sections of prefabricated internals. Their main advantage is a low pressure drop per theoretical stage, which is important in vacuum distillation. They can also be operated efficiently with a large vapour to liquid volume ratio.

Impulse packing has a structure composed of a plurality of hollow prism-shaped elements with narrow necks, giving constricted and widened channels as the fluids pass through in countercurrent flow. The manufacturer gives liquid flow rates in the range from 0.5 to 200 m³/m² column cross-section per hour, *F*-factor in equation (2.40) up to 5, and *HETP* ranging from 0.4 to 0.67 m.

Koch-Sulzer packing consists of parallel layers of corrugated wire gauze

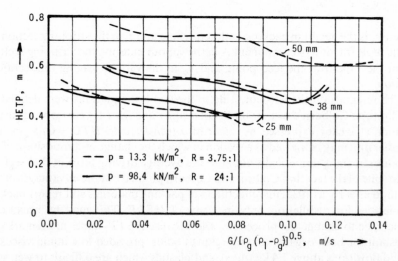

Figure 2.57 Height per theoretical stage for Hy-Pak rings in a 380 mm diameter column as a function of the capacity factor $G/[\rho_g(\rho_l - \rho_g)]^{0.5}$ m/s.

arranged in a sloping pattern. Owing to the capillary action of the wire gauze, this packing can be operated with a liquid mass flow as low as 0.35 kg/(m² s) and still have a reasonable efficiency. Together with the low pressure drop, this makes the packing well suited for vacuum distillation. The column can be designed for an F-factor in equation (2.40) between 2.1 and 2.4 and an *HETP* between 0.25 and 0.3 m.

The packing is-more expensive than any type of dumped packing.

Knit-mesh packings are available from several manufacturers. They are made from woven wire gauze wire diameter 0.15 to 0.18 mm. *HETP* ranges from 0.15 m in small columns to 0.6 m in large columns.

Kloss and *Neo-Kloss packings* are constructed, respectively, with an assembly of coiled springs, or with concentric cyclinders. The pressure drop is extremely low, but their-*HETP* range from 0.6 to 1.2 m.

Glitsch grid consists of an assembly of stamped panels with a high open area, stacked on top of each other. It has a low pressure drop and a high liquid capacity, and is better suited for use in absorbers and strippers than in distillation columns where the *HETP* can be as high as 1.8 m.

Example 2.1 RAOULT'S LAW

In the pressure range 13 to 1750 Pa (0.1 to 13 mm mercury), the vapour pressures of di-*n*-butylphtalate (component A) and di-*n*-butylsebacate (component B) are given by the equations[48]

$$\ln P_A = 19.719 - 2328/T - 1\,658\,000/T^2 \tag{a}$$

$$\ln P_B = 19.765 - 2267/T - 1\,851\,000/T^2 \tag{b}$$

where P_A and P_B are in Pa $= N/m^2$ and T in K.

(a) Assuming Raoult's law is valid, plot the mole fraction y_A of component A in the vapour phase as a function of the mole fraction x_A of the same component in a liquid mixture of A and B boiling at constant total pressure $P = 600$ Pa (4.5 mm mercury).

(b) Derive an expression for the relative volatility α_{AB} as a function of temperature.

Solution

(a) Equations (a) and (b) with pressure 600 Pa give the saturation temperature of the two components as 450.8 and 466.45 K. Table a gives compositions at temperatures selected between the saturation temperatures of the two components, and the calculations are explained under notes.

Table a   Calculations in Example 2.1

T	K	453	456	458	460	462	464	Notes
P_A	Pa	665	765	839	918	1004	1097	eq.(a)
P_B	Pa	311	362	400	441	486	535	eq.(b)
x_A		0.815	0.590	0.456	0.333	0.220	0.116	eq.(2.5)
y_A		0.904	0.753	0.638	0.510	0.368	0.212	eq.(2.4)
α_{AB}		2.138	2.114	2.098	2.082	2.067	2.052	eq.(c)

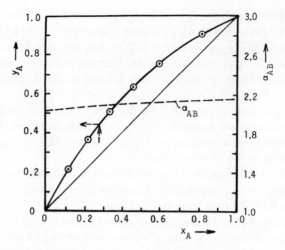

Figure 2.58 Mole fraction y_A of di-n-butylphtalate in the vapour as a function of mole fraction x_A of the same compound in a liquid mixture of di-n-butylphtalate and di-n-butylsebacate boiling at 600 Pa

The mole fractions x_A and y_A from Table a are plotted in Figure 2.58 together with the relative volatility α_{AB}.

(b) The relative volatility of an ideal mixture obeying Raoult's law is given by equation (2.8),

$$\alpha_{AB} = P_A/P_B$$

or with P_A and P_B from equations (a) and (b),

$$\alpha_{AB} = \exp\left[\left(19.719 - \frac{2328}{T} - \frac{1\,658\,000}{T^2}\right) - \left(19.765 - \frac{2267}{T} - \frac{1\,851\,000}{T^2}\right)\right]$$

$$= \exp\left(-0.046 - \frac{61}{T} + \frac{193\,000}{T^2}\right) \tag{c}$$

Values of α_{AB} calculated by equation (c) are given in Table a and plotted in Figure 2.58.

Example 2.2 VAN LAAR'S EQUATIONS

At pressure $P = 760$ mm mercury $= 101.32$ kPa a mixture of ethanol and water forms an azeotrope with mole fraction ethanol $y = x = 0.894$, boiling at 78.15 °C. These data will be used in this example to demonstrate the possibility of estimating the whole vapour–liquid equilibrium curve from only one experimental point and vapour pressure data for the pure compounds.

The vapour pressure of the pure components are given by the Antoine equation

$$\ln P = A - B/(t + C) \tag{a}$$

where t is the temperature in °C, and P the pressure in Pa. Hála et al.[49] give the constants in Table a for the two compounds.

(a) Using the data for the azeotrope, calculate the constants in Van Laar's equations.
(b) Estimate points on the vapour–liquid equilibrium curve at 101.32 kPa pressure.

Table a Constants in equation (a)

	A	B	C
Ethanol (compound 1)	23.6885	3737.6	228.98
Water (compound 2)	23.2370	3841.2	228.00

Solution

(a) At temperature 78.15 °C, equation (a) gives the vapour pressure of ethanol,

$$P_1 = \exp\left(23.6885 - \frac{3737.6}{78.15 + 228.98}\right) = 100\,615\,\text{Pa}$$

and of water,

$$P_2 = \exp\left(23.2370 - \frac{3841.2}{78.15 + 228}\right) = 43\,923\,\text{Pa}$$

Equation (2.9) gives the corresponding activity coefficients as

$$\gamma_1 = \frac{y_1}{x_1}\frac{P}{P_1} = \frac{0.894}{0.894} \times \frac{101\,320}{100\,615} = 1.0070$$

or

$$\ln\gamma_1 = 0.006\,98$$

and

$$\gamma_2 = \frac{y_2}{x_2}\frac{P}{P_2} = \frac{0.106}{0.106} \times \frac{101\,320}{43.923} = 2.3068$$

or

$$\ln\gamma_2 = 0.8358$$

For convenience, equations (2.11) and (2.12) are rewritten as

$$\ln\gamma_1 = \frac{B}{(1 + Cx_1/x_2)^2} \tag{b}$$

and

$$\ln\gamma_2 = \frac{BC}{(C + x_2/x_1)^2} \tag{c}$$

where $B = A_{12}$ and $C = A_{12}/A_{21}$.
Equation (b) with $x_1 = 0.894$, $x_2 = 0.106$ and $\ln\gamma_1 = 0.006\,98$ gives

$$B = 0.006\,98(1 + 8.434C)^2$$

This value is inserted in equation (c) together with $\ln\gamma_2 = 0.8358$, giving

$$C = 1.683$$

and

$$B = 0.006\,98(1 + 8.434 \times 1.683)^2 = 1.611$$

(b) Equation (2.9) with γ_1 from equation (b) and P_1 from equation (a), gives the vapour mole fraction of ethanol,

$$y_1 = x_1\exp\left[\frac{1.611}{(1 + 1.683x_1/x_2)^2} + 23.6885 - \frac{3737.6}{t + 228.98}\right]\bigg/ 101\,320 \tag{d}$$

The corresponding expression for the vapour mole fraction of water is

$$y_2 = x_2 \exp\left[\frac{1.611 \times 1.683}{(1.683x_2/x_1)^2} + 23.2370 - \frac{3841.2}{t+228}\right] \Big/ 101\,320 \qquad (e)$$

Equations (d) and (e) are solved for selected values of x_1 with $x_2 = 1 - x_1$. The temperature t in °C is determined as the value that gives

$$y_1 + y_2 = 1.00 \qquad (f)$$

The result of this calculation is given in Table b together with $y_{1\,exp}$ read from the curve Figure 2.2 which is based on experimental data by several investigators.

Table b. Calculated and experimental vapour–liquid equilibrium data

x_1	0.1	0.2	0.3	0.4	0.6	0.8	0.9	Notes
t °C	86.8	83.2	81.6	80.6	79.1	78.3	78.15	Satisfies
y_1	0.435	0.537	0.587	0.626	0.704	0.821	0.899	eq.(d) eq.(f)
$y_{1\,exp}$	0.425	0.525	0.575	0.615	0.70	0.82	0.90	Fig.2.2

Taking into account that y_1 in Table b is calculated on the basis of only one experimental point, the values coincide remarkably well with the experimental data.
Note: It is always better to calculate the constants of Van Laar's equations (or other equations for the activity coefficients) from more than one experimental point.

Example 2.3   DIFFERENTIAL DISTILLATION

Ethanol is recovered from an ethanol–water mixture by differential distillation at atmospheric pressure. The mole fraction of ethanol in the still is reduced from the initial value $x_0 = 0.20$ to $x = 0.01$.
(a) Calculate the number of kmoles left in the still per kmole initial charge.
(b) What percentage of the ethanol in the initial mixture is recovered in the distillate?
(c) What is the fraction of ethanol in the distillate?

Solution

(a) Table a gives differences $y - x$ read from an enlarged plot of Figure 2.2.

Table a   Vapour-liquid equilibrium data for ethanol–water mixtures

x	0.01	0.02	0.04	0.08	0.12	0.16	0.20
$y - x$	0.10	0.16	0.23	0.31	0.34	0.34	0.325
$\dfrac{1}{y-x}$	10.0	6.25	4.35	3.23	2.94	2.94	3.08

Figure 2.59 gives $1/(y-x)$ from Table a, as a function of the mole fraction x in the still.
The hatched area under the curve in Figure 2.56 between $x = 0.01$ and $x_0 = 0.20$ is the integral in equation (2.13). It is approximately 17.2 squares. Each square is $2 \times 0.02 = 0.04$, giving the area $17.2 \times 0.04 = 0.69$. Equation (2.13) gives

$$\ln(L/L_0) = -0.69$$

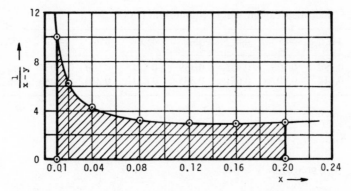

Figure 2.59 Integration of equation (2.13)

or

$$L/L_0 = 0.502$$

(b) The initial charge, 1 kmol, contained 0.2 kmol ethanol. Left in the still after the distillation, $0.01 \times 0.502 = 0.005$ kmol ethanol. Percentage recovered in distillate, $100(0.2 - 0.005)/0.2 = 97.5$ per cent.

(c) The mole fraction of ethanol in the distillate is

$$(0.2 - 0.005)/(1 - 0.502) = 0.39$$

or weight fraction

$$0.39 \times 46/(0.39 \times 46 + 0.61 \times 18) = 0.62$$

Example 2.4 McCABLE–THIELE METHOD

A mixture of 60 mol per cent lauric acid (component 1) and 40 mol per cent myristic acid (component 2) is to be separated by distillation at 7.5 mm mercury = 1000 Pa pressure, to give overhead and bottom products of 90 and 15 mol per cent lauric acid. The feed is at its boiling-point at 1000 Pa. The vapour pressures of the pure components are given by the equations[48]

$$\ln P_1 = 20.8375 - 2210/T - 1\,715\,400/T^2 \tag{a}$$

and

$$\ln P_2 = 19.9356 - 1467/T - 2\,072\,300/T^2 \tag{b}$$

where P is in Pa and T in K.

Assume Raoult's law is valid.

(a) Plot the vapour–liquid equilibrium curve.

(b) Determine the minimum reflux ratio, R_{min}.

(c) Determine the reflux ratio R that gives a total of five theoretical stages including the reboiler.

Solution

(a) At 100 Pa pressure equations (a) and (b) give the saturation temperatures of the two components, $T_1 = 439.1$ K and $T_2 = 459.1$ K. Selected temperatures between these limits give the data on the equilibrium curve listed in Table a and plotted in Figure 2.60.

(b) A straight line through the points ($x_1 = y_1 = 0.90$) at the diagonal and ($x_1 = 0.60$, $y_1 = 0.81$) at the equilibrium curve in Figure 2.60 intersects the ordinate at $y_1 = 0.625$.

82

Table a   Calculated vapour–liquid equilibrium data

TK	442	445	448	450	453	456	458	Notes
x_1	0.788	0.598	0.435	0.339	0.211	0.100	0.034	eq.(2.5)
y_1	0.914	0.807	0.681	0.585	0.421	0.230	0.086	eq.(2.4)

From Figure 2.17,

$$x_D/(R_{min} + 1) = 0.625$$

or

$$R_{min} = 0.90/0.625 - 1 = 0.44$$

(c) Different operating lines are plotted in Figure 2.60 until the one is found that gives five steps from $x_1 = 0.15$ to $x_1 = 0.9$ (five steps = four theoretical stages + reboiler). This is obtained for the operating lines through the point $x = 0.60$, $y = 0.703$. An extension of the upper operating line to $x = 0$ gives $y = 0.31$.

$$0.90/(R + 1) = 0.31$$

or

$$R = 1.90$$

and

$$R/R_{min} = 1.90/0.44 = 4.32$$

Note: The low number of theoretical stages is selected to keep the pressure drop in the column at an acceptable level. A partial condenser may also be used in order to improve

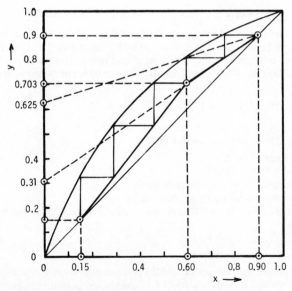

Figure 2.60   McCabe–Thiele diagram for a mixture of lauric acid and myristic acid

*the separation without increasing the number of theoretical stages. Four theoretical stages plus reboiler and partial condenser, both giving one theoretical stage, give the reflux ratio $R = 1.143$ or $R/R_{min} = 2.60$.

Example 2.5 DIFFERENT MOLAR HEATS OF VAPORIZATION

Figure 2.61 shows schematically a regenerator for triethylene glycol (TEG) used in an absorption column for drying of natural gas.

The feed F is 1540 kg liquid per hour consisting of 1414 kg TEG with molecular weight 150.2 and 126 kg water. The bottoms B contains 98.2 wt per cent TEG and the distillate D less than 0.25 wt per cent.

For the temperatures in the rectifier (100–110 °C), data in reference 50 and the vapour pressure curve in reference 51 both correspond to a molar heat of vaporization of TEG approximately 1.7–1.8 times the molar heat of vaporization of water.

Table a Equilibrium data for a triethylene glycol–water mixture at atmospheric pressure with X and Y as weight fractions of water in the liquid and in the vapour [52]

X	0.01	0.02	0.04	0.06	0.10	0.20	0.30	0.35	0.535
Y	0.225	0.625	0.79	0.865	0.925	0.97	0.985	0.99	0.995

(a) Recalculate Table a with fractions x' and y' based on fictitious molecular weights ('latent heat units'), assuming the molar heat of vaporization of TEG as 1.7 time the molar heat of vaporization of water.
(b) Estimate the number of theoretical stages needed with reflux ratio $R' = 0.6$.
(c) Select tower internals and determine tower diameter and height.

Solution

(a) The TEG is assigned the fictitious molecular weight or latent heat unit $M' = 150.2/1.7 = 88.4$. A mixture of 1 kg contains $X/18$ kmol water and $(1 - X)/88.4$ 'kmol' TEG, giving

$$x' = \frac{X/18}{X/18 + (1 - X)/88.4} \tag{a}$$

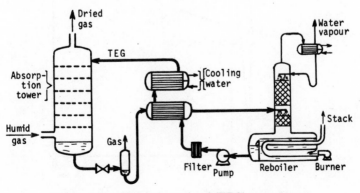

Figure 2.61 Triethylene glycol (TEG) regenerator

and for the vapour

$$y' = \frac{Y/18}{Y/18 + (1 - Y)/88.4}$$

(b)

which give the equilibrium data in Table b.

Table b

x'	0.047	0.091	0.170	0.239	0.353	0.551	0.678	0.726	0.850
y'	0.588	0.891	0.949	0.969	0.984	0.994	0.997	0.998	0.999

(b) The composition of feed and products expressed in x'-units are

$$x'_F = \frac{126/18}{126/18 + 1414/88.4} = 0.304$$

$$x'_D = \frac{0.9975/18}{0.9975/18 + (1 - 0.9975)/88.4} = 0.9995$$

$$x'_B = \frac{(1 - 0.982)/18}{(1 - 0.982)/18 + 0.982/88.4} = 0.083$$

These points are plotted in Figure 2.62 together with the equilibrium curve and the operating line that intersects the ordinate at

$$\frac{x'_D}{R' + 1} = \frac{0.9995}{0.6 + 1} = 0.625$$

Figure 2.62 gives $y' \approx 0.875$ for vapour from the reboiler. The following steps can be determined either in a large-scale diagram or by step-by-step calculation. With $D' = L'/0.6$ and $V' = 1.6\,L'/0.6$ (equation (2.19)) gives the operating line.

$$y' = \frac{L'}{1.6L'/0.6}x' + \frac{L'/0.6}{1.6L'/0.6}0.9995$$

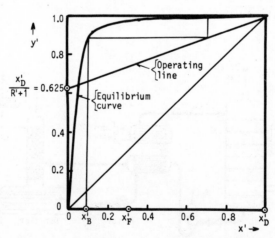

Figure 2.62 McCabe–Thiele diagram for TEG–water with a fictitious molecular weight for TEG of 88.4

or

$$x' = \frac{1.6}{0.6}y' - \frac{0.9995}{0.6} \tag{c}$$

The top part of the equilibrium curve may be approximated by a straight line through the point $x' = 0.678$ and $y' = 0.997$ (Table b),

$$y' \approx 0.997 + (1 - 0.997)\frac{x' - 0.678}{1 - 0.678} \tag{d}$$

Table c   Step-by-step calculation

	x' eq.(c)	y' eq.(d)	Notes
Reboiler		0.875	From equilibrium diagram
Ist stage	0.6673	0.9969	
2nd stage	0.9924	0.9999	

Table c shows that the reboiler plus two theoretical stages fulfil the requirement $x'_D > 0.9995$.

(c) Table 2.4 indicates moderate foaming in glycol regenerators. Probably it is also a small-diameter column, and Table 2.3. gives a randomly packed column as the best choice for a small column operated with a foaming system.

Material balances on weight basis give

$$1540 = B + D \text{ kg/h}$$

and

$$126 = 0.018B + 0.9975D \text{ kg/h}$$

giving $D = 100 \text{ kg/h}$ or $0.9975 \times 100/18 + 0.0025 \times 100/150.2 = 5.54 \text{ kmol/h}$ and $V = (R + 1)D = 1.6 \times 5.54 = 8.87 \text{ kmol/h}$.

The highest value of the abscissa (Figure 2.56) is at the bottom of the packing where $x' = 0.667$ and $y' = 0.875$. Here 1 'kmol' of the liquid contains $0.667 \times 18 \text{ kg}$ water and $(1 - 0.667)88.4 \text{ kg}$ TEG, giving weight fractions

$$X = \frac{0.667 \times 18}{0.667 \times 18 + (1 - 0.667)88.4} = 0.290 \text{ kg/kg}$$

and

$$Y = \frac{0.875 \times 18}{0.875 \times 18 + (1 - 0.875)88.4} = 0.588 \text{ kg/kg}$$

or ratio L/G in Figure 2.56,

$$\frac{L}{G} = \frac{0.6 \times 5.54/[0.290/18 + (1 - 0.290)/150.2]}{1.6 \times 5.54/[0.588/18 + (1 - 0.588)/150.2]} = 0.637$$

The temperature of the liquid is only slightly above 100 °C, and the vapour density is estimated from the ideal gas law with $T = 373 \text{ K}$ and

$$y = \frac{0.588/18}{0.588/18 + (1 - 0.588)/150.2} = 0.923$$

$$\rho_g = \frac{10^5}{8314 \times 373}[0.923 \times 18 + (1 - 0.923)150.2] = 0.91 \text{ kg/m}^3$$

The liquid density is close to 1000 kg/m^3, and the abscissa in Figure 2.56 is

$$\frac{L}{G}\sqrt{\frac{\rho_g}{\rho_1}} = 0.637 \sqrt{\frac{0.91}{1000}} = 0.019$$

which is slightly to the left of the curves. It indicates that the liquid may not give complete wetting of the packing. With $(L/G)\sqrt{\rho_g/\rho_1} = 0.019$, the curve for a pressure drop of 40 mm H_2O/m gives

$$\frac{G^2 F_p(0.003/1000)^{0.1}}{0.91(1000-0.91)} \approx 0.04 \tag{e}$$

where $\mu = 0.003 \text{ N s/m}^2$ is the viscosity estimated from data in reference 51, and G is the mass flux of vapour in the bottom of the column, referred to the total column cross-section, $\text{kg/(m}^2\text{ s)}$,

$$G = \frac{1.6 \times 5.54/[0.588/18 + (1-0.588)/150.2]}{3600(\pi/4)D^2} = \frac{0.0885}{D^2} \tag{f}$$

Equations (e) and (f) give the column diameter

$$D = 0.088 F_p^{1/4} \tag{g}$$

Using ceramic Intalox saddles gives the results listed in Table d.

Table d   Calculated tower diameters and packing heights

Nominal size (mm)	25	40	50	Notes
Packing factor, F_p	92	52	40	Table 2.6
Tower diameter, D (m)	0.27	0.24	0.22	eq. (g)
Packing height[†] (m)	0.6	0.8	1.05	One theoretical stage

[†] Packing height listed in Table d is the height given on p. 76 for one theoretical stage with 0.15 m added for low irrigation.

It seems reasonable to select a 25 mm nominal packing size and a column diameter of 300 mm, which gives the possibility of operating with a somewhat higher reflux ratio. The column to packing ratio is $300/25 = 12$, which is within the acceptable range.

The calculations under (b) gave three theoretical stages including reboiler and partial condenser (dephlegmator). With the reboiler corresponding to one stage and the partial condenser close to one stage, the packing should correspond to between one and two theoretical stages. A packing giving two theoretical stages corresponds to a packing height of $2 \times 0.6 = 1.2$ m.

Comments: The feed contains less water than the liquid at any point in the packing, and it is reasonable to use the simple arrangement with feeds directly into the reboiler.

The regenerator for TEG at the West Sole field in the North Sea is reported to have a packing height of 1.4m.[53]

Example 2.6   VARYING RELATIVE VOLATILITY

For the propylene–n-propane binary system, Schweitzer[54] refers to A. B. Hill's data for volatility as a polynominal where the constants are functions of the pressure. At 20 bar pressure, the polynominal is

$$\alpha = 1.1506 - 0.0548x - 0.0174x^2 \tag{a}$$

where x is the mole fraction of propylene.

Find for a propylene–propane splitter operated at 20 bar pressure with feed at the boiling-point and the mole fraction of propylene, $x_F = 0.6$ and propane $(1 - x_F) = 0.4$, with the mole fraction of propylene in distillate $x_D = 0.98$ and in bottoms $x_B = 0.05$:

(a) minimum reflux ratio;

(b) minimum number of theoretical stages;

(c) number of theoretical stages below and above the feed point with reflux ratio $R = 1.5 R_{min}$.

Solution

(a) Equation (a) gives for $x_F = 0.6$,

$$\alpha = 1.1506 - 0.0548 \times 0.6 - 0.0174 \times 0.6^2 = 1.1115$$

Equation (2.15),

$$y_F = \frac{1.1115 \times 0.60}{1 + (1.1115 - 1)0.6} = 0.625$$

From Figure 2.17 is seen that

$$\frac{y_F - x_D/(R_{min} + 1)}{- x_F} = \frac{x_D - y_F}{x_D - x_F}$$

$$\frac{0.625 - 0.980/(R_{min} + 1)}{0.60} = \frac{0.980 - 0.625}{0.980 - 0.60}, \qquad R_{min} = 14.2$$

(b) A graphical solution must be carried out on a very large scale in order to give reasonable accuracy.

Equations (2.25) and (2.27) give an approximate solution.

Equation (a) with $x = 0.98$, $(0.98 + 0.050)/2$, and 0.050 gives the relative volatility at the top, in the middle, and at the bottom, and equation (2.27) the average volatility

$$\alpha = (1.080 \times 1.029 \times 1.148)^{1/3} = 1.106$$

This value inserted in the Fenske equation (2.25) gives the minimum number of theoretical stages (total reflux),

$$N_{min} = \frac{\ln[(0.98/0.02)(0.95/0.05)]}{\ln 1.106} = 67.6 \text{ theoretical stages}$$

Table a.   Program for pocket calculators HP 19C, 25, 29C, and 33

Place the following in storage	Step no.	Step no.	Step no.	Step no.	Step no.
x_B STO 0	01 LBL 1	10 RCL 3	19 1	28 $x \leqslant y$	37 —
x_D STO 1	02 1	11 ×	20 —	29 GTO 2	38 ÷
1.1506 STO 2	03 STO + 5	12 +	21 RCL 0	30 GTO 1	39 RCL 5
− 0.0548 STO 3	04 RCL 0	13 RCL 2	22 ×	31 LBL 2	40 $x \rightleftarrows y$
− 0.0174 STO 4	05 STO 6	14 +	23 1	32 RCL 0	41 −
0 STO 5	06 x^2	15 STO 7	24 +	33 RCL 1	42 R/S
	07 RCL 4	16 RCL 0	25 ÷	34 —	
	08 ×	17 ×	26 STO 0	35 RCL 0	
	09 RCL 0	18 RCL 7	27 RCL 1	36 RCL 6	

An accurate calculation may be carried out with a programmable pocket calculator using the flow chart (Figure 2.14) with $V/L = 1.0$ and $D = 0$. This gives the program in Table a.

This program with $x_B = 0.05$ and $x_D = 0.98$ gives $N_{min} = 67.0$ theoretical stages.

(c) With the basis 1 kmol feed, a material balance for propylene gives $0.60 = 0.98D + 0.05B = 0.98D + 0.05(1 - D)$
or distillate per kmol feed,

$$D = 0.591$$

and bottoms per kmol feed,

$$B = 0.409$$

$$R = 1.5R_{min} = 21.3 = L/D, \qquad L = 12.59$$
$$L' = L + 1 = 13.59$$

Equation (2.19) gives the operating line for the rectifying section,

$$y_{n+1} = \frac{12.59}{12.59 + 0.591}x_n + \frac{0.591}{12.59 + 0.591}0.98$$

or

$$x_{n-1} = 1.0469\,y_n - 0.0460 \tag{b}$$

and equation (2.21) for the stripping section,

$$y_m = \frac{13.59}{13.59 - 0.409}x_{m+1} - \frac{0.409}{13.59 - 0.409}0.05$$

or

$$x_{m+1} = 0.9699\,y_m + 0.0015 \tag{c}$$

With equation (a) giving the equilibrium curve and equations (b) and (c) the operating lines, the number of theoretical stages may be determined graphically, as shown in Figure 2.16, or analytically. For the rectifying section, the program under b may be used with the following modifications: $x_B = 0.05$ is replaced by $x_F = 0.6$ in storage 0, the constant 1.0469 is placed in STO 8 and -0.0460 in STO 9, and steps 26 and 27 are replaced by the steps in Table b.

Table b   Steps 26–31 replacing steps 26 and 27 in Table a

Step no.		Step no.		Step no.	
26	RCL 8	28	RCL 9	30	STO 0
27	×	29	+	31	RCL 1

This gives 76.8 theoretical stages in the rectifying section.

The number of theoretical stages in the stripping section is calculated with the same program with $x_B = 0.05$, $x_D = 0.98$ substituted by $x_F = 0.6$ in STO 1, the constant 0.9699 in STO 8, and 0.0015 in STO 9, giving 40.0 theoretical stages in the stripping section. This gives a total of $76.8 + 40.0 \approx 117$ theoretical stages including the reboiler.

Comments: One hundred and seventeen theoretical stages is not exact. Heat losses from the column, and to a lesser extent temperature gradients in the column, may give some deviation from the assumption of constant molar overflow.

Example 2.7 STAGE-TO-STAGE CALCULATION, TERNARY MIXTURE

According to German patent No. 1 301 533 a solution of formaldehyde in butanol can be obtained from an aqueous formaldehyde solution by distillation with butanol.

The data in Table a were obtained for the ternary mixture at atmospheric pressure in an equilibrium still in the laboratory.[55] The subscripts are for formaldehyde, 2 for water, 3 for n-butanol.

Table a   Laboratory measurements at atmospheric pressure

Temperature (°C)	Mole fraction in liquid			Mole fraction in vapour		
	x_1	x_2	x_3	y_1	y_2	y_3
117.0	0.603	0.043	0.354	0.311	0.528	0.161
113.5	0.541	0.040	0.419	0.306	0.488	0.206
107.0	0.482	0.156	0.362	0.144	0.733	0.123
98.0	0.356	0.428	0.216	0.137	0.761	0.101
	0.265	0.450	0.285	0.091	0.730	0.179
99.0	0.261	0.309	0.430	0.060	0.749	0.191
98.5	0.200	0.473	0.327	0.054	0.756	0.190
95.3	0.113	0.498	0.389	0.026	0.751	0.223
93.0	0.044	0.500	0.456	0.010	0.750	0.240
95.0	0.052	0.897	0.051	0.036	0.770	0.195

Water and n-butanol have an azeotrope with $z_2 = y_2 = 0.771$ and $x_3 = y_3 = 0.229$. The condensed azeotrope separates in a bottom liquid layer with 98.2 mol per cent water and 1.8 mol per cent n-butanol, and an upper liquid layer with 50.7 mol per cent water and 49.3 mol per cent n-butanol.

A distillation column will be used to obtain a bottom product with $x_1 = 0.603$, $x_2 = 0.043$, and $x_3 = 0.354$, and a top product with the mole fraction of formaldehyde less than 0.001. The fresh feeds consist of an aqueous solution of formaldehyde with $x_1 = 0.26$ and $x_2 = 0.74$, and pure liquid n-butanol. Estimate the number of theoretical stages needed.

Solution

Figure 2.63 shows the column with the known concentrations.

With 1 mol aqueous feed F_a as basis, a material balance for each component taken around the whole column, gives

$$0.260 = 0.603B \qquad\qquad B = 0.431 \text{ mol}$$
$$0.740 = 0.043B + 0.982D \qquad D = 0.735 \text{ mol}$$
$$1.00F_b = 0.354B + 0.018D \qquad F_b = 0.166 \text{ mol}$$

Assuming azeotropic vapour from the top stage, a material balance for water around the decanter gives

$$0.771V = 0.982D + 0.507(V - D), \qquad V = 1.322 \text{ mol}$$

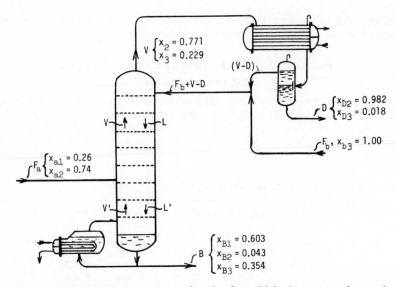

Figure 2.63 Distillation column for the formaldehyde–water–n-butanol ternary mixture. Here F_a is the aqueous feed and F_b the feed of n-butanol

Reflux in the rectifying section,

$$L = V + F_b - D = 0.753 \text{ mol}$$

Reflux in the stripping section,

$$L' = L + 1 = 1.753 \text{ mol}$$

A material balance around the top of the column, as shown in Figure 2.64, gives

$$0.753(x_i)_{n+1} = 1.322(y_i)_{n+1} + 0.166(x_i)_{F_b} - 0.735(x_i)_D$$

or with the data for each component,

$$(x_1)_{n+1} = 1.756(y_1)_n \tag{a}$$
$$(x_2)_{n+1} = 1.756(y_2)_n - 0.959 \tag{b}$$
$$(x_3)_{n+1} = 1.756(y_3)_n + 0.203 \tag{c}$$

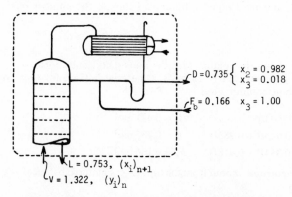

Figure 2.64 Material balance at top of column

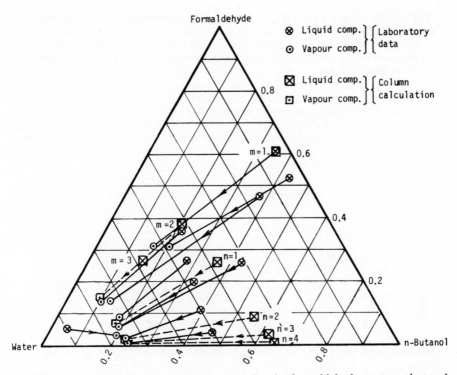

Figure 2.65 Vapour–liquid equilibrium data for the formaldehyde–water–n-butanol ternary system

Equation (2.32) gives the material balance for the stripping section with $m = 1$ being the reboiler,

$$(y_i)_m = \frac{1.753}{1.322}(x_i)_{m+1} - \frac{0.431}{1.322}(x_i)_B$$

or with the given $(x_i)_B$,

$$(x_1)_{m+1} = 0.754(y_1)_m + 0.148 \tag{d}$$
$$(x_2)_{m+1} = 0.754(y_2)_m + 0.011 \tag{e}$$
$$(x_3)_{m+1} = 0.754(y_3)_m + 0.087 \tag{f}$$

As no other data for this system are available in the open literature, the vapour–liquid equilibria have to be estimated from the data given in Table a.

Table b. Compositions in the stripping section

	From eqs. (d), (e), and (f)			From Figure 2.65		
	x_1	x_2	x_3	y_1	y_2	y_3
Reboiler, $m = 1$	0.603	0.043	0.354	0.311	0.528	0.161
$m = 2$	0.383	0.409	0.208	0.15	0.76	0.09
$m = 3$	0.261	0.584	0.155	0.15	0.76	0.09

The first line in Table b gives $(x_i)_{m=1}$ and $(y_i)_{m=1}$ for the reboiler, which are the data from the first line in Table a. The $(x_i)_2$-values for the second stage are calculated from equations (d), (e) and (f), and plotted in Figure 2.65 together with all the data from Table a. Comparison with the laboratory data gives the dotted tie-line to the vapour composition y_i given in Table b for $m = 2$. This procedure is repeated for stage $m = 3$.

From Table b is seen that the vapour is unchanged from $m = 2$ to $m = 3$. Hence, it is of no use with more than two to three theoretical stages below the feed point.

The vapour ascending from the feed point into the rectifying section is given in the last line in Table b and transferred to the first line in Table c. In the rectifying section the calculation procedure is the same as for the stripping section. The result is given in Table c.

Table c Compositions in the rectifying section

	From eqs.(a), (b), and (c)			From Figure 2.65		
	x_1	x_2	x_3	y_1	y_2	y_3
Feed point				0.15	0.76	0.09
$n = 1$	0.263	0.376	0.361	0.07	0.755	0.175
$n = 2$	0.123	0.326	0.510	0.02	0.75	0.23
$n = 3$[†]	0.035	0.358	0.607	0.007	0.75	0.243
$n = 4$	0.012	0.358	0.630	0.002	0.75	0.248
$n = 5$	0.004	0.358	0.638	0.001	0.75	0.249
$n = 6$	0.002	0.358	0.640	0.000	0.75	0.25

[†]From this point the vapour composition is estimated to be $y_1 \approx 0.2\, x_1$ and $y_2 = 0.75$.

The liquid composition calculated in Tables b and c is plotted in Figure 2.66.

The estimate gives nine theoretical stages including the reboiler, and the feed point at the third theoretical stage.

Note: Three special circumstances increase the uncertainty of the calculations:

1. Different molal heats of vaporization for formaldehyde have been reported in the literature,[56] ranging from 1.0 to 1.3 times the molal heat of vaporization for water and butanol.
2. Formaldehyde in alcoholic solution forms simple hemiacetale, $H_2C\diagup{}^{OH}_{OR}$, and in aqueous solutions forms methyleneglycol, $CH_2(OH)_2$, and polyoxymethyleneglycol

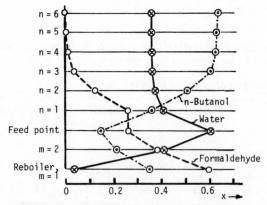

Figure 2.66 Liquid composition in the column

(paraformaldehyde), $HO(CH_2O)_nOH$. Even in concentrated solutions, monomeric formaldehyde is present only in a small amount. The vaporization of the volatile components is followed by reactions in the liquid phase, and it is reasonable to assume that the measured 'equilibrium' data depend somewhat on the rate of vaporization. Such an influence has been shown by Iliceto[57] for mixtures of formaldehyde–water at reduced pressure.

3. Commercial grade formaldehyde includes some methanol that may influence the equilibrium.

Example 2.8 MULTICOMPONENT DISTILLATION

A mixture of four fatty acids is to be separated in two fractions by fractional distillation under vacuum. The mole fractions of each of the acids in the feed are:

Myristic acid (C_{14}), $x_{14} = 0.05$

Palmitic acid (C_{16}), $x_{16} = 0.35$

Stearic acid (C_{18}), $x_{18} = 0.50$

Eicosanoic acid (C_{20}), $x_{20} = 0.10$

The column will be operated with steam injection and at a vapour pressure of fatty acids of 500 Pa = 3.75 mm mercury.

In the temperature range 150 to 250°C, the vapour pressure curves given by Jantzen and Erdmann[57] are approximated by the equation

$$\log P = A - B/T - C/T^2 \tag{a}$$

with P in Pa, T in K, and the constants given in Table a.

Table a Constants in equation (a)

		A	B	C
Myristic acid	(C_{14})	8.6579	637.2	900 000
Palmitic acid	(C_{16})	7.1860	− 755.3	1 317 024
Stearic acid	(C_{18})	7.9783	127.7	1 160 900
Eiscosanoic acid	(C_{20})	10.7539	2674.9	666 158

In the distillate the mole fraction of palmitic acid plus myristic acid shall be 0.85 and the mole fraction of stearic acid plus eicosanoic acid in the bottoms 0.80. The feed is at its boiling-point, 179.9°C, and Raoult's and Dalton's laws are assumed valid.

(a) Calculate the amount of distillate and of bottoms.

(b) Estimate the composition of the products.

(c) Estimate the minimum reflux ratio, R_{min}, and the minimum number of theoretical stages, N_{min}.

(d) With reflux ratio $R = L/D = 1.8$, estimate the number of theoretical stages.

(e) Determine the number of theoretical stages by a stage-by-stage calculation, assuming reflux ratio $R = 1.8$.

Solution

(a) The amount of distillate and bottoms per mole feed are determined by a material balance for the sum of myristic and palmitic acid:

$$0.4 = 0.85D + 0.20(1 - D)$$

$$D = 0.308 \, \text{kmol/kmol feed}$$

$$B = 1 - D = 0.692 \, \text{kmol/kmol feed}$$

(b) As a first approximation, the distillate is assumed to be without eicosanoic acid and the bottoms without myristic acid, i.e.

$(x_{14})_D = 0.05/0.308 = 0.162$

$(x_{16})_D = 0.85 - 0.162 = 0.688$

$(x_{18})_D = 1.00 - 0.85 = 0.15$

$(x_{20})_B = 0.10/0.692 = 0.145$

$(x_{18})_B = 0.80 - 0.145 = 0.655$

$(x_{16})_B = 1.00 - 0.80 = 0.200$

The volatility of component i relative to stearic acid as the heavy key, is

$$\alpha_{i-18} = \exp_{10}(\log P_i - \log P_{18}) \tag{b}$$

where $\log P_i$ and $\log P_{18}$ are calculated from equation (a).

The temperatures at the top and at the bottom of the column are the temperatures where

$$\sum [(x_i)_D P_i] = 500 \, \text{Pa} \tag{c}$$

and

$$\sum [(x_i)_B P_i] = 500 \, \text{Pa} \tag{d}$$

giving the temperatures 461.2 K and 477.8 K, and the relative volatilities in Table b.

Table b   Relative volatilities

Temperature (K)	α_{14-18}	α_{16-18}	α_{20-18}	Notes
461.2	6.33	2.45	0.379	eqs. (a) and (b)
477.8	5.70	2.35	0.409	eqs. (a) and (b)
Average	6.01	2.40	0.394	eq. (3.23)

The Fenske equation (2.35),

$$N_{min} = \frac{\text{Log}\left[\left(\dfrac{0.688}{0.15}\right)\left(\dfrac{0.655}{0.20}\right)\right]}{\log 2.40} = 3.09$$

Geddes's equation, equation (2.36), gives the mole fraction of the non-key components as

$\log[(x_{14})_D/(x_{14})_B] = 3.09 \log 6.01 + \log(0.15/0.655)$

$\log[(x_{20})_D/(x_{20})_B] = 3.09 \log 0.394 + \log(0.15/0.655)$

or

$(x_{14})_D/(x_{14})_B = 58.4$ and $(x_{20})_D/(x_{20})_B = 0.0129$

$0.05 = 0.308 \times 58.4(x_{14})_B + 0.692(x_{14})_B$

$(x_{14})_B = 0.0027$

$0.10 = 0.308(x_{20})_D + 0.692(x_{20})_D/0.0129$

$(x_{20})_D = 0.0019$

The following balances give the composition of the products:

$(x_{18})_D = 1.00 - (0.85 + 0.0019) = 0.148$

$(x_{16})_B = 1.00 - (0.80 + 0.0027) = 0.197$

$(x_i)_F = 0.308(x_i)_D + 0.692(x_i)_B$

$\sum(x_i)_D = \sum(x_i)_B = 1.000$

The four equations give the compositions in Table c.

Table c   Composition of distillate and bottoms

Component	C_{14}	C_{16}	C_{18}	C_{20}
Distillate, $(x_i)_D$	0.157	0.693	0.148	0.0019
Bottoms, $(x_i)_B$	0.0027	0.197	0.657	0.143

(c) The saturation temperatures of the feed and of the distillate are calculated by equations (c) and (d), giving 471.1 K and 461.3 K. The corresponding relative volatilities are given in Table d.

Table d   Relative volatilities

Temperature (K)	α_{14-18}	α_{16-18}	α_{20-18}	Notes
461.3	6.33	2.44	0.38	eqs. (a) and (b)
471.1	5.94	2.39	0.40	eqs. (a) and (b)
Average	6.13	2.42	0.39	eq. (2.26)

Equation (2.38) with the average relative volatilities for the rectifying section gives the Underwood constant θ.

$$0 = \frac{6.13 \times 0.05}{6.13 - \theta} + \frac{2.42 \times 0.35}{2.42 - \theta} + \frac{1.00 \times 0.50}{1.00 - \theta} + \frac{0.39 \times 0.10}{0.39 - \theta}$$

$\theta = 1.5161$

Equation (2.37),

$$R_{min} + 1 = \frac{6.13 \times 0.157}{6.13 - 1.516} + \frac{2.45 \times 0.693}{2.45 - 1.516} + \frac{1.00 \times 0.148}{1.00 - 1.1516} + \frac{0.39 \times 0.0019}{0.39 - 1.516}$$

$R_{min} = 0.74$

Equation (2.35),

$$N_{min} = \frac{\log\left[\left(\frac{0.693}{0.148}\right)\left(\frac{0.657}{0.197}\right)\right]}{\log 2.40} = 3.14$$

(d) $\dfrac{R - R_{min}}{R + 1} = \dfrac{1.8 - 0.74}{1.8 + 1} = 0.38$

Figure 2.32 gives

$$\frac{N - N_{min}}{N + 1} = \frac{N - 3.14}{N + 1} = 0.32$$

$N = 5.1$

(e) The composition of vapour leaving a theoretical stage is calculated from equation (2.7),

$$y_i = \alpha_{i-18} \frac{x_i}{x_{18}} y_{18} \tag{e}$$

where α_{i-18} is calculated from equation (b) with the temperature from equations (c) and (d), and y_{18} from the balance

$$\sum y_i = 1.00 = \left(\alpha_{14-18} \frac{x_{14}}{x_{18}} + \alpha_{16-18} \frac{x_{16}}{x_{18}} + 1.00 + \alpha_{20-18} \frac{x_{20}}{x_{18}} \right) y_{18} \tag{f}$$

The reflux above and below the feed point and the flow rate of vapour are

$$L = RD = 1.8 \times 0.308 = 0.554 \text{ kmol/kmol feed}$$
$$L' = RD + 1 = 0.554 + 1 = 1.554 \text{ kmol/kmol feed}$$
$$V = V' = L + D = 0.554 + 0.308 = 0.862 \text{ kmol/kmol feed}$$

These data inserted in equation (2.34) for the stripping section and (2.33) for the rectifying section of the column, give

$$(x_i)_{m+1} = 0.555(y_i)_m + 0.445(x_i)_B \tag{g}$$

and

$$(x_i)_n = 1.556(y_i)_{n+1} - 0.556(x_i)_D \tag{h}$$

The calculations are carried out as follows:

1. The composition of the vapour leaving the reboiler is calculated from equations (e) and (f) with the relative volatility from equation (b) and $T = 477.6$ K.
2. The liquid composition at the next theoretical stage is calculated from equation (g) where $(y_i)_m$ is the mole fraction in the vapour from below.

Table e   Stage-by-stage calculation

Component		C_{14}	C_{16}	C_{18}	C_{20}	Notes
1st stage (reboiler)	$(x_i)_B$	0.0027	0.197	0.657	0.143	Sat.temp. 477.6 K
	α_{i-18}	5.71	2.37	1.00	0.41	eq. (b)
	y_i	0.013	0.390	0.548	0.049	eqs. (e) and (f)
2nd stage	x_i	0.008	0.304	0.597	0.091	eq. (g), 474.5 K
	α_{i-18}	5.81	2.37	1.00	0.40	eq. (b)
	y_i	0.033	0.514	0.426	0.026	eqs. (e) and (f)
3rd stage, feed point	x_i	0.0195	0.373	0.529	0.078	eq. (g), 472.2 K
	α_{i-18}	5.90	2.38	1.00	0.40	eq. (b)
	y_i	0.0736	0.568	0.338	0.020	eqs. (e) and (f)
4th stage	x_i	0.0272	0.488	0.444	0.030	eq. (h), 469.9 K
	α_{i-18}	5.98	2.40	1.00	0.39	eq. (b)
	y_i	0.0909	0.654	0.248	0.007	eqs. (e) and (f)
5th stage	x_i	0.054	0.632	0.304	0.010	eq. (h), 466.2 K
	α_{i-18}	6.13	2.42	1.00	0.39	eq. (b)
	y_i	0.153	0.705	0.140	0.002	eqs. (e) and (f)

3. Equations (c) and (d) give the temperature of the liquid and equation (b) the relative volatilities.
4. The composition of the vapour leaving this stage is calculated by the same procedure as for the reboiler.

This procedure is repeated until the feed point, where the mole fraction of the light key in the liquid has passed the mole fraction of the light key (x_{16}) in the feed.

For stages above the feed point the procedure is the same as for stages below the feed point, except that equation (h) is used instead of equation (g). The calculations are shown in Table e.

The mole fraction of myristic acid and palmitic acid in the vapour from the fifth stage, $0.153 + 0.705 = 0.858$, is only slightly above the 0.85 in the distillate, i.e. close to five theoretical stages are needed to obtain the given compositions with a reflux ratio $R = 1.8$.

Note: The calculations should start at the end where the composition is best known. Starting with the reboiler, even a small error in the mole fraction of component(s) lighter than he light key gives a large error in the composition at the top. If the calculations are started at the top, the same applies to the mole fraction of component(s) heavier than the heavy key.

It is sometimes best to work the calculations from both ends.

Problems

2.1 Assuming Raoult's law valid, calculate the boiling-point for a mixture with mole fraction of benzene $x_A = 0.4$ and mole fraction of toluene $x_B = 0.6$, and the dew-point for a vapour with the same mole fractions. The total pressure is 760 mm Hg and the vapour pressure of the two components are given by the equations[59]

$$\ln P_A = 20.7638 - 2771.233/(219.888 + \theta)$$

and

$$\ln P_B = 20.914\,28 - 3101.059/(219.693 + \theta)$$

where the pressures are in $Pa = N/m^2$ and the temperatures θ in °C.

2.2 Using the vapour pressure equations given in Problem 2.1, calculate the relative volatility α_{AB} (benzene relative to toluene) as a function of the mole fraction x_A of benzene at total pressure $P = 760$ mm Hg.

2.3 Reference 51 gives the constants $A_{12} = 0.3625$ and $A_{21} = 0.2418$ in the van Laar equations for mixtures of methanol (1) and water (2) at 760 mm Hg pressure, and the constants in Table a for the Antoine vapour pressure equation,

$$\log_{10} p = A - B/(C + \theta)$$

Table a. Constants in the Antoine equation where p is in mm Hg and θ in °C

	A	B	C
Methanol (1)	8.072 46	1574.99	238.86
Water (2)	7.966 81	1668.21	228.00

Plot the vapour–liquid equilibrium diagram $y = f(x)$.

2.4 A mixture of 100 kmol containing mole fractions of benzene $x_A = 0.45$ and of toluene $x_B = 0.55$ is vaporized at atmospheric pressure (101.3 kPa) under differential conditions until 35 kmol is left in the still. Using Raoult's law, the vapour pressure equations in Problem 2.1, and an average value of the relative volatility, calculate the average composition of the total amount of vapour distilled and of the liquid left.

2.5 A mixture containing 0.6 kmol ethanol and 9.4 kmol water is vaporized at 101.3 kPa pressure until the mole fraction of ethanol in the still is $x_{A1} = 0.004$. The van Laar constants for this system are[49] $A_{12} = 0.7715$ and $A_{21} = 0.3848$. Using vapour pressure data from Example 2.2, calculate the number of kmoles evaporated and the average composition of the total amount of vapour expressed in mole fractions and in per cent by weight.

2.6 A mixture of 0.45 kmol benzene and 0.55 kmol water is boiling at 101.32 kPa pressure. The liquids are immiscible, and the vapour pressure equations for the two components are given in Problems 2.1 and 2.3.
(a) Calculate the composition of the vapour when the boiling liquid consists of two phases.
(b) Calculate the amount of liquid left in the still at the moment when one of the phases has disappeared completely.

2.7 A continuous distillation column for methanol–water has two feeds. Feed $F_1 = 200$ kmol/h is saturated vapour with mole fraction methanol $y_{F1} = 0.5$, and feed $F_2 = 100$ kmol/h saturated liquid with $x_{F1} = 0.3$. The column is operated with direct steam (S kmol steam/h). Five per cent of the total methanol is lost in the bottoms, and the rest recovered in the distillate with $x_D = 0.95$. The reflux ratio at the top of the column is $R = L/D = 1.4$.
(a) Calculate the amount of distillate, D kg/h.
(b) Calculate the steam consumption, S kg/h.
(c) Use the data in Problem 2, and determine number of theoretical stages needed and where the feeds should be introduced.
(d) Determine minimum reflux ratio, R_{min}.

2.8 The feed to the column to the left in Figure 2.28 is liquid at its boiling point with mole fraction water $x_F = 0.25$ and mole fraction n-butanol $(1 - x_F) = 0.75$. The mole fraction of water in the bottoms is $x_B = 0.006$, giving vapour from the reboiler with mole fraction $y =$ appr. 0.05. The reflux from the decanter has mole fraction $x_0 =$ appr. 0.50.
Table a gives vapour–liquid equilibrium data for this system,[13] which has two liquid phases when $0.44 < x < 0.98$.

Table a. Vapour–liquid equilibrium data

x	0.050	0.092	0.097	0.181	0.291	0.303	0.417	0.546
y	0.253	0.388	0.402	0.556	0.660	0.666	0.724	0.750

Table a (continued)

x	0.550	0.752	0.901	0.902	0.980	0.981	0.991	0.992
y	0.753	0.754	0.754	0.754	0.760	0.763	0.839	0.850

(a) Calculate the number of theoretical stages needed when the internal reflux ratio $R_{int} = L/V = 0.90$.

(b) Determine minimum internal reflux ratio $(L/V)_{min}$.

2.9 Calculate the internal reflux ratio $R_{int} = L/V$ and the number of theoretical stages needed in the column to the right in Figure 2.28, based on the equilibrium data given in problem 2.8 and the following specifications. The liquid feed to the column is at its boiling point with mole fraction water 0.98. The mole fraction water vapour in the vapour from the top of the column is 0.80 and the mole fraction water in the bottom product is 0.999.

Symbols

A	Area, m^2
A_a	Active area, i.e. column area minus area of inlet downcomer and area outside weir, m^2
A_h	Hole area (sieve tray), m^2
a	System factor, Table 2.4
a	Active interfacial area per unit volume, m^2/m^3
B	Bottoms, kmol/s
D	Distillate, kmol/s
F	Feed, kmol/s
F_p	Packing factor, Table 2.6
f	Vaporized fraction of feed, kmol/kmol
G	Mass flow rate of vapour or gas, $kg/(m^2 s)$
G'	Mass flow rate of vapour or gas in the stripping section, $kg/(m^2 s)$
g	Local acceleration due to gravity, m/s^2
H	Height, m
$HTEP$	Height equivalent to a theoretical plate, m
HTU	Height of a transfer unit, m
h	Liquid head, m
K_i	Equilibrium constant for component i, $K_i = y_i/x_i$
L	Amount of liquid, kmol
L	Liquid flow rate, kmol/s or $kg/(m^2 s)$
L'	Liquid flow rate in the stripping section, kmol/s
l	Length, m
M	Molecular weight, kg/kmol
m	Slope of equilibrium line
m	Stage number in the stripping section
N	Total number of theoretical stages
N	Mass flux, $kmol/(m^2 s)$
NTU	Number of transfer units
n	Number of kmoles
n	Stage number in the rectifying section
P	Total pressure, $Pa = N/m^2$
p	Partial pressure, $Pa = N/m^2$
Δp	Pressure difference, $Pa = N/m^2$
q	Energy to convert 1 mol of feed to saturated vapour divided by the molar heat of vaporization, J/J
R	Gas constant, 8314.3 N s/(kmol K)
R	Reflux ratio, $R = L/D$
S	Sidestream, kmol/s
S	Direct steam, kmol/s
S	Cross-section, m^2
T	Absolute temperature, K

V	Velocity, m/s
V	Molar vapour flow rate, kmol/s
V'	Molar vapour flow rate in the stripping section, kmol/s
x	Mole fraction in the liquid phase
y	Mole fraction in the vapour phase
z	Height, m
α	Angle, radians
α_{AB}	Volatility of A relative to B, $\alpha_{AB} = (y_A/x_A)/(y_B/x_B)$
β	Aeration factor, Figure 2.45
γ	Activity coefficient
γ	Surface tension, N/m
μ	Dynamic viscosity, $N\,s/m^2 = kg/(s\,m)$
η_M	Murphree efficiency, equation (2.51)
η_{Mp}	Murphree local or point efficiency, equation (2.52)
η_o	Overall plate efficiency, equation (2.50)
ρ	Density, kg/m^3

Subscripts

A, B, C	Components A, B and C
B	Bottoms
D	Distillate
F	Feed
g	Vapour or gas phase
h	Refers to hole area (sieve plate)
i	Component i
i	Interface
l	Liquid phase
m	Stage number in stripping section
n	Stage number in rectifying section
S	Sidestream
V	Vapour phase
W	Weir

Superscripts

*	Equilibrium
'	Stripping section

References

1. Forbes, R., *A Short History of the Art of Distillation*, E. J. Brill, Leiden, 1970.
2. Ewell, R. H., J. M. Harrison, and L. Berg, Azeotropic distillation, *Ind. Eng. Chem.*, **36**, 871–875 (1944).
3. Udovenko, V. V., and L. G. Fatkulina, *Zh. Fiz. Khim.*, **26**, 1438 (1952).
4. Reinders, W., and C. H. deMinjer, *Rec. Trav. Chim. Pays Bas*, **59**, 369 (1940).
5. Van Laar J. J., Über Dampfspannungen von binären Gemischen, *Z. Phys. Chemie*, **72**, 723–751 (1910).
6. Margules, M., *Sitzungsberichte der Math.-Naturw. Klasse der Kaiserlichen Akad. der Wissenschaften (Wien)*, **104**, 1243 (1895).
7. Wilson, G. M., Vapor-liquid equilibrium. XI. A new expression for the excess energy of Mixing, *J. AM. Chem. Soc.*, **86**, 127–130 (1964).
8. Abrams, D. S., and J. M. Prausnitz, Statistical thermodynamics of liquid mixtures: a

new expression for the excess free energy of partly or completely miscible systems, *A.I.Ch.E.J.*, **21** (No. 1), 116–128 (1975).

9. NGPSA, *Engineering Data Book*, Natural Gas Processers Suppliers Ass., Tulsa, Oklahoma, 1974.

10. Perry, R. H., and C. H. Chilton, *Chemical Engineers' Handbook*, 5th edn, McGraw-Hill, New York, 1973.

11. Karr, A. E., E. G. Scheibel, W. M. Bowes, and D. F. Othmer, Composition of vapors from boiling solutions, *Ind. Eng. Chem.*, **43**, 961–968 (1951).

12. Robinson, C. S., and E. R. Gilliland, *Elements of Fractional Distillation*, 4th edn, McGraw-Hill, New York, 1950.

13. Smith, Th. E., and R. F. Bonner, Vapor-liquid equilibrium still for partly miscible liquids, *Ind. Eng. Chem.*, **41**, 2867–2871 (1949).

14. McCabe, W. L., and E. W. Thiele, Graphical design of fractionating columns, *Ind. Eng. Chem.*, **17**, 605–611 (1925).

15. Benenati, R. F., Solving engineering problems on programmable pocket calculators, *Chem. Eng.*, **84** (14 March), 129–132 (1977).

16. Fenske, M. R., Fractionation of straight-run Pennsylvania gasoline, *Ind. Eng. Chem.*, **24**, 482–485 (1932).

17. Van Winkle, M., and W. G. Todd, Optimum fractionation design by simple graphical methods, *Chem. Eng.*, **78**, (20 Sept.), 136–148 (1971).

18. Ponchon, M., *Technique Moderne*, **13**, 20, 25 (1921).

19. Savarit, R., *Arts et Métiers*, **65**, 142, 178, 241, 266, 307 (1922).

20. Lewis, W. K., and G. L. Matheson, Studies in distillation, *Ind. Eng. Chem.*, **24**, 494–498 (1932).

21. Hegner, B., and U. Block, Development and application of simulation model for three-phase distillation, *A.I.Ch.E. Journal*, **22**, 582–589 (1976).

22. Torres-Marchal, C., Graphical design for ternary distillation systems, *Chem. Eng.*, **88** (No. 21), 134–141 (1981).

23. Torres-Marchal, J., Calculating vapor–liquid eqilibria for ternary systems, *Chem. Eng.*, **88** (No. 21), 141–155 (1981).

24. Frank, O., Shortcuts for distillation design, *Chem. Eng.*, **84**, (14 March), 111–128 (1977).

25. Geddes, R. L., A general index of fractional distillation power for hydrocarbon mixtures, *A.I.Ch.E.J.*, **4**, 389–392 (1958).

26. Underwood, A. J. V., The theory and practice of testing stills, *Trans. Inst. Chem. Engrs* (*London*), **10**, 111–152 (1932).

27. Underwood, A. J. V., Fractional distillation of multicomponent mixtures, *Chem. Eng. Progr.*, **44**, 603–613 (1948).

28. Gilliland, E. R., Multicomponent rectification. Estimation of the number of theoretical plates as a function of the reflux ratio, *Ind. Eng. Chem.*, **32**, 1220–1223 (1949).

29. Kirkbride, C. G., *Petrol. Refiner*, **23**, 32 (1945).

30 Kesler, M., Shortcut program for multicomponent distillation *Chem. Eng.*, **88** (No. 9), 85–88 (1981).

31. Glitsch, *Ballast Tray Design Manual*, Bulletin No. 4900, 3rd Edn., Glitsch Inc., Dallas, Texas, 1979.

32. Huang, C. J., and J. R. Hodson, Perforated trays, *Petrol Refiner*, **37**, No. 2, 104–118 (1958).

33. Kister, H. Z., Column internals, *Chem. Eng.*, **87** (19 May), 138–142: (28 July), 79–83; (8 Sept. 8), 119–123; (17 Nov.), 283–285 (1980).

34. Kister, H. Z., Inspection assures troublefree operation, *Chem. Eng.*, **88**, 9 Feb., 107–109 (1981).

35. Stichelmair, J., Dimensionierung des Gas/Flüssigkeit-Kontaktapparates Bodenkolonne, *Chem.-Ing.-Techn.*, **50**, 281–284, 383–387, 453–456 (1978).

36. Blecher, H. C., and T. M. Nichols, Capital and operating costs of pollution control

102

equipment modules, *Data Manual*, Vol. 2, EPA-R5-73-023b (July), PB-224356 (1973).
37. Leibson, I., R. E. Kelly, and L. A. Bullington, How to design perforated trays, *Petrol Refiner*, **36** (2), 127–133 (1957).
38. Klein, G. F., Simplified model calculates valve-tray pressure drop, *Chem. Eng.*, **89**, 3 May, 81–85 (1982).
39. Krishna, R., H. F. Martinez, R. Shreedar, and G. L. Standart, Murphree point efficiencies in multicomponent systems, *Trans. I. Chem. E.*, **55**, 178–183 (1977).
40. A.I.Ch.E., *Bubble Tray Design Manual*, New York, 1958.
41. Drickamer, H. G., and J. R. Bradford, Overall plate efficiency of commercial hydrocarbon fractionating columns as a function of viscosity, *Trans. A.I.Ch.E.*, **39**, 319–357 (1943).
42. O'Connell, H. E., Plate efficiency of fractionating columns and absorbers, *Trans. A.I.Ch.E.*, **42**, 741–755 (1946).
43. Eckert, J. S., Selecting the proper distillation column packing, *Chem. Eng. Progress*, **66** (3), 39–44 (1970).
44. Norton, *Tower Packings*, Bulletin TP-78, 1974.
45. Norton, *Intalox Metal Tower Packing*, Bulletin M1–81, 1977.
46. Chilton, T. H., and A. P. Colburn, Distillation and absorption in packed columns, *Ind. Eng. Chem.*, **27**, 255–260, 904 (1935).
47. Norton, *High Performance Metal Tower Packing Hy-Pak*, Bulletin Hy-40, 1978.
48. Hammer, E., and A. L. Lydersen, The vapour pressure of di-*n*-butylphtalate, di-*n*-butylsebacate, lauric acid and myristic acid, *Chem. Eng. Sci.*, **7**, 66–72 (1958).
49. Hala, E., I. Wichterle, J. Polak, and T. Boublik, *Vapour–Liquid Equilibrium Data at Normal Pressures*, Pergamon Press, Oxford, 1968.
50. Union Carbide, *Gas Treating Chemicals*, Union Carbide Corporation, New York, 1969.
51. Dow, *Gas Conditioning Fact Book*, The Dow Chemical Company, Midland, Michigan, 1962.
52. Campbell, J. M., *Gas Conditioning and Processing*, Vol. II, Campbell Petroleum Series, Norman, Oklahoma, 1978.
53. Baldwin, A. H., Offshore production facilities in the North Sea, *9th Commonwealth Mining and Metallurgy Congress*, **1**, 521–535 (1969).
54. Schweitzer, P. A., *Handbook of Separation Techniques for Chemical Engineers*, McGraw-Hill, New York, 1979.
55. Egeland, I., Personal communication, 1979.
56. Walker, J. F., *Formaldehyde*, 3rd edn., Reinhold, New York, 1964.
57. Iliceto, A., Sul sistema acqua-formaldeide, *Chim. Ind.* (Milano), **36**, 197–201 (1954).
58. Jantzen, E., and W. Erdmann, Siedepunkte der Fettsäuren, *Fette u. Seifen*, **54**, 197–201 (1952).
59. Boublik, T., V. Fried, and E. Hala, *The Vapour Pressure of Pure Substances*, Elsevier, Amsterdam, 1973.

CHAPTER 3

Gas Absorption and Desorption

Gas absorption is the unit operation in which one or more components of a gas mixture are dissolved in a liquid. The absorption may be either a purely physical phenomenon, or it may involve reactions in the liquid phase.

Gas desorption or *stripping* is the reverse of absorption, with transfer of one or more components from the liquid to the gas.

An absorbed component or solute is usually recovered from the liquid by processes such as distillation or stripping, and the absorbing liquid or solvent can be either discarded or recycled to the absorption process. An example of this is the absorption of H_2S from natural gas by intimate contact with a monoethanolamine solution in an absorption tower, followed by removal of concentrated H_2S from a stripping tower before the amine solution is recycled to the absorption tower.

Gas–liquid equilibria

Calculation of absorption and desorption is based on material and enthalpy balances and gas–liquid equilibrium data.

In absorption the solute is usually removed from relatively large amounts in inert carrier gas. If the solute concentration is also low in the liquid, the partial pressure at equilibrium may be expressed by *Henry's law*

$$p = Hx \tag{3.1}$$

where x = mole fraction of A in the liquid, kmol A/kmol liquid, and H = Henry's law constant for A, Pa/mol fraction.

Except for substances such as electrolytes which dissociate in solution, Henry's law expresses a general behaviour of dilute solutions. Henry's law constant for a certain compound depends on temperature and on the solvent. The value of H has to be determined experimentally. Figure 3.1 shows the equilibrium pressure of SO_2 over aqueous solutions at different temperatures.[1] The dotted lines represent Henry's law with the constant H given in Figure 3.2 as a function of temperature.

The partial pressure p is, by definition, equal to the product of the gas-phase mole fraction and the total pressure,

$$p \equiv yP \tag{3.2}$$

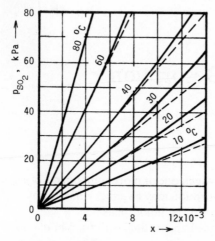

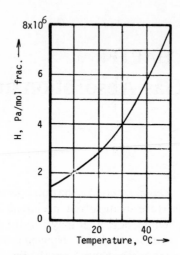

Figure 3.1 Partial pressure of SO_2 in aqueous solution as a function of the mole fraction of SO_2 in the liquid

Figure 3.2 Henry's law constant for SO_2 in aqueous solution as a function of temperature

or with Henry's law,

$$y^* = \frac{H}{P} x \tag{3.3}$$

where P = the total pressure, N/m^2 = Pa, and y^* = mole fraction of A in gas in equilibrium with the liquid

Absorption

Calculation of theoretical stages

A solute A in a gas mixture of A and B shall be absorbed in a liquid C. In the following it is assumed that B is practically insoluble in C, and that evaporation of C into the gas is negligible.

The following symbols will be used:

G = kmol/s of inert gas B

G' = kmol/s of inert gas B and solute A in gas = $G/(1 - y)$

y = mole fraction of A in gas

L = kmol/s of liquid C

L' = kmol/s of C and A in the liquid = $L/(1 - x)$

x = mole fraction of A in liquid.

The G kmol inert gas B is mixed with a kmol of A, giving the mole fraction $y = a/(G + a)$, or $a/G = y/(1 - y)$ kmol A/kmol B in the gas. The same reasoning applied to the liquid phase gives $x/(1 - x)$ kmol A/kmol C.

Figure 3.3 is a schematic drawing of an absorption tower where the gas comes into intimate contact with the liquid that flows over perforated plates.

A material balance for component A, as shown by the dotted line in Figure 3.3,

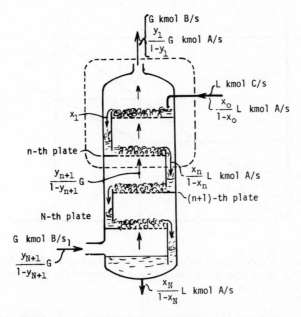

Figure 3.3   Absorption tower where gas is contacted with liquid in countercurrent flow. The gas flows through perforated plates and bubbles through liquid flowing over the plates

gives

$$\frac{y_{n+1}}{1-y_{n+1}}G + \frac{x_0}{1-x_0}L = \frac{x_n}{1-x_n}L + \frac{y_1}{1-y_1}G$$

or

$$\frac{y_{n+1}}{1-y_{n+1}} = \frac{L}{G}\frac{x_n}{1-x_n} + \left(\frac{y_1}{1-y_1} - \frac{L}{G}\frac{x_0}{1-x_0}\right) \tag{3.4}$$

The denominators $1-y$ and $1-x$ are often close to 1.0, and the approximation $1-y \approx 1-x$ gives

$$y_{n+1} = \frac{L}{G}x_n + \left(y_1 - \frac{L}{G}x_0\right) \tag{3.5}$$

Equations (3.4) and (3.5) are important mass balances. They are the equations of the *operating line*. The approximate equation corresponds to a straight line in an x–y diagram.

A *theoretical stage* can be visualized as a plate where the liquid is completely mixed, and the gas leaving the plate is in equilibrium with the liquid. In this case the mole fraction in the gas from the nth plate is given by equation (3.3),

$$y_n = \frac{H}{P}x_n \tag{3.6}$$

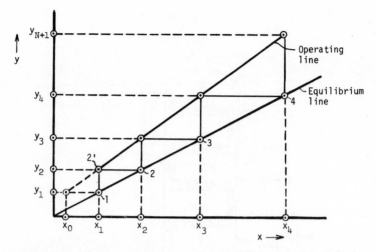

Figure 3.4   McCabe–Thiele method. Mole fraction of solute A in the gas phase in an absorption tower, y, as a function of the mole fraction x of A in the liquid phase

which is the equation of the *equilibrium line* in the x–y diagram (Figure 3.4). In equation (3.6) H is constant only for non-dissociating compounds in dilute solutions. At pressures above 10–15 bar, a correction term due to pressure may also be taken into account. The conditions may also be complicated by reactions in the liquid phase.

The mole fraction of A in the entering gas, which is the gas entering the last plate N, y_{N+1}, will usually be known, and the mole fraction y_1 of the gas from the top plate will be specified. The last number is plotted in Figure 3.4, and the corresponding composition of the liquid at an ideal top plate, x_1, is located at the equilibrium line. The composition of the gas entering the top plate is determined by insertion of x_1 in the equation for the operating line,—or graphically as shown in Figure 3.4, point 2'. The gas with mole fraction y_2 is in equilibrium with liquid with composition x_2, point 2 on the equilibrium line. This is continued step by step until the mole fraction y_{N+1} of the entering gas is reached. The number of steps involved is the number of theoretical stages required for a reduction of y from y_{N+1} to y_1.

The Kremser equation

Figure 3.5 shows the nth stage of an absorption tower. In this figure G' is the number of moles of inert gas and solute A in the gas, and L' the number of moles of liquid, including absorbed A.

A material balance around the nth stage as shown with the dotted line in Figure 3.5, and with x substituted by $(P/H)y$ (equation (3.6)), gives

$$G'_{n+1}y_{n+1} + L'_{n-1}\frac{P}{H}y_{n-1} = G'_n y_n + L'_n\frac{P}{H}y_n$$

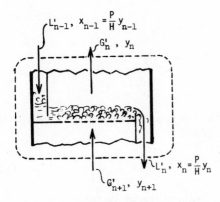

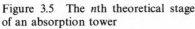

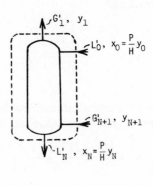

Figure 3.5 The nth theoretical stage of an absorption tower

Figure 3.6 Material balance over the whole absorber

or solved for y_n with the approximations $G'_n/G'_{n+1} \approx 1.0$ and $L'_n/L'_{n-1} \approx 1.0$,

$$y_n = \frac{y_{n+1} + Ay_{n-1}}{1 + A} \tag{3.7}$$

where

$$A = \frac{P}{H}\frac{L'}{G'} = \text{absorption factor} \tag{3.8}$$

One single theoretical stage gives

$$y_1 = \frac{y_2 + Ay_0}{1 + A} \tag{3.9}$$

where y_1 = mole fraction in gas leaving stage 1, y_2 = mole fraction in gas entering stage 1, and y_0 = mole fraction in gas in equilibrium with entering liquid (liquid from stage 0, hypothetical stage above top plate).

Equation (3.7) applied to two theoretical stages gives

$$y_2 = \frac{y_3 + Ay_1}{1 + A} \tag{3.10}$$

Substituting the value of y_1 and solving for y_2,

$$y_2 = \frac{(A + 1)y_3 + A^2 y_0}{A^2 + A + 1}$$

and for an absorber with three theoretical stages,

$$y_3 = \frac{(A^2 + A + 1)y_4 + A^3 y_0}{A^3 + A^2 + A + 1}$$

Rewriting,

$$y_3 = \frac{\left(\dfrac{A^3 - 1}{A - 1}\right)y_4 + A^3 y_0}{\dfrac{A^4 - 1}{A - 1}} = \frac{(A^3 - 1)y_4 + A^3(A - 1)y_0}{A^4 - 1}$$

or for N theoretical stages with mole fraction y_{N+1} in the entering rich gas (from a hypothetical stage below the bottom plate),

$$y_N = \frac{(A^N - 1)y_{N+1} + A^N(A - 1)y_0}{A^{N+1} - 1} \tag{3.11}$$

Figure 3.6 shows a material balance around the whole tower,

$$G'_{N+1}y_{N+1} + L'_0 \frac{P}{H}y_0 = G'_1 y_1 + L'_N \frac{P}{H}y_N$$

or solved for y_N with the approximations $G'_{N+1}/G'_1 \approx 1.0$ and

$$\frac{L'_0}{G'_1}\frac{P}{H} \approx \frac{L'_N}{G'_1}\frac{P}{H} \approx A$$

$$y_N = (y_{N+1} - y_1)/A + y_0$$

Substituting this value in equation (3.11) and rearranging, gives the mole fraction in the exit gas,

$$y_1 = \left(\frac{A - 1}{A^{N+1} - 1}\right)y_{N+1} + \left(\frac{A^{N+1} - A}{A^{N+1} - 1}\right)y_0 \tag{3.12}$$

This is the original Kremser equation.[2] Souders and Brown[43] substituted $(A - 1)/(A^{N+1} - 1)$ by $[1 - (A^{N+1} - A)/(A^{N+1} - 1)]$ and rearranged the equation to the more convenient form.

$$\frac{y_{N+1} - y_1}{y_{N+1} - y_0} = \frac{A^{N+1} - A}{A^{N+1} - 1} \tag{3.13}$$

This equation can be used to calculate the exit mole fraction y_1 when the number of theoretical stages, N_1 is known. Rearranging to a form explicit in N, gives

$$N = \frac{\ln\left[\dfrac{y_{N+1} - y_0}{y_1 - y_0}\left(1 - \dfrac{1}{A}\right) + \dfrac{1}{A}\right]}{\ln A} \tag{3.14}$$

When the absorption factor $A = 1$, the right-hand side of equation (3.13) reduces to $N/(N + 1)$.

Calculation by iteration A rigorous calculation must be based on both equilibrium data and mass and energy balances for all components. This means that

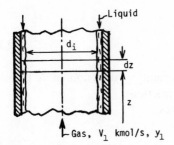

Figure 3.7 Liquid in counter-current contact with gas in a vertical tube

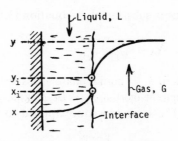

Figure 3.8 Liquid L in contact with gas G. Mole fraction y_i of A in the gas and x_i in the liquid at the interface

enthalpy data has also to be known. An iteration method for solution of the equations for absorbers with a known number of theoretical stages is given by Khoury.[4,5]

Calculation with individual mass transfer coefficients

Stefan diffusion. Figure 3.7 shows a vertical tube with liquid flowing down along the walls in contact with a rising gas, and Figure 3.8 an enlarged picture of a section of the wall and the mole fractions of the solute in the gas and the liquid phase.

The gas in contact with the liquid surface is assumed to be in equilibrium with the liquid at the interface, i.e.

$$y_i = \frac{H}{P} x_i \tag{3.3}$$

for a gas following Henry's law. If A diffuses through a non-diffusing liquid film, the flux of A is given by equation (1.27),

$$N_A = -\frac{Dc}{\Delta z} \ln \frac{1 - x_i}{1 - x} \tag{3.15}$$

where Δz = thickness of the film, and $1 - x$ = mole fraction of components other than A.

The logarithmic mean mole fraction of components other than A in the film is defined by the equation

$$(1 - x)_{lm} = \frac{(1 - x) - (1 - x_i)}{\ln \dfrac{1 - x}{1 - x_i}} = \frac{x_i - x}{\ln \dfrac{1 - x}{1 - x_i}} \tag{3.16}$$

or

$$\ln \frac{1 - x}{1 - x_i} = \frac{x_i - x}{(1 - x)_{lm}}$$

110

This is substituted in equation (3.15),

$$N_A = k_{xs} \frac{x_i - x}{(1 - x)_{lm}} \tag{3.17}$$

where k_{xs} = liquid phase mass transfer coefficient by diffusion of A through stagnant inert(s), $k_{xs} = Dc/\Delta z \, \text{kmol}/(\text{m}^2 \, \text{s mol fraction})$.

The same procedure applied to the gas gives

$$N_A = k_{ys} \frac{y - y_i}{(1 - y)_{lm}} \tag{3.18}$$

where

$$(1 - y)_{lm} = \frac{(1 - y_i) - (1 - y)}{\ln \dfrac{1 - y_i}{1 - y}} = \frac{y - y_i}{\ln \dfrac{1 - y_i}{1 - y}} \tag{3.19}$$

At steady state the flux of A from the gas equals the flux of A in the liquid,

$$N_A = k_{ys} \frac{y - y_i}{(1 - y)_{lm}} = k_{xs} \frac{x_i - x}{(1 - x)_{lm}} \tag{3.20}$$

or

$$\frac{y - y_i}{x - x_i} = - \frac{k_{xs}(1 - y)_{lm}}{k_{ys}(1 - x)_{lm}} \tag{3.21}$$

The right-hand side of equation (3.21) gives the slope of a straight line from point 1 on the operating line (Figure 3.9) to point 2 which gives the equilibrium mole fractions at the interface.

The tie-line between points 1 and 2 in Figure 3.9 is found by a trial-and-error method, as x_i and y_i are included in the logarithmic means $(1 - x)_{lm}$ and $(1 - y)_{lm}$. For the first trial, the ratio $(1 - y)_{lm}/(1 - x)_{lm}$ can be assumed to be 1.0, and equation (3.21) is used to find x_i and y_i. These values are used in equations (3.16)

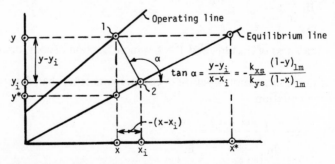

Figure 3.9   Line with slope $-(k_{xs}/k_{ys})[(1 - y)_{lm}/(1 - x)_{lm}]$ that connects a point 1 on the operating line (x, y), with corresponding point 2 on the equilibrium line (x_i, y_i) by diffusion through non-diffusing films (Stefan diffusion)

and (3.19) to obtain $(1 - x)_{lm}$ and $(1 - y)_{lm}$ which is inserted in equation (3.21) to obtain the new slope of the tie-line and a new point 2 in Figure 3.9. Two or three such trials are usually enough.

This method of obtaining equilibrium mole fractions is also valid for curved operating and equilibrium lines.

In dilute mixtures $(1 - y)_{lm}/(1 - x)_{lm}$ may be approximated by 1.0, avoiding the trial procedure.

The number of kmoles of A transferred from the gas to the liquid per unit time in height dz in Figure 3.7, is the molar flux at that point times the area of interface. In the tube (Figure 3.7) and with a liquid film without ripples, the interfacial area is $\pi d_i dz$. Applied to packed towers, however, it is convenient to express the area of the interface as $aS\,dz$, where a = area of interface per unit volume of the tower, m^2/m^3, S = cross-sectional area of the tower, m^2, and dz = height, m.

By diffusion of A through films of non-diffusing components, equation (3.20) gives the mass transfer per unit time and per unit interfacial area,—or over height dz with interfacial area $aS\,dz$,

$$dN_A = N_A(aS\,dz) = \frac{k_{ys}a}{(1-y)_{lm}}(y - y_i)S\,dz = \frac{k_{xs}a}{(1-x)_{lm}}(x_i - x)S\,dz \qquad (3.22)$$

where $k_{ys}a$ and $k_{xs}a$ are volumetric mass transfer coefficients, $kmol/(s \cdot m^3$ packing·mol fraction), for the gas and the liquid side respectively.

Moles of A transferred equals the reduction in number of moles in the gas,

$$dN_A = -d(G'y) = -d\left(\frac{G}{1-y}y\right) = \frac{-G}{(1-y)^2}dy \qquad (3.23)$$

where $G = G'(1 - y)$ = moles of non-diffusing gas/s.

Equations (3.22) and (3.23) give

$$dz = \frac{-G}{k_{ys}aS}\frac{(1-y)_{lm}}{(y-y_i)(1-y)^2}dy$$

Integration of this equation gives the total height of the tube or tower needed to reduce the mole fraction of A in the gas from y_0 to y_1,

$$z = \frac{G}{k_{ys}aS}\int_{y=y_1}^{y_0}\frac{(1-y)_{lm}}{(y-y_i)(1-y)^2}dy \qquad (3.24)$$

where $G/(k_{ys}aS)$ is assumed to be constant. If G is substituted by $G'(1 - y)$, equation (3.24) becomes

$$z = \frac{1}{k_{ys}aS}\int_{y=y_1}^{y_0}\frac{G'(1-y)_{lm}}{(y-y_i)(1-y)}dy \qquad (3.25)$$

The integral in equation (3.24) is the hatched area of Figure 3.10. This plot can be made after calculation of a few points on the curve.

112

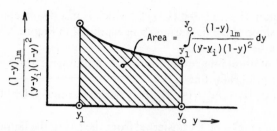

Figure 3.10 Graphical integration of equation (3.24)

The same procedure applied to the liquid side, gives

$$z = \frac{L}{k_{ys}aS} \int_{x=x_1}^{x_0} \frac{(1-x)_{lm}}{(x_i - x)(1-x)^2} \, dx \qquad (3.26)$$

or

$$z = \frac{1}{k_{xs}aS} \int_{x=x_1}^{x_0} \frac{L'(1-x)_{lm}}{(x_i - x)(1-x)} \, dx \qquad (3.27)$$

where L' = total moles of liquid per second, and $L = L'(1 - x)$ = moles of non-diffusing liquid components, kmol/s.

In dilute gases the ratios $(1 - y)_{lm}/(1 - y)^2$ and $(1 - y)_{lm}/(1 - y)$ in equations (3.24) and (3.25) may be approximated by 1.0. Similarly, with dilute liquid solutions, the ratios $(1 - x)_{lm}/(1 - x)^2$ and $(1 - x)_{lm}/(1 - x)$ in equations (3.26) and (3.27) may be approximated by 1.0.

With an average value of G' in equation (3.25) and of L' in equation (3.27), the two equations may be written

$$z = \frac{G'}{k_{ys}aS} \int_{y_1}^{y_0} \frac{(1-y)_{lm}}{(y - y_i)(1-y)} \, dy \qquad (3.25a)$$

and

$$z = \frac{L'}{k_{xs}aS} \int_{x_1}^{x_0} \frac{(1-x)_{lm}}{(x_i - x)(1-x)} \, dx \qquad (3.27a)$$

As in distillation (p. 75), the two integrals represent the number of transfer units (NTU) and the fractions $G'/(k_{ys}aS)$ and $L'/(k_{xs}aS)$ the height per transfer unit (HTU).

Empirical mass transfer coefficients

Mass transfer coefficients reported in the literature refer to different driving forces. For the gas phase it is concentration, partial pressure, or mole fraction,

Table 3.1 Driving force and mass transfer coefficient. The mass transfer coefficient $k_{g,c}$ can be substituted by $RTk_{g,p}$.

Driving force (subscript i at interface)	Units	Mass transfer coefficient				
		Equimolar counter diffusion	Stefan diffusion (through stagnant film)	Units	Conversion to mole fraction as driving force, units kmol/(m² s mol frac.)	
$c_g - c_{g,i}$	$\dfrac{kmol}{m^3}$	$k_{g,c}$	$k_{g,c}\dfrac{c_t}{(c_{inert})_{lm}}$	$\dfrac{m}{s}$	$k_{ye} = c_t k_{g,c}$	$k_{ys} = k_{g,c}\dfrac{c_t}{(1-y)_{lm}}$
$p_g - p_{g,i}$	$Pa = \dfrac{N}{m^2}$	$k_{g,p}$	$k_{g,p}\dfrac{P}{(p_{inert})_{lm}}$	$\dfrac{kmol}{m^2\,s\,Pa}$	$k_{ye} = Pk_{g,p}$	$k_{ys} = k_{g,p}\dfrac{P}{(1-y)_{lm}}$
$y - y_i$	mole frac.				k_{ye}	$k_{ys} = k_{ye}\dfrac{1}{(1-y)_{lm}}$
$c_{l,i} - c_l$	$\dfrac{kmol}{m^3}$	k_l	$k_l\dfrac{c_t}{(c_{inert})_{lm}}$	$\dfrac{m}{s}$	$k_{xe} = c_t k_l$	$k_{xs} = k_l\dfrac{c_t}{(1-x)_{lm}}$
$x_i - x$	mole frac.				k_{xe}	$k_{xs} = k_{xe}\dfrac{1}{(1-x)_{lm}}$

Table 3.2 Properties of tower packings. Reproduced from ref. 9 by permission of Butterworths, London

Item no	Packing	Material	Dimensions (mm)			Number per m³	Void-age ε	Surface area a, (m²/m³)	Hydraulic diameter $d_h = 4\varepsilon/a$ (m)
			Pitch or dia.	Height	Thick-ness				
1	Plain grids	Wood	25	25	6	—	0.75	99	0.030
2	grids	Wood	25	50	6	—	0.75	90	0.033
3	Serrated grids	Wood	50	50	10	—	0.83	43	0.077
4	Stacked	Ceramic	100	100	10	950	0.73	60	0.049
5	Raschig	Ceramic	75	75	10	2 300	0.65	80	0.033
6	rings	Ceramic	50	50	5	7 400	0.74	115	0.026
7		Metal	50	50	1.6	6 180	0.92	97	0.038
8	Random	Ceramic	75	75	10	1 840	0.72	64	0.045
9	Raschig	Ceramic	50	50	6	5 820	0.76	90	0.034
10	rings	Ceramic	38	38	5	14 100	0.72	126	0.023
11		Ceramic	25	25	2.4	46 000	0.80	180	0.018
12	Solid	Coke	25–50				0.40	115	0.014

1	11		12	13	14		15	16	17
	Constants in eq.(3.34)		Minimum liquid velocity V_{lmin} (m/s)	Optimum gas velocity, $V_{go}^\dagger$ (m/s)	Loading velocity, $V_g^\dagger$ (m/s)		Relative cost installed packing, per m³	Liquid side factor, eq.(3.29), C_l	Gas side factor, eq.(3.31), C_g
Item no.	n_0	$b \times 10^{-4}$			$V_l = V_{lmin}$	$V_l = 2V_{lmin}$			
1	6.7	0.97	0.0022	1.5–2.4	2.8	1.3	100	1100	0.07
2	4.2	0.56	0.0020	1.7–2.5	2.8	1.6	80	880	0.06
3	2.9	−1.8	0.0010	2.1–3.3	4.2	2.0	100	900	0.095
4	3.7	0.54	0.0013	1.6–2.4	2.6	1.3	40–150	760	0.055
5	10	2.0	0.0018	1.1–1.5	1.7	0.8	60	860	0.055
6	28	2.4	0.0026	0.9	1.1	0.5	200–290	880	0.055
7	60	3.0	0.0022	0.7–0.9	1.2	0.5	165	850	0.13
8	50	2.4	0.0014	0.7–1.25	2.1	1.0	45–160	760	0.10
9	115	3.3	0.0020	0.55–0.85	1.1	0.5	90	820	0.11
10	130	4.0	0.0028	0.5–0.8	—	—	150–290	950	0.11
11	200	5.3	0.0040	0.4–0.6	0.8	0.4	230–360	1100	0.11
12	—	—	0.0026	0.25	—	—	30	690	0.085

†Optimum velocity and loading velocity are valid for air with density 1.2 kg/m³. For gases with density ρ_g, the numbers in column 14 are to be multiplied by $(1.2/\rho_g)^{1/3}$ and in column 15 by $(1.2/\rho_g)^{1/2}$. Above the loading velocity, pressure drops rise steeply with increasing gas velocities. Loading velocities are roughly 60 per cent of flooding velocities for random packings and 35 per cent for stacked packings.

and for the liquid phase concentration or mole fraction. Table 3.1 gives units and conversion factors for mass transfer coefficients given for equimolar counterdiffusion.

Table 3.1 is given for SI units. Data are often reported in other units, for instance $k_{g,c}$ and k_1 in cm/s (reference 6) and $k_{g,p}$ in kmol/(h m² kPa) or 1b-moles/(ft² hr atm) (reference 7).

Correlations, proposed in the literature, are often rather complicated, while the values of the exponents may be given with an accuracy of three decimals!

Liquid side mass transfer coefficients One of the simpler correlations for the liquid side mass transfer coefficients in packed towers is given by Norman.[8] It covers within ± 20 per cent his measurements with Raschig rings from 10 to 50 mm and Berl saddles from 13 to 38 mm nominal size,

$$\frac{k_1 a}{D} = \beta \left(\frac{V_1 \rho_1}{\mu_1} \right)^n \left(\frac{\mu_1}{\rho_1 D} \right)^{0.5} \tag{3.28}$$

where k_1 = liquid side mass transfer coefficient, m/s (for conversion see Table 3.1), a = specific area of packing material, m²/m³, D = diffusivity of dilute in liquid, m²/s, V_1 = superficial liquid velocity, m/s, ρ_1 = liquid density, kg/m³, μ_1 = dynamic viscosity of liquid, Ns/m² = kg/(s m), $\beta \sim 0.044$ and $n = 0.78$ for the Raschig rings, and $\beta \approx 0.086$ and $n = 0.72$ for the Berl saddles.

Morris and Jackson[9] proposed the correlation

$$k_1 = C_1 (\rho_1/\mu_1)^{0.2} D^{0.5} l^{0.7} \tag{3.29}$$

where C_1 = liquid-film packing factor, given in Table 3.2, column 16, and l = wetting rate, for packed towers defined as the volumetric liquid flow rate divided by the specific area of the packing material, $l = V_1/a$, m³/(s m). To give complete wetting with aqueous solutions, $l > 0.000\,02$ m³/(m s).

Gas side mass transfer coefficients For packing materials with nominal size larger than 15 mm, Charpentier[6] proposed the correlation

$$\frac{k_{g,p} P}{G} = 5.3 (ad)^{-1.7} \left(\frac{Gd}{\mu_g} \right)^{-0.3} \left(\frac{\mu_g}{\rho_g D_g} \right)^{-0.5} \pm 30 \text{ per cent} \tag{3.30}$$

where $k_{g,p}$ = gas side mass transfer coefficient, kmol/(m² s Pa), G = molar superficial gas flow rate, kmol/(m²s), d = packing nominal diameter, m, μ_g = dynamic viscosity of gas, Ns/m² = kg/(s m), ρ_g = density of gas, kg/m³, and D_g = solute gas diffusivity, m²/s.

The following correlation was used by Morris and Jackson[9] for $Re > 5000$:

$$\frac{k_{g,p} R T d_h}{D_g} = C_g \left(\frac{\rho_g V_r d_h}{\mu_g} \right)^{0.75} \left(\frac{\mu_g}{\rho_g D_g} \right)^{0.5} \tag{3.31}$$

where C_g = gas film factor (Table 3.2, column 17), d_h = hydraulic diameter, m (Table 3.2, column 10), D_g = gas phase diffusivity, m²/s, V_r = gas velocity relative to the liquid surface velocity, m/s,

$$V_r \approx V_g/\varepsilon + V_{ls} \tag{3.32}$$

V_g = gas velocity referred to empty tower, m/s, ε = voidage of packing (Table 3.2, column 8), and V_{ls} = liquid surface velocity, m/s,

$$V_{ls} \approx 0.02 + 150l \tag{3.33}$$

for wetting rates $10^{-4} < l < 6 \times 10^{-4} \, \text{m}^3/(\text{s m})$.

Table 3.2 also includes the constants n_0 and b in the equation for the pressure drop through the packing,

$$\Delta P = (n_0 + e^{bl})\rho_g \frac{V_g^2}{2} h \tag{3.34}$$

where h = height of packing, m.

The Norton Company reports products $k_{g,p}a$ that, for some packings, may be reproduced by the equation

$$k_{g,p}a = AL^n \tag{3.35}$$

where L = liquid flow rate, $\text{kg}/(\text{m}^2 \, \text{s})$, and the constants A and n are given in Table 3.3 for absorption in water of carbon dioxide in air at atmospheric pressure and temperature 24 °C, molar gas flow rate $G = 1.22 \, \text{kg}/(\text{m}^2 \, \text{s})$ and liquid rates L between 2.7 and 40 $\text{kg}/(\text{m}^2 \, \text{s})$. The product $k_{g,p}a$ for other gas mixtures and other molar gas flow rates may be estimated from the equation

$$k_{g,p}a = 15.4\left[\left(\frac{GD_g}{\rho_g}\right)^{0.5}\left(\frac{G}{\mu_g}\right)^{0.25} AL^n\right] \tag{3.36}$$

Equation (3.35) gives mass transfer coefficients obtained in closely controlled pilot plant tests. For commercial towers it is recommended to add 50 per cent to packing heights calculated from data obtained in pilot plant tests.

The mass transfer coefficients given in this chapter are all for countercurrent flow in packed towers. Data for mass transfer in concurrent packed towers,

Table 3.3 Constants in equations (3.35) and (3.36) for superficial liquid velocities $7.5 \times 10^{-4} < V_1 < 0.012 \, \text{m/s}$ based on data in references 7, 10, and 11

Packing	Hy-Pak, metal				Metal Intalox saddle			Super Intalox saddle, plastic		
Nominal								Number		
size (mm)	25	38	50	75	25	40	50	1	2	3
Pieces per m³	29 900	9350	3670	1090	158 000	47 000	14 700			
Voidage, ε	0.96	0.96	0.97	0.97	0.967	0.973	0.978			
$A \times 10^6$	1.00	0.70	0.61	0.41	0.87	0.73	0.62	0.98	0.63	0.30
n	0.18	0.25	0.28	0.23	0.33	0.33	0.33	0.17	0.18	0.25

packed and open bubble columns, plate columns, spray towers, and venturis are collected by Laurent and Charpentier,[6,12,13] and for stirred vessels by Sridhar and Potter.[14] However, much of this material is obtained with small-scale equipment and with model systems with chemical reaction in the liquid phase. When designing full-scale absorption equipment, great care must be taken when using data obtained under different conditions. Important conditions may change without being easily recognized. For instance, it has been found that the active interfacial area depends on the reaction rate and the absorption capacity of the liquid phase.[15] Also, the wetting characteristics of the packing will frequently vary with the system and deposits formed during operation.

Overall mass transfer coefficients

The individual products $k_y a$ and $k_x a$ are often not known for the system and equipment of interest, and it is common to make use of the product $K_y a$ or $K_x a$ with the overall mass transfer coefficient K_y or K_x. Expressions for the overall mass transfer coefficients are derived in Chapter 1, equations (1.53) and (1.55),

$$\frac{1}{K_y} = \frac{1}{k_y} + \frac{m}{k_x} \tag{3.37}$$

and

$$\frac{1}{K_x} = \frac{1}{mk_y} + \frac{1}{k_x} \tag{3.38}$$

where $m =$ the slope of the equilibrium line.

In a similar manner as for equations (3.25) and (3.27), the final equations for tower height can be derived in terms of overall coefficients,

$$z = \frac{1}{K_{ys} aS} \int_{y=y_1}^{y_0} \frac{G'(1-y)^*_{lm}}{(y-y^*)(1-y)} \, dy \tag{3.39}$$

and

$$z = \frac{1}{K_{xs} aS} \int_{x=x_1}^{x_0} \frac{L'(1-x)^*_{lm}}{(x^*-x)(1-x)} \, dx \tag{3.40}$$

where

$$(1-y)^*_{lm} = \frac{(1-y^*)-(1-y)}{\ln[(1-y^*)/(1-y)]} \tag{3.41}$$

and

$$(1-x)^*_{lm} = \frac{(1-x)-(1-x^*)}{\ln[(1-x)/(1-x^*)]} \tag{3.42}$$

Equations (3.39) and (3.40) are strictly valid only when $K_{ys}a$ and $K_{xs}a$ are constants, and the integral may be solved graphically.

Experimental data are sometimes reported as K_Ga-values with the difference in partial pressure as the driving force. As an example, in Norton's Bulletin HY-40[7], K_Ga is given in $kmol/(h\,m^3\,kPa)$ and in $1b\text{-}mol/(ft^3\,h\,atm)$. Since the partial pressure p equals yP,

$$K_Ga\Delta p = K_GaP\Delta y = K_{ys}a\Delta y$$

or

$$K_{ys}a = PK_Ga \tag{3.43}$$

where $K_{ys}a$ is in $kmol/(s\,m^3\,mol\,frac.)$, P in $Pa = N/m^2$, and K_Ga in $kmol/(s\,m^3\,Pa)$.

With K_Ga in $kmol/(h\,m^3\,kPa)$, equation (3.43) becomes

$$K_{ys}a = \frac{PK_Ga}{3600 \times 10^3} = 2.78 \times 10^{-7}\,PK_Ga \tag{3.44}$$

and with K_Ga in $1b\text{-}mol/(h\,ft^3\,atm)$,

$$K_{ys}a = \frac{0.4536 \times 35.315}{3600 \times 101\,325}PK_Ga = 4.39 \times 10^{-8}PK_Ga \tag{3.45}$$

With mixtures with mole fractions y and x in the gas and liquid stream less than about 0.1, all terms in equations (3.39) and (3.40) except $(y - y^*)$ and $(x^* - x)$ can be taken outside the integral, using average values at the top and bottom of the tower,

$$z \approx \left[\frac{G'}{K_{ys}aS}\frac{(1-y)^*_{lm}}{1-y}\right]\int_{y_1}^{y_0}\frac{dy}{y-y^*} \tag{3.46}$$

and

$$z \approx \left[\frac{L'}{K_{xs}aS}\frac{(1-x)^*_{lm}}{1-x}\right]\int_{x_1}^{x_0}\frac{dx}{x^*-x} \tag{3.47}$$

The equilibrium concentrations y^* and x^* are determined as shown in Figure 3.11, with operating line as given by equation (3.4), or (3.5) for dilute mixtures. For dilute mixtures following Henry's law, equations (3.3) and (3.5) give

$$y - y^* = \frac{L}{G}x + \left(y_1 - \frac{L}{G}x_0\right) - \frac{H}{P}x = \left(\frac{L}{G} - \frac{H}{P}\right)x + \left(y_1 - \frac{L}{G}x_0\right)$$

or with y^* from equation (3.5),

$$x = \frac{G}{L}y - \left(\frac{G}{L}y_1 - x_0\right)$$

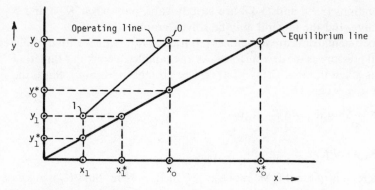

Figure 3.11   Operating line, equilibrium line, and equilibrium compositions in a packed absorption tower

The last two equations give

$$y - y^* = by + c \tag{3.48}$$

where

$$b = 1 - \frac{GH}{LP} \tag{3.49}$$

$$c = \frac{H}{P}\left(\frac{G}{L}y_1 - x_0\right) \tag{3.50}$$

Substituting equation (3.48) into (3.39) gives the final result,

$$z = \left[\frac{G'}{K_{ys}aS}\frac{(1-y)^*_{\text{lm}}}{1-y}\right]\ln\frac{by_0 + c}{by_1 + c} \tag{3.51}$$

The same procedure applied to the liquid side,
gives

$$z = \left[\frac{L'}{K_{xs}aS}\frac{(1-x)^*_{\text{lm}}}{1-x}\right]\ln\frac{dx_0 + e}{dx_1 + e} \tag{3.52}$$

where

$$d = \frac{P}{H}\frac{L}{G} - 1 \tag{3.53}$$

$$e = \frac{P}{H}\left(y_1 - \frac{L}{G}x_0\right) \tag{3.54}$$

Equation (3.51) is preferred if most of the resistance to mass transfer is in the gas, and equation (3.52) if it is in the liquid. Table 3.4 indicates the source of the primary resistance for some common systems. But, under certain circumstances, these systems may fall into categories other than those indicated. For instance,

Table 3.4 Source of major resistance to mass transfer for some common systems.[16] Reproduced from ref. 16. Copyright © 1973 McGraw-Hill. Used with the permission of McGraw-Hill Book Company

Liquid phase	Both phases	Gas phase
O_2–H_2O	SO_2–H_2O	NH_3–H_2O
CO_2–H_2O	NO_2–H_2O	NH_3–acid
Cl_2–H_2O	Acetone–H_2O	SO_2–alkali
CO_2–NaOH	H_2S–alkali	HCl–H_2O
CO_2–amines	H_2S–amines	H_2O–acid

H_2S–alkali can be gas-phase controlled if the H_2S is dilute. Also, both phases are important for CO_2–H_2O if the liquid-to-gas ratio is high.

Absorption with chemical reaction

Absorption of gases as oxygen and nitrogen in water are purely physical absorptions without chemical reaction. But most commercially important absorption processes are accompanied by chemical reactions in the liquid phase, which enlarges the liquid capacity and increases the absorption rate. Table 3.5 shows examples of reagents used by absorption of various gases in aqueous solutions.

The mechanism of the chemical reactions may be complex, as in the case of absorption of H_2S in monoethanol or diethanolamine solutions. The following is limited to the case of an irreversible reaction, and is based on an idealized film model, together with the assumption of equilibrium between the gas phase and the liquid phase at the interface.

Figure 3.12 shows the real concentration profile by reaction between component A from the gas and component B in the liquid when the reaction takes place in a stagnant liquid film. Figure 3.13 gives the idealized concentration profiles. Based on this model, Van Krevelen and Hoftijzer[18,19] derived the

Table 3.5 Absorption systems with chemical reaction[17]

Solute gas, A	Reagent, B	Solute gas, A	Reagent, B
CO_2	Carbonates	H_2S	Ethanolamines
CO_2	Hydroxides	H_2S	$Fe(OH)_3$
CO_2	Ethanolamines	SO_3	H_2SO_4
CO	Cuprous amine complexes	C_2H_4	KOH
CO	Cuprous ammonium chloride	C_2H_4	Trialkyl phosphates
SO_2	$Ca(OH)_2$	Olefins	Cuprous ammonium complexes
SO_2	$HCrO_4$	NO	$FeSO_4$
SO_2	KOH	NO	$Ca(OH)_2$
Cl_2	H_2O	NO	H_2SO_4
Cl_2	$FeCl_2$	NO_2	H_2O

122

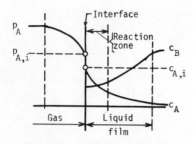

Figure 3.12 Partial pressure of A in the gas and concentration c_A of A and c_B of B in the liquid film

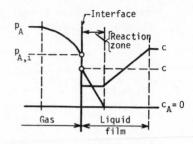

Figure 3.13 Idealized model for the Krevelen–Hoftijzer correlation

chemical acceleration factor, also called the chemical *enhancement factor*,

$$E = \frac{Ha[1-(E-1)Z]^{1/2}}{\tanh\{Ha[1-(E-1)Z]^{1/2}\}} \tag{3.55}$$

given graphically in Figure 3.14. The symbols in equation (3.55) are as follows:

$$Z = \frac{(m/n)D_A c_{A,i}}{D_B c_B} \tag{3.56}$$

where m/n = stoichiometric constant, m moles of B react with n moles of A, D_A = Diffusion coefficient of A in liquid, m²/s, D_B = diffusion coefficient of B in liquid, m²/s, $c_{A,i}$ = concentration of A at the liquid interface, kmol/m³, c_B = concentration of B in bulk of liquid, kmol/m³, Ha = Hatta number, modified by Brian[20] to

$$Ha = \frac{\left(\frac{2}{n+1}k_{AB}c_B^m c_{A,i}^{n-1}D_A\right)^{1/2}}{k_l} \tag{3.57}$$

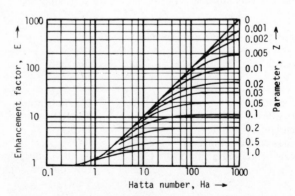

Figure 3.14 The chemical enhancement factor E as a function of the Hatta number Ha and the concentration diffusion parameter Z[17]

where k_{AB} = reaction rate constant, $m^3/(s \, kmol)$, in the equation for a reaction of general order,

$$\frac{dc_A}{dt} = - k_{AB} c_A^n c_B^m \tag{3.58}$$

where c_A = concentration of A in the liquid, and k_1 = liquid phase mass transfer coefficient by physical absorption, m/s.

A first-order reaction where $m = n = 1$ gives the Hatta number as

$$Ha = \frac{(k_{AB} c_B D_A)^{1/2}}{k_1} \tag{3.59}$$

Using the chemical enhancement factor the calculation procedure is the same as without chemical reaction, but with the liquid phase mass transfer coefficient multiplied by the enhancement factor E. Thus, the expression for $\tan \alpha$ in Figure 3.9 becomes

$$\tan \alpha = \frac{y - y_i}{x - x_i} = - \frac{E k_{xs} (1 - y)_{lm}}{k_{ys} (1 - x)_{lm}} \tag{3.60}$$

in the case of Stefan diffusion in both phases and

$$\tan \alpha = \frac{y - y_i}{x - x_i} = \frac{E k_{xe}}{k_{ys}} (1 - y)_{lm} \tag{3.61}$$

with Stefan diffusion in the gas and equimolar counterdiffusion in the liquid. Equations (3.37) and (3.38) become

$$\frac{1}{k_y} = \frac{1}{K_y} + \frac{m}{E k_x} \tag{3.62}$$

and

$$\frac{1}{K_x} = \frac{1}{m k_y} + \frac{1}{E k_x} \tag{3.63}$$

Simple equations for the molar flux may be used in special cases, as

$$N_A = c_{A,i} (k_{AB} c_A D_A)^{1/2} \tag{3.64}$$

for a relatively rapid pseudo-first-order reaction with $E > 2$ and $Z < 0.1/E$, and

$$N_A = c_{A,i} k_1 (1 + 1/Z) \tag{3.65}$$

for an infinitely fast reaction that takes place in a plane parallel to the interface, where $E > 10/Z$.

Figure 3.15 is a packed tower where a component A in a gas stream containing G kmol/s inert gas is absorbed in L kmol/s liquid entering with mole fraction x_{B0} of component B.

A material balance for component A around the top of the tower yields

$$\frac{y}{1 - y} G + L x_{B0} = L x_B + \frac{y_1}{1 - y_1} G$$

124

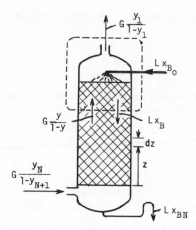

Figure 3.15 Absorption tower;
y = mole fraction A in the gas and
x_B = mole fraction B in the liquid

or

$$\frac{y}{1-y} = \frac{L}{G}x_B + \left(\frac{y_1}{1-y_1} - \frac{L}{G}x_{B0}\right) \tag{3.66}$$

which is the equation for the operating line, assuming L kmol/s liquid to be constant.

The total height of packing needed to reduce the mole fraction of A from y_N to y_1 is given by equation (3.24),

$$z = \frac{G}{k_{ys}aS} \int_{y_1}^{y_N} \frac{(1-y)_{lm}}{(y-y_i)(1-y)^2}dy \tag{3.67}$$

Figure 3.16 shows to the left the equilibrium curve for component A, $y = f_A(x)$, and to the right the operating line, $y = f_B(x_B)$, as given by equation (3.66). In the case of a rapid reaction completed within the liquid film, the mole fraction x in the bulk of the liquid is zero, and equation (3.61) simplifies to

$$\tan\alpha = \frac{y_i - y}{x_i} = -\frac{Ek_{xe}}{k_{ys}}(1-y)_{lm} \tag{3.68}$$

The mole fraction y_i at the interface as a function of y in the bulk of the gas, is determined as shown in Figure 3.16, following the arrows from an arbitrarily chosen point x_B between x_{BN} and x_{B0}. This makes it possible to plot the curve for the expression in the integral equation (3.66) and integrate graphically as shown in Figure 3.10.

Summary of calculations

1. Using equation (3.66) for the operating line, plot y as a function of x_B as shown in Figure 3.16.

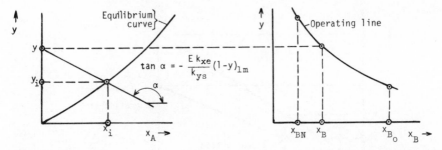

Figure 3.16 Graphical determination of corresponding values of y in the bulk of the gas and y_i at the interface

2. Plot the equilibrium curve for component A, using the same scale for y as for the operating line (see Figure 3.16).
3. Select a value of x_B within the range of interest. The corresponding value of y given by the operating line is plotted at the ordinate of the equilibrium diagram.
4. Estimate the enhancement factor E, plot a line as shown in Figure 3.16 with $\tan \alpha$ as given by equation (3.68), and read x_i and y_i from the intersection with the equilibrium curve.
5. Calculate $c_{A,i} = x_i c$ and the enhancement factor E from equations (3.56) and (3.57) or (3.59), and Figure 3.14 or equation (3.55). If this value of E deviates significantly from the estimated value under 4, recalculate $\tan \alpha$.
6. Repeat the procedure under 3, 4, and 5 for other values of x_B.
7. Plot $[(1 - y)_{lm}]/[(y - y_i)(1 - y)^2]$ versus y as shown in Figure 3.10 and carry out the graphical integration between y_1 and y_N. Use this value in equation (3.67) giving the height of packing.

Recently, promising methods for sizing of commercial absorption columns with chemical reaction from laboratory data have been recommended.[21-24] In one method, the operating conditions of the column are reproduced stepwise in a stirred vessel which is provided with separate stirrers for the gas and the liquid phase, and with baffles, to give well-defined interfacial area. In separate experiments with suitable absorption systems, the relation between k_g and the gas phase revolutions per second and k_l and the liquid phase revolutions per second, is determined. Based on this information k_g and k_l can be set at the same values as in the absorption column. The latter follows from known correlations. The measurements are performed with compositions in the gas and liquid phase set at the same values as at different heights in the column. The absorption rates will then be the same as locally in the column. Simple numerical integration of the measured rate data will yield the necessary height of the packing.

Desorption

Gas desorption or *stripping* is employed when it is desired to transfer a volatile component from a liquid mixture into a gas. It is the reverse of absorption and

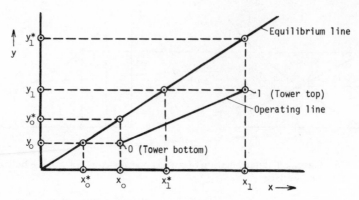

Figure 3.17 Operating line, equilibrium line, and equilibrium compositions in desorption

treated by the same methods. The only differences are that the operating line is located below the equilibrium line as shown in Figure 3.17, and the driving forces listed in Table 3.1 are reversed.

Gas absorption equipment

Figure 3.18 is a schematic drawing of the main types of equipment used for gas absorption.

Which type to be preferred depends on the circumstances, such as liquid-to-gas ratio, gas volume, rate-controlling resistance, corrosivity of the fluids, etc. Except for the venturi, the absorber may also be equipped with a wire mesh demister.

Plate columns The description of plate columns given in Chapter 2 and most of the criteria given in Table 2.3, does also apply to absorption columns. The plate columns are well suited for large gas-to-liquid ratios, large gas volumes (diameter > 1.0 m), processes where the composition of the exit gas approaches the equilibrium composition in contact with the entering liquid, and if intermediate cooling is required in absorption or intermediate heating in desorption.

Systematically packed towers are mainly used for absorption from relatively large gas volumes where it is important to have a low pressure drop, and where the ratio of gas rate to liquid rate is high.

Figures 3.19 and 3.20 show a wooden grid packing. Data for this packing are given in Table 3.2.

Randomly packed towers are usually cheaper than plate columns for smaller diameters (diameter < 1.0 m), and for processes with highly corrosive fluids. The pressure drop is less than in plate columns, and the liquid hold-up is low.

Data for randomly packed towers are given in Chapter 2 and in Table 3.2.

Countercurrent flow is used in most cases with purely physical absorption. However, when an irreversible reaction occurs between the dissolved solute and

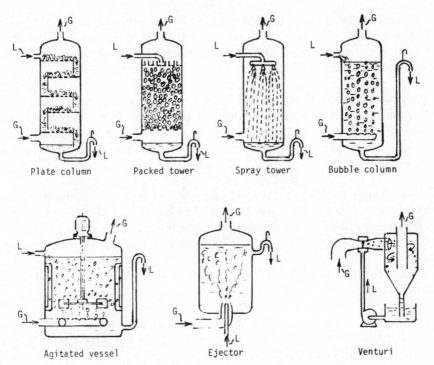

Figure 3.18   Schematic drawings of main types of gas absorption equipment

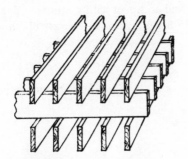

Figure 3.19   Plain wooden grid packing. In large towers vertical rods may be placed with intervals to keep the grids in position

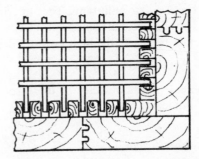

Figure 3.20   Cross-section of corner of square wooden tower with the grid packing in Figure 3.19

the reactive species in the liquid, then there is no difference in the mean driving force in concurrent and in countercurrent flow. Towers with concurrent flow can be operated with higher gas and liquid velocities than with countercurrent flow, and without the limitation by flooding. The packing may even be a catalyst for liquid phase reactions. This is used in petroleum industries for hydro-treatment and in chemical processing for hydrogenation or liquid phase oxidation[25].

Irrigation should be sufficient to ensure complete wetting of the packing, and there is a danger of maldistribution of the fluids. High towers need intermediate support of the packing, and redistribution of the liquid is also recommended. For further details refer to Chapter 2 and to the special literature.[9,26]

Spray towers Spray towers are of interest when only one, or at most two, theoretical stages are required, or when the gas pressure drop must be kept at a minimum. They can handle difficult fluids, as dust- or soot-containing gases. They are also used when large amounts of liquid are required to avoid excessive rise in temperature.

Bubble columns Bubble columns are commonly used in practice as absorbers and desorbers requiring large liquid hold-up. The liquid may also contain solids.

The major disadvantages are back-mixing in the liquid, high gas side-pressure drop, and coalescence of bubbles. Coalescence is reduced in packed bubble columns.

The physical properties of the liquid have a marked effect on the performance[27].

Agitated vessels A mechanically agitated vessel may be preferred in cases with a slow chemical reaction in the liquid phase, especially for low rates of gas flow. The agitator helps to break up the bubbles.

Venturi scrubbers Venturi scrubbers are used for simultaneous removal of gaseous and particulate pollutants. They are simple and compact with no moving parts and low first cost of equipment. Their main disadvantages are the relatively short contact time and the high pressure drop[28],

$$\Delta P = 1000 V_g^2 (V_l/V_g) \tag{3.69}$$

where V_g = gas velocity in the throat, m/s, and V_l/V_g = liquid to gas ratio, m³ liquid/m³ gas.

Absorber design

The first step in absorber design is the selection of type of absorber. Some of the advantages and disadvantages of the different types are given on pp. 126–128.

Plate columns If a plate column is selected, the number of theoretical stages needed may be calculated by a stage-by-stage calculation as given on p. 104, or by the Kremser equation, p. 106.

For *column diameter, tray spacing*, etc. refer to p. 60–69 in Chapter 2.

Plate efficiencies are often low. They should preferably be based on full-scale experimental data with the system in question. Figure 3.21 give a correlation reported by O'Connell[29]. It should be used with care, as it does not include factors such as liquid depth, gas and liquid velocity, and type of plate. At high total pressures it also seems to give too high efficiencies.

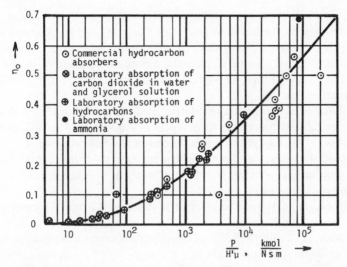

Figure 3.21   Correlation of overall plate efficiencies for absorbers. H' is Henry's law constant with the units Pa/(kmol/m³), P total pressure, Pa, μ liquid dynamic viscosity, N s/m²

Plate efficiencies may be estimated from individual mass transfer coefficients as shown in Example 3.3. For bubble cap plates, Sharma et al.[30] recommend the equations

$$k_{g,p}a = 3.6 \times 10^{-4} V_g^{0.75} S^{-0.67} D_g^{0.5} \tag{370}$$

and

$$k_l a = 1000 \, V_g^{0.75} S^{-0.67} D_l^{0.5} \tag{3.71}$$

where $k_{g,p}$ = gas phase mass transfer coefficient for diffusion in dilute solutions, kmol/(m²s Pa), k_l = liquid phase mass transfer coefficient for diffusion in dilute solutions with concentration as driving force (see Table 3.1), m/s, a = effective interfacial area per unit volume of dispersion, m²/m³ V_g = superficial gas velocity, m/s, S = submergence measured as the distance from half of the slot height to the top of the dispersion, m. In the experiments S varied between 0.07 and 0.22 m, D_g = gas phase diffusion coefficient, m²/s, and D_l = liquid phase diffusion coefficient, m²/s.

Charpentier[6] reports the interfacial area a' in m² per m² plate area as

$$a' = 245 \, V_g^{0.5} S^{0.83} \tag{3.72}$$

for bubble cap plates, and

$$a' = 20 \text{ to } 32 \text{ m}^2/\text{m}^2 \tag{3.73}$$

for sieve plates with dispersion heights in the range 0.08–0.16 m.

Data reported for sieve-plate liquid phase mass transfer coefficients show some discrepancy due to the influence of the presence of electrolyte, antifoaming agents, solids, or ionic strength.[6]

For *column diameter, tray spacing*, etc. refer to pp. 60–69 in Chapter 2.

Packed columns If a packed column is selected, the possible types of packing depend strongly on the liquid to gas ratio. For grids and stacked and dumped Rasching rings, column 13 in Table 3.2, gives optimum gas velocity which gives the cross-section, and column 12 gives the minimum liquid velocity. For several types of dumped packings, Figure 2.53 indicates $(L/G)(\rho_g/\rho_l)^{1/2} > 0.02$ to give sufficient irrigation.

The *column cross-section* may be determined from column 13 in Table 3.2, or from Figure 2.56 with pressure drop $\Delta p/H$ in the range 15 to 30 mm H_2O/m.

The *height of packing* may be determined in the following steps:

— The liquid side mass transfer coefficient is estimated from equations (3.28) or (3.29), and the gas side mass transfer coefficient from equations (3.30) or (3.31), or from equations (3.35) and (3.36).

— The overall mass transfer coefficient is calculated from equations (3.37) or (3.38).

— Integration of equations (3.39) or (3.40) (for dilute mixtures equations (3.46) or (3.47)) gives the packing height z.

The *pressure drop* is calculated from equation (3.34) with the constants in Table 3.3.

Other types of absorbers If another type of absorber is selected, refer to references 4 or 11 for mass transfer data.

Example 3.1 HENRY'S LAW

Owing to inertia of the water in the pipeline to a hydroelectric power station, the pressure in the pipeline increases while the controller reduces the water flow. At a hydroelectric power station located in a tunnel and operating with 60 bar water pressure in front of the turbines, it was considered using an adjacent tunnel as an air cushion (Figure 3.22) in order to reduce pressure fluctuations in the pipeline.

By shut-down, the air in the air cushion could be released through the engine room, and a too low oxygen content could be a hazard to the operators.

At 5 °C, Henry's law constant for oxygen is $H = 2.95 \times 10^9$ Pa/mole fraction and for nitrogen $H = 6.05 \times 10^9$ Pa/mole fraction.[16] The oxygen and nitrogen content of the water in the pipeline is assumed to be in equilibrium with air at 0.94 bar pressure and 5 °C temperature.

Estimate the mole fraction of oxygen in the air cushion at steady state.

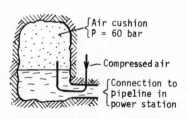

Figure 3.22 Tunnel used as air cushion

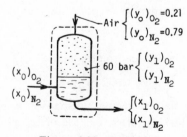

Figure 3.23 Symbols

Solution

Basis 1 kmol water passing into and out of the tunnel, symbols in Figure 3.23. Equation (3.1) gives the gas content in the water from the pipeline,

$$0.21 \times 0.94 \times 10^5 = 2.95 \times 10^9 (x_0)_{O_2}, \qquad (x_0)_{O_2} = 6.7 \times 10^{-6} \, \text{kmol O}_2/\text{kmol water}$$

$$0.79 \times 0.94 \times 10^5 = 6.05 \times 10^9 (x_0)_{N_2}, \qquad \underline{(x_0)_{N_2} = 12.3 \times 10^{-6} \, \text{kmol N}_2/\text{kmol water}}$$

Gas in incoming water $\qquad\qquad\qquad 19.0 \times 10^{-6} \, \text{kmol gas/kmol water}$

An oxygen balance as shown with the dotted cage in Figure 3.23, gives

$$6.7 \times 10^{-6} + 0.21[(x_1)_{O_2} + (x_1)_{N_2} - 19.0 \times 10^{-6}] = (x_1)_{O_2} \tag{a}$$

where the term in brackets is the absorbed gas in kmol/kmol water. The total pressure in the air cushion is the sum of the partial pressures,

$$60 \times 10^5 = 2.95 \times 10^9 (x_1)_{O_2} + 6.05 \times 10^9 (x_1)_{N_2} \tag{b}$$

Equations (a) and (b) are two equations with two unknowns, giving

$$(x_1)_{O_2} = 0.000\,449.$$
$$p_{O_2} = 2.95 \times 10^9 \times 0.000\,449 = 7.23 \times 10^5 \, \text{Pa}$$
$$(y_1)_{O_2} = \frac{7.23 \times 10^5}{60 \times 10^5} = 0.12$$

Example 3.2 GAS ABSORPTION IN PACKED COLUMN

Figure 3.24 shows schematically part of the Fläkt-Hydro flue gas desulphurization process[31] where SO_2 is absorbed in sea-water in countercurrent flow in a packed column.

Uncontaminated sea-water is alkaline with a pH of about 8.3, and it has a considerable buffer capacity. For absorption calculations equilibrium data for SO_2 given by Bromley[32] in the temperature range $\theta = 10\text{--}25\,°C$ may be approximated by the equation

$$p \approx (3320 + 1230\theta)(x - x_0)^{1.9} \tag{a}$$

where p = partial pressure of SO_2 in Pa for mole fractions in the liquid $x > x_0 = 3.5 \times 10^{-5}$; $p = 0$ for $x \leqslant x_0$, and θ = temperature, $°C$.

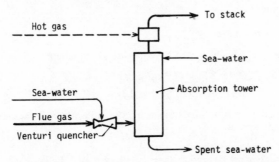

Figure 3.24 The Fläkt-Hydro FGD system where SO_2 is absorbed in cold sea-water. Under unfavourable weather conditions a small amount of hot gas is added to the exit gas giving a temperature rise up to 25 °C

The density of normal sea-water with salinity $S = 34.5\,\text{kg}/1000\,\text{kg}$ sea-water and temperature $10\,^\circ\text{C}$ is $\rho = 1026\,\text{kg/m}^3$, and the viscosity is $\mu = 0.0013\,\text{Ns/m}^2$.

(a) Determine a reasonable diameter for a column packed with no. 3 plastic saddles, operating isothermally at $10\,^\circ\text{C}$, and with $2380\,\text{kmol/h}$ flue gas with average molecular weight $M = 31$ and pressure $P = 1.03 \times 10^5\,\text{Pa}$ (1.03 bar) in countercurrent flow with $535\,\text{m}^3$ sea-water/h.

(b) Estimate the gas phase volumetric mass transfer coefficient $k_{ys}a\,\text{kmol}/(\text{m}^3\text{s}$ mol fraction) when the dynamic viscosity of the gas is $\mu_g = 1.77 \times 10^{-5}\,\text{Ns/m}^2$.

(c) The venturi quencher system and the column together shall be designed for 99 per cent efficiency, reducing the SO_2-content in the gas from $y = 0.0014$ to $y_1 = 0.000\,014\,\text{kmol}$ SO_2/kmol flue gas. Calculate the necessary height of packing in the column, assuming that 10 per cent of the SO_2 is absorbed in $55\,\text{m}^3/\text{h}$ sea-water in the venturi quenching system, and that the major resistance to mass transfer is in the gas phase, i.e. $K_y \approx k_y$.

Solution

(a) Density of flue gas,

$$\rho_g = \frac{MP}{RT} = \frac{31 \times 1.03 \times 10^5}{8314 \times 283} = 1.36\,\text{kg/m}^3$$

Figure 2.56 with abscissa,

$$\frac{535 \times 1026}{2380 \times 31}\sqrt{\frac{1.36}{1026}} = 0.27$$

pressure drop $30\,\text{mm}\,H_2O/\text{m}$ (p. 74), and $F_p = 16$ (Table 2.6), give

$$\frac{G^2 F_p(\mu/\rho_1)^{0.1}}{\rho_g(\rho_1 - \rho_g)} = \frac{G^2 16(0.0013/1026)^{0.1}}{1.36(1026 - 1.36)} = 0.0175$$

or

$$G = 2.43\,\text{kg}/(\text{m}^2\,\text{s})$$

Column cross-section,

$$A = \frac{2380 \times 31}{3600 \times 2.43} = 8.4\,\text{m}^2$$

and column diamter

$$D = \sqrt{4 \times 8.4/\pi} = 3.3\,\text{m}$$

(b) The product $k_{g,p}a$ is calculated by equation (3.36) where

$$L = 535 \times 1026/(3600 \times 8.5) = 17.9\,\text{kg}/(\text{m}^2\,\text{s})$$
$$A = 0.30 \times 10^{-6} \quad \text{and} \quad n = 0.25 \text{ (Table 3.3)}.$$

and D_g is calculated from equation (1.62), using the molecular volumes given in Table 1.1 for sulphur dioxide and air,

$$D_g = \frac{0.0101 \times 283^{1.75}(1/64.06 + 1/29)^{1/2}}{1.03 \times 10^5(41.1^{1/3} + 20.1^{1/3})^2} = 1.13 \times 10^{-5}\,\text{m}^2/\text{s}$$

Equation (3.36),

$$k_{g,p}a = 15.4\left(\frac{2.43 \times 1.13 \times 10^{-5}}{1.36}\right)^{0.5}\left(\frac{2.43}{1.77 \times 10^{-5}}\right)^{0.25} 0.30 \times 10^{-6} \times 17.9^{0.25}$$
$$= 8.2 \times 10^{-7}\,\text{kmol}/(\text{m}^3\,\text{s}\,\text{Pa})$$

or with the conversion factor (Table 3.1)

$$k_{ys}a = k_{g,p}a\frac{P}{(1-y)_{lm}} \approx 8.2 \times 10^{-7} \times 1.03 \times 10^5 = 0.084\frac{kmol}{m^3\,s\,mol\,frac.}$$

(c) $535\,m^3/h$ sea-water contains $535(1026 - 34.5)/18.02 = 29\,440\,kmol\,H_2O/h$. The mole fraction SO_2 at the inlet to the column is

$$y_0 = 0.9 \times 0.0014 = 0.00126$$

Absorbed,

$$(y_0 - y_1)2380 = (0.00126 - 0.000014)2380 = 29.65\,kmol\,SO_2/h,$$

giving the mole fraction in the sea-water from the column,

$$x_n = 29.65/29\,440 = 0.0010$$

Equation (a) gives the corresponding equilibrium pressure of SO_2,

$$p = (3320 + 1230 \times 10)(0.0010 - 3.5 \times 10^{-5})^{1.9} = 0.030\,Pa$$

which is negligible. With $(1-y)_{lm}/(1-x)_{lm} \approx 1.0$, equation (3.46) gives the height of packing,

$$z = \frac{G'}{K_{ys}aS}\int_{y_1}^{y_0}\frac{dy}{y} = \frac{2380/3600}{0.084 \times 8.5}\int_{0.000014}^{0.00126}\frac{dy}{y} = 4.2\,m$$

Comments: In the solution some oxidation of the bisulphite/sulphite system will occur. This has been neglected in the present calculations.

References 31 and 33 report 99.4 per cent efficiency of a similar system being used for industrial boiler flue gas of the same composition as in this example.

Example 3.3 GAS ABSORPTION IN A PLATE COLUMN

The water content of $2.4\,kmol/s$ natural gas at 69 bar pressure and 30 °C temperature is to be reduced from $0.014\,kg$ water vapour/kmol gas to $0.0026\,kg$ water vapour/kmol gas by absorption in triethylene glycol (TEG) in countercurrent flow in a plate column. The TEG solution entering the top plate is $0.40\,kg/s$ containing 98.2 wt per cent TEG, and at 30 °C temperature.

Campbell[34] reports an activity coefficient for water in TEG corresponding to the van Laar equation

$$\ln\gamma_1 = \frac{-0.94}{(1 + 1.13x_1/x_2)^2} \tag{a}$$

giving the partial pressure of water in a TEG–water solution at 30 °C,

$$p_1 = \exp\left[\frac{-0.94}{(1 + 1.13x_1/x_2)^2}\right]4241x_1 \quad N/m^2 \tag{b}$$

The natural gas has an average molecular weight of 20. At 30 °C temperature and 69 bar pressure, the compressibility factor of the gas is $z \approx 0.90$, i.e.

$$Pv = 0.90nRT \tag{c}$$

The water vapour is assumed to be an ideal gas, and the process is assumed to be isothermal. Figure 3.25 gives the surface tension, density, and viscosity of TEG solutions at 30 °C.

The molar volume of liquid TEG at its normal boiling-point, 278 °C, is estimated to approximately $0.17\,m^3/kmol$. The molecular weight of TEG is 150.2.

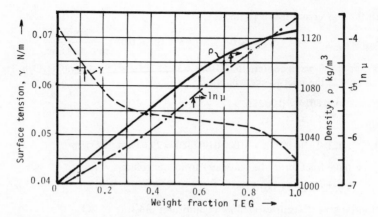

Figure 3.25 Properties of TEG–water solutions at 30 °C. The viscosity μ is in N s/m². Reproduced by permission of Campbell Petroleum Series, Norman, Oklahoma

(a) Calculate the number of theoretical stages needed.
(b) Calculate column cross-section.
(c) Estimate the plate efficiency.
(d) Select a reasonable number of actual plates.

Solution

(a) Absorbed, $2.4(0.014 - 0.0026) = 0.0274$ kg water/s
With the symbols in Figure 3.6:

$$\text{TEG in } L_0', \quad 0.4 \times 0.982 = 0.3928 \text{ kg/s}$$
$$\text{Water in } L_0', \quad 0.4 - 0.3928 = 0.0072 \text{ kg/s}$$

Water in L_N', $0.0274 + 0.0072 = 0.0346$ kg/s

$$x_0 = \frac{0.0072/18}{0.0072/18 + 0.3928/150.2} = 0.133$$

$$x_N = \frac{0.0346/18}{0.0346/18 + 0.3928/150.2} = 0.424$$

Equation (b) gives Henry's law constant for the feed,

$$H = \left[\exp\frac{-0.94}{(1 + 1.13 \cdot 0.133/0.867)^2} \right] 4241 = 2143 \text{ Pa/mol frac.}$$

and at the bottom,

$$H = \left[\exp\frac{-0.94}{(1 + 1.13 \cdot 0.424/0.576)^2} \right] 4241 = 3205 \text{ Pa/mol frac.}$$

With the deviation from ideal gas behaviour given by equation (c), the number of kmoles natural gas in a volume v is $n = Pv/0.90RT$, while the number of kmoles water vapour in the same volume is $n_1 = p_1 v/RT$, or $y_1 = n_1/(n + n_1) \approx n_1/n = p_1/(P/0.90)$. This is taken into account by replacement of P with $P/0.90$ in the expression for the absorption factor A, equation (3.8).

Absorption factor of the liquid feed,

$$A = \frac{P/0.90}{H}(L'/G') = \frac{69 \times 10^5}{2143}(0.0072/18 + 0.3928/150.2)/2.4 = 4.49$$

and at the bottom,

$$A = \frac{69 \times 10^5/0.90}{3205}(0.0346/18 + 0.3928/150.2)/2.4 = 4.52$$

or average, $A = 4.51$.

The gas phase mole fractions in equation (3.14) are (symbols in Figure 3.6),

$$y_{N+1} = \frac{0.014/18}{1 + 0.014/18} = 0.000\,777$$

$$y_1 = \frac{0.0026/18}{1 + 0.0026/18} = 0.000\,144$$

$$y_0 = \frac{H}{P/0.90}x_0 = \frac{2143}{69 \times 10^5/0.90}0.133 = 0.000\,037$$

The Kremser equation (3.14) gives the number of theoretical stages,

$$N = \frac{\ln\left[\frac{0.000\,777 - 0.000\,037}{0.000\,144 - 0.000\,037}\left(1 - \frac{1}{4.51}\right) + \frac{1}{4.51}\right]}{\ln 4.51} = 1.14$$

(b) The weight fraction of TEG in the liquid from the bottom tray is $0.3928/(0.3928 + 0.0346) = 0.919$, and Figure 3.25 gives surface tension $\gamma = 0.049\,\text{N/m}$ and liquid density $\rho_1 = 1123\,\text{kg/m}^3$. Equation (c) gives the vapour density,

$$\rho_g = \frac{20 \times 69 \times 10^5}{0.90 \times 8314.3 \times 303.2} = 61\,\text{kg/m}^3$$

The abscissa of Figure 2.41,

$$\frac{0.3928 + 0.0346}{20 \times 2.4}\left(\frac{61}{1123}\right)^{0.5} = 0.0021$$

or $C_f = 0.26$ for tray spacing 610 mm.

Equation (2.41) gives the flooding velocity,

$$V_f = 0.26 \times 0.049^{0.2}\left(\frac{1123 - 61}{61}\right)^{0.5} = 0.59\,\text{m/s}$$

Equation (2.42) with $\varphi = 0.8$ and $a = 0.75$ (Table 2.4),

$$V_g = 0.75 \times 0.8 \times 0.59 = 0.36\,\text{m/s}$$

and column area minus downcomer area,

$$A = \frac{20 \times 2.4/61}{0.36} = 2.19\,m^2$$

Equation (2.43) gives liquid velocity in downcomer,

$V_d = 0.17 \times 0.75 = 0.13 \, \text{m/s}$ or downcomer area

$$A_d = \frac{0.4/1123}{0.13} = 0.003 \, \text{m}^2$$

Column diameter 1.7 m gives total cross-section $(\pi/4) \, 1.7^2 = 2.27 \, \text{m}^2$, and Figure 2.40 gives reversed liquid flow.

(c) The following data are inserted in equation (3.70):

$V_g = 0.36 \, \text{m/s}$

$S = 0.12 \, \text{m}$ (average value)

D_g is estimated. Appendix 1 gives $PD_{AB} = 3.61 \, \text{Pa}(\text{m}^2/\text{s})$ for the methane–water vapour binary system at 352.3 K, or at 303.2 K, (equation (1.62)),

$$PD_{AB} = 3.61(303.2/352.3)^{1.75} = 2.78 \, \text{Pa}(\text{m}^2/\text{s})$$

Equation (3.70).

$$k_{g,p}a = 3.6 \times 10^{-4} \times 0.36^{0.75} \times 0.12^{-0.67}[2.78/(69 \times 10^5)]^{0.5}$$
$$= 4.4 \times 10^{-7} \, \text{kmol}/(\text{m}^3 \, \text{s} \, \text{Pa})$$

Equation (1.64) gives for water in TEG in infinite dilution,

$$D_{AB}^0 = 1.17 \times 10^{-16} \times 150.2^{0.5} \frac{303.2}{0.028 \times 0.0189^{0.6}} = 1.7 \times 10^{-10} \, \text{m}^2/\text{s}$$

where 0.028 is the viscosity of TEG (Figure 3.25). TEG in water in infinite dilution.

$$D_{AB}^0 = 1.17 \times 10^{-16} \times 18^{0.5} \frac{303.2}{0.00083 \times 0.17^{0.6}} = 1.7 \times 10^{-10} \, \text{m}^2/\text{s}$$

Equation (1.66) for the liquid from the bottom tray,

$$D_{AB} = (1.7 \times 10^{-10})^{0.576}(5.3 \times 10^{-10})^{0.424} = 2.7 \times 10^{-10} \, \text{m}^2/\text{s}$$

Equation (3.71),

$$k_l a = 1000 \times 0.36^{0.75} \times 0.12^{-0.67}(2.7 \times 10^{-10})^{0.5} = 0.032 \, \text{m/s}$$

Using the conversion factors for Stefan diffusion (Table 3.1), the mass transfer in kmol per second and per m^2 tray surface is

$$w = k_l aS \frac{c_t}{(1-x)_{lm}}(x_i - x) = k_{g,p}aS \frac{P}{(p_{inert})_{lm}}(p - p_i)_{lm} \tag{d}$$

or with the data for the bottom tray,

$$0.032 \times 0.12 \frac{c_t}{(1-x)_{lm}}(x_i - 0.424) = 4.4 \times 10^{-7} \times 0.12 \frac{P}{(p_{inert})_{lm}}(p - p_i)_{lm} \tag{e}$$

where as a first approximation $c_1 \approx (0.919/150.2 + 0.081/18)1126 = 11.96 \, \text{kmol/m}^3$, $(1-x)_{lm} \approx 1 - 0.424 = 0.576$, and $P \approx p_{(inert)_{lm}}$,

$$(x_i - 0.424) = 6.6 \times 10^{-7}\left\{p - \exp\left[\frac{-0.94}{[1 - 1.13x_i/(1 - x_i)]^2}\right]4241 x_i\right\}_{lm} \tag{f}$$

The volume of 1 kmol gas is

$$v = 0.90RT/P = 0.90 \times 8314.3 \times 303.2/(69 \times 10^5) = 0.329\,\text{m}^3$$

giving the partial pressure of water vapour in the entering gas,

$$p_1 = n_1 RT/v = (0.014/18)8314.3 \times 303.2/0.329 = 5960\,\text{Pa}$$

To obtain the order of magnitude of x_i, this pressure is inserted for p in equation (f), giving $x_i = 0.427$. That is practically all resistance to mass transfer is in the gas phase. Hence, it is sufficient with the approximation $x_i = 0.427$, giving

$$p_i \approx \left[\exp\frac{-0.94}{(1 + 1.13 \times 0.427/0.573)^2}\right]4241 \times 0.427 = 1373\,\text{Pa}$$

$$w = 4.4 \times 10^{-7} \times 0.12(p - 1373)_{\text{lm}}\,\text{kmol}/(\text{m}^2\ \text{trays})$$

The partial pressure of water vapour in the gas leaving the bottom tray is

$$p_N = \frac{2.4 \times 0.014 - 2.19 \times 18w}{2.4 \times 0.014}5960\,\text{Pa}$$

$$(p - 1373)_{\text{lm}} = \frac{(5960 - 1373) - (p_N - 1373)}{\ln\dfrac{5960 - 1373}{p_N - 1373}}\,\text{Pa}$$

There three last equations give $p_N = 4544$ Pa. The efficiency of the last (Nth) tray,

$$\eta = \frac{5960 - 4544}{5960 - 1373} = 0.31$$

(d) Adding a safety factor of at least 20 per cent gives

$$1.2 \times 1.14/0.31 = 4.4, \quad \text{i.e. five actual trays.}$$

Comment: Campbell[34] uses 25 per cent plate efficiency for the absorption in TEG of water vapour from natural gas.

Actual installations have from four to eight trays.

The heat released by condensation of 0.0274 kg water vapour per second is only enough to increase the gas temperature by 3 °C, and it is not worth while carrying out the more complicated calculation where temperature changes are taken into account.

It is possible that two parallel columns with diamter 1.2 m would be preferred, so that half the capacity is available if one column is shut down for repair.

Example 3.4 DESORPTION OF OXYGEN FROM WATER

To reduce corrosion and bacteria growth, dissolved oxygen in water is removed before the water is used for injection in an oil reservoir for water drive. The oxygen content may be reduced to 0.03–0.05 ppm by desorption, and the rest removed by chemical scavenging with sodium sulphite.

Within a limited temperature range, Henry's law constants for oxygen and nitrogen in water reported by Freshour[35] and Shelton[36] can be reproduced by the equation

$$H = (A + B\theta)10^9 \qquad \text{Pa/mole fraction} \tag{a}$$

where θ = temperature in °C

A and B = constants given in Table a.

138

Table a   Constants A and B in equation (a)

	Oxygen	Nitrogen	Note
Temp. range, °C	0–40	0–32	Fresh
A	2.61	5.41	water
B	0.071	0.131	
Temp. range, °C	0–20	0–25	Sea-
A	3.43	7.70	water
B	0.092	0.182	

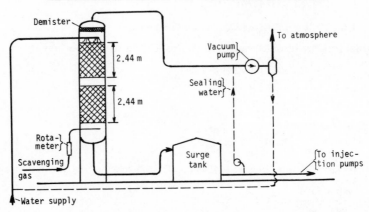

Figure 3.26   Vacuum deaerator with two sections 1676 mm diameter packed with 50 mm Intalox plastic saddles, each section 2.44 m high. Reproduced by permission; copyright Society of Petroleum Engineers of AIME, SPE 4064

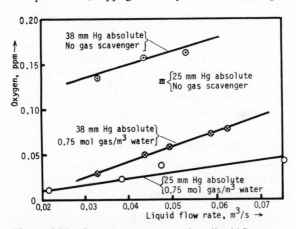
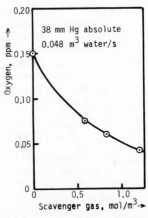

Figure 3.27   Oxygen content at various liquid flow rates without and with 0.000 75 kmol scavenging gas per m³ water. Reproduced by permission; copyright Society of Petroleum Engineers of AIME, SPE, 4064

Figure 3.28   Effect of gas scavenging on oxygen content of water. Reproduced by permission; copyright Society of Petroleum Engineers of AIME, SPE 4064.

Below 57 °C the saturation pressure of water vapour is,

$$p = \exp[23.7093 - 4111/(237.7 + \theta)] \, \text{Pa} \qquad (b)$$

for fresh water, and

$$p = \exp[23.6907 - 4111/(237.7 + \theta)] \, \text{Pa} \qquad (c)$$

for sea-water.

Figure 3.26 shows a one-stage vacuum deaeration process and Figures 3.27 and 3.28 results obtained with this installation.[37]

(a) Calculate the content of oxygen and nitrogen in water at 21 °C (70 °F) in equilibrium with air with 0.21 atm partial pressure of oxygen and 0.79 atm partial pressure of nitrogen.

(b) The experimental data reported in Figure 3.27 with 0.75 mol scavenging gas per m^3 water are given in Tables b and c.

Table b Pressure 38 mm Hg absolute, 0.75 mol scavenging gas/m^3 water

m^3 water/s	0.033	0.044	0.0495	0.0585	0.0625
Oxygen (ppm)	0.028	0.050	0.058	0.073	0.079

Table c Pressure 25 mm Hg absolute, 0.75 mol scavenging gas/m^3 water

m^3 water/s	0.022	0.0385	0.0475	0.0755
Oxygen (ppm)	0.011	0.0225	0.038	0.044

Assuming water temperature 21 °C and all resistance to mass transfer being in the liquid phase, calculate the products ak_{xs} where a is the contact area in m^2 per m^3 packing and k_{xs} the mass transfer coefficient in kmol/(m^2s mole fraction).

(c) Estimate the capacity (water flow rate) for the same column when operated with sea-water at 10 °C, total pressure 15 mm Hg = 2000 Pa, and when 0.3 mol scavenging gas per m^3 water yields water with 0.05 ppm oxygen. The sea-water has a density of 1026 kg/m^3 and contains approximately 55 kmol water/m^3.

(d) Calculate the volumetric flow from the top of the column.

Solution

(a) Equation (3.1) solved for x and with the constants in equation (a), gives the mole fractions,

$$x_{O_2} = \frac{0.21 \times 1.01\,325 \times 10^5}{(2.61 + 0.071 \times 21)10^9} = 5.19 \times 10^{-6} \, \text{kmol } O_2/\text{kmol water}$$

and

$$x_{N_2} = \frac{0.79 \times 1.01\,325 \times 10^5}{(5.41 + 0.131 \times 21)10^9} = 9.81 \times 10^{-6} \, \text{kmol } N_2/\text{kmol water}$$

or

$$32 \times 5.19 \times 10^{-6}/18 = 9.2 \times 10^{-6} \, \text{kg } O_2/\text{kg water or } 9.2 \, \text{ppm}$$

and

$$28 \times 9.81 \times 10^{-6}/18 = 15.3 \times 10^{-.6} \, \text{kg } N_2/\text{kg water or } 15.3 \, \text{ppm}$$

(b) The following calculations refer to the first point in Table b.

Total pressure, $\qquad P = 38 \times 133.32 = 5066 \text{ Pa}$

Pressure of water vapour, $p_{H_2O} = \exp[23.7093 - 4111/(237.7 + 21)] = \underline{2486 \text{ Pa}}$
Partial pressure of oxygen, nitrogen, and scavenging gas $\qquad \underline{2580 \text{ Pa}}$

Exit mole fraction of oxygen,

$$x_1 = (0.028/9.2)5.19 \times 10^{-6} = 1.58 \times 10^{-8} \text{ kmol O}_2/\text{kmol water}$$

The amount of oxygen released below a point with mole fraction x of oxygen is

$$x - 1.58 \times 10^{-8} \text{ kmol O}_2/\text{kmol water}$$

The amount of nitrogen released below the same point is (see comment at end of solution)

$$\frac{9.81 \times 10^{-6}}{(5.19 \times 10^{-6})^2}[x^2 - (1.58 \times 10^{-8})^2] = 3.64 \times 10^5 x^2 - 9.1 \times 10^{-11} \frac{\text{kmol N}_2}{\text{kmol water}}$$

Total gas at this point,

$$(x - 1.58 \times 10^{-8}) + (3.64 \times 10^5 x^2 - 9.1 \times 10^{-11}) + \frac{0.00075}{1000/18}$$

$$= 3.64 \times 10^5 x^2 + x + 1.348 \times 10^{-5} \text{ kmol gas/kmol water}$$

The equilibrium mole fraction oxygen at the interface, x_i, is the partial pressure of oxygen divided by Henry's law constant for oxygen, $H = (2.61 + 0.071 \times 21)10^9 = 4.1 \times 10^9$, giving

$$x_i = \frac{2580(x - 1.58 \times 10^{-8})}{(3.64 \times 10^5 x^2 + x + 1.348 \times 10^{-5})4.1 \times 10^9}$$

$$x - x_i = \frac{x^3 + 2.75 \times 10^{-6} x^2 + 3.53 \times 10^{-11} x + 2.7 \times 10^{-20}}{x^2 + 2.75 \times 10^{-6} x + 3.7 \times 10^{-11}} \qquad (d)$$

Rearranging equation (3.27) for desorption with $(x - x_i)$ instead of $(x - x_i)$ and $(1 - x)_{lm}$

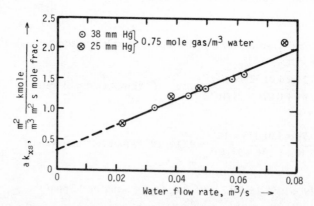

Figure 3.29 Mass transfer coefficient per unit volume of packing as a function of liquid flow rate in the column of Figure 3.26 with cross-section 2.2 m^2

and $(1 - x)^2 \approx 1.00$, gives

$$ak_{xs} = \frac{L}{Sz} \int_{x_1}^{x_0} \frac{dx}{x - x_i} = \frac{0.033 \times 1000/18}{(\pi/4)1.676^2 \times 4.88} \int_{x_1}^{x_0} \frac{dx}{x - x_i}$$

$$= 0.17 \int_{1.58 \times 10^{-8}}^{5.19 \times 10^{-6}} \frac{dx}{x - x_i} \tag{e}$$

From equation (d), $x - x_i$ inserted in equation (e) gives

$$ak_{xs} = 0.17 \times 6.01 = 1.02 \frac{m^2}{m^3} \frac{kmol}{m^2 \, s \, mole \, fraction}$$

The same calculations applied to the rest of the data in Tables b and c give the results listed in Tables d and e and plotted in Figure 3.29.

Table d Pressure 38 mm Hg, 0.75 mol scavenging gas/m^3 water

Water (m^3/s)	0.033	0.044	0.0495	0.0585	0.0625
$ak_{xs} \dfrac{m^2}{m^3} \dfrac{kmol}{m^2 \, s \, mole \, frac.}$	1.02	1.23	1.34	1.51	1.59

Table e Pressure 25 mm Hg, 0.75 mol scavenging gas/m^3 water.

(water (m^3/s)	0.022	0.0385	0.0475	0.0755
$ak_{xs} \dfrac{m^2}{m^3} \dfrac{kmol}{m^2 \, s \, mole \, frac.}$	0.77	1.21	1.36	2.11

The increase in the product ak_{xs} with increasing flow rate is probably due to both increased interfacial area a and increased mixing in the liquid.

(c) Mole fractions in the sea-water,

$$x_{O_2} = \frac{0.21 \times 1.01 \, 325 \times 10^5}{(3.43 + 0.092 \times 10)10^9} = 4.89 \times 10^{-6} \, kmol \, O_2/kmol \, water$$

and

$$x_{N_2} = \frac{0.79 \times 1.013 \, 25 \times 10^5}{(7.7 + 0.182 \times 10)10^9} = 8.41 \times 10^{-6} \, kmol \, N_2/kmol \, water$$

Here, 0.05 oxygen corresponds to the mole fraction

$$x_1 = (0.05 \times 10^{-6} \times 1026/32)55 = 2.9 \times 10^{-8} \, kmol \, O_2/kmol \, water$$

The partial pressure of oxygen, nitrogen, and scavenging gas, is

$$2000 - \exp[23.6907 - 4111/(237.7 + 10)] = 795 \, Pa$$

The same procedure as under (b) gives total gas

$$x - 2.9 \times 10^{-8} + 3.517 \times 10^5 [x^2 - (2.9 \times 10^{-8})^2] + 0.0003/55$$
$$= 3.517 \times 10^5 x^2 + x + 5.43 \times 10^{-6} \, \text{kmol gas/kmol water}$$

and the equilibrium mole fraction of oxygen at the interface,

$$x_i = \frac{795(x - 2.9 \times 10^{-8})}{(3.517 \times 10^5 x^2 + x + 5.43 \times 10^{-6})(3.43 + 0.092 \times 10)10^9}$$

$$x - x_i = \frac{x^3 + 2.843 \times 10^{-6} x^2 + 1.49 \times 10^{-11} x + 1.5 \times 10^{-20}}{x^2 + 2.843 x + 1.54 \times 10^{-11}}$$

The straight line in Figure 3.29 corresponds to

$$ak_{xs} = 0.32 + 0.38 L \; \frac{\text{m}^2}{\text{m}^3} \frac{\text{kmol}}{\text{m}^2 \, \text{s mole fraction}}$$

where L = water flow rate, kmol/s.

Equation (3.27) with $x_1 = 2.9 \times 10^{-8}$ and $x_0 = 4.89 \times 10^{-6}$, gives

$$4.88 = \frac{L}{(0.32 + 0.38L)(\pi/4)1.676^2} \int_{x_1}^{x_0} \frac{x^2 + 2.84 \times 10^{-6} x + 1.54 \times 10^{-11}}{x^3 + 2.84 \times 10^{-6} x^2 + 1.49 \times 10^{-11} x + 1.5 \times 10^{-20}} dx$$

$$= \frac{L}{2.2(0.32 + 0.38L)} 5.51$$

$$L = 3.0 \, \text{kmol/s}$$

or

$$3.0/55 = 0.055 \, \text{m}^3 \, \text{sea-water/s}$$

(d) $4.89 \times 10^{-6} - 2.9 \times 10^{-8}$ $= 4.86 \times 10^{-6} \, \text{kmol O}_2/\text{kmol water}$
$8.41 \times 10^{-6} - 3.517 \times 10^5 (2.9 \times 10^{-8})^2$ $= 8.41 \times 10^{-6} \, \text{kmol N}_2/\text{kmol water}$
$0.0003/55$ $= 5.45 \times 10^{-6} \, \text{kmol gas/kmol water}$

Oxygen, nitrogen, and scavenging gas $= 18.72 \times 10^{-6} \, \text{kmol/kmol water}$

Water vapour, $\dfrac{2000 - 795}{795} 18.72 \times 10^{-6} = 28.38 \times 10^{-6} \, \text{kmol/kmol water}$

Total gas and water vapour $= 47.10 \times 10^{-6} \, \text{kmol/kmol water}$

or

$$47.10 \times 10^{-6} \times 3.0 = 1.41 \times 10^{-4} \, \text{kmol/s}.$$

Volume,

$$v = \frac{1.41 \times 10^{-4} \times 8314 \times 283}{2000} = 0.17 \, \text{m}^3/\text{s}.$$

Comment: The calculations are based on the assumptions of entering water in equilibrium with atmospheric air at the water temperature in the column, that the experimental data were obtained with water temperature 21 °C, and no air leakage.

Supersaturation has only a small influence of the performance of the column, but considerable influence on the size of the vacuum pump. Reference 37 does not give the water temperature, but only the design temperature, 70 °F. In industrial installations, it is also good engineering practice to make allowance for some leakage in vacuum equipment.[38]

The equation for mole fraction nitrogen at a point with mole fraction x of oxygen comes from the following reasoning.

Taking the water, water vapour, and scavenging gas as inerts, the stripping of oxygen and nitrogen is similar to a differential distillation of a binary mixture of the two components. Referring the differential distillation to the oxygen and nitrogen contained in $1/[(5.19 + 9.81)10^{-6}] = 6.67 \times 10^4$ kmol water, L_0 in equation (2.16) is 1.0, and this equation gives the amount of oxygen plus nitrogen left in the water as

$$L = \exp\left(\ln\frac{x_N'}{(x_N')_0} - \alpha \ln\frac{1 - x_N'}{(x_O')_0} \right) \bigg/ (\alpha - 1) \text{ kmol } O_2 + N_2 \tag{f}$$

where $x_N' = $ mole fraction nitrogen in L, $(x_N')_0 = $ mole fraction nitrogen in L_0, $(x_N')_0 = 9.81/(9.81 + 5.19) = 0.654$, $(x_O')_0 = $ mole fraction oxygen in L_0, $(x_O')_0 = 1 - 0.654 = 0.346$, and $\alpha = $ relative volatility of nitrogen $= $ ratio of Henry's law constant for the two components, $\alpha = 8.16 \times 10^9/(4.1 \times 10^9) = 2.0$.

The corresponding mole fraction oxygen in the water is

$$x = (1 - x_N')L/(6.67 \times 10^4) \text{ kmol } O_2/\text{kmol water} \tag{g}$$

and of nitrogen

$$x_N = xx_N'/(1 - x_N') \text{ kmol } N_2/\text{kmol water} \tag{h}$$

For the special case of $\alpha = 2.0$, the three equations give

$$x_N = [(x_N)_0/x_0^2]x^2 \text{ kmol } N_2/\text{kmol water}$$

where $(x_N)_0 = $ kmol N_2/kmol water before stripping, $(x_N)_0 = 9.81 \times 10^{-6}$, and $x_0 = $ kmol O_2/kmol water before stripping, $x_0 = 5.19 \times 10^{-6}$.

Problems

3.1 Sea-water with temperature 10 °C is in contact with air at pressure 101.5 kPa and with mole fraction oxygen 0.21. Using the Henry's law constants and the vapour pressure equation given in Example 3.4, calculate
(a) the oxygen content in the sea-water in equilibrium with the ambient air, and
(b) the equilibrium oxygen content in the sea-water after evacuation to absolute pressure 1.5 kPa.

3.2 Exit gas from a process contains 30 per cent by volume inert gases with molecular weight 29 and 70 per cent H_2S and water vapour (per cent given on dry basis). The H_2S-content is to be reduced to 1 per cent by volume (dry basis) by countercurrent absorption in water at 19 °C in a tower with bubble cap trays.

The total pressure of the gas from the process is 1.00 bar, the partial pressure of water vapour 0.022 bar, and the temperature 19 °C. Henry's law constant for H_2S in water is $H = 5 \times 10^7$ Pa/mole fraction.
(a) Neglecting heats of absorption, calculate the amount of water vapour that condenses in the absorber in kmol per kmol inert gas, assuming equilibrium at all trays. *Ans.* 5.2 $\times 10^{-2}$ kmol H_2O/kmol inerts.
(b) Plot the operating line and the equilibrium line, using $X = $ kmol H_2S/kmol H_2O and $Y = $ kmol H_2S/kmol inert gases. Determine the number of equilibrium stages needed if the amount of water is 1.2 times minimum (minimum gives infinite number of stages). *Ans.* Five stages.

3.3 The ammonium content in 980 kg/h air has to be reduced from 5 per cent by volume to

0.01 per cent by absorption in water. For this purpose it is considered that an existing tower should be used, with diameter 0.5 m and 4 m high packing of 50 mm ceramic Raschig rings, item no. 9 in Table 3.2.

Pressure and temperature in the tower will be 1.00 bar and 20 °C. The absorption equilibrium can be approximated as $y = 1.14x$, and the resistance to mass transfer is almost entirely in the gas phase (Table 3.4).

What is the maximum obtainable ammonia concentration in the water from the column if the requirement maximum 0.01 per cent NH_3 in the exit gas has to be met? How does the liquid velocity compare with minimum liquid velocity given in Table 3.2?

3.4 50 kmol/h exit gas from a process contains 98 per cent by volume air and 2 per cent by volume benzene. 95 per cent of the benzene is recovered by absorption in an oil with average molecular weight 260. The benzene-rich oil from the absorption tower is pumped to a stripping tower where benzene is stripped off with superheated water vapour. The vapours from the stripping tower are condensed to give two liquid phases, and the oil from the bottom of the stripping tower is cooled and returned to the absorption tower.

The absorption tower is operated at total pressure 106.7 kPa and temperature 27 °C. The liquid feed has mole fraction benzene $x = 0.005$ and temperature 20 °C. The liquid flow rate is 1.5 times minimum.

The stripping tower is operated at total pressure 101.3 kPa and temperature 120 °C, and the water vapour entering the tower has the same temperature and pressure. The stripping tower is an existing tower with five theoretical stages.

Both towers are operated with countercurrent flow.

The oil–benzene system is ideal and follows Henry's law, $p_A = Hx$. The vapour pressure of pure benzene is 13.9 kPa at 27 °C and 300 kPa at 120 °C. The vapour pressure of the oil is negligible.

(a) Calculate points on the equilibrium curve and plot the curve.
(b) Determine the minimum liquid flow rate in the absorption tower and the number of theoretical stages needed. *Ans.* Eight stages.
(c) Determine the amount of stripping vapour needed. *Ans.* 4.0 kmol/h.

3.5 Figure 3.30 is an absorption tower where a component A with molecular weight 20 is recovered by absorption in pure water at 50 °C, giving a product containing 40 kg A/m³.

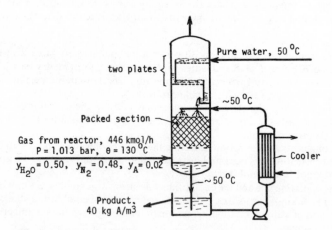

Figure 3.30 Absorption tower with two plates and a packed section with recirculation

With c_A in kmol A/m^3 and Henry's law written as $p_A = H'c_A$, Henry's law constant is

$$H' = 26.7 \frac{Pa}{kmol/m^3}$$

The vapour pressure of water over the aqueous solution is practically the same as for pure water. The heat of absorption (condensation) of water is 2365 kJ/kg, and of component A, 2300 kJ/kg. The heat capacity of nitrogen is 1.00 kJ/(kg °C) and of water vapour 1.88 kJ/(kg °C).

(a) Assuming that recirculation in the packed section is large enough to give almost constant liquid temperature and concentration, and also equilibrium conditions, calculate the amount of A absorbed in the packed section. *Ans.* 176 kg/h.

(b) Calculate the amount of heat removed from the system. *Ans.* 9.83 GJ/h = 2730 kW.

(c) Assuming a plate efficiency of 60 per cent for both plates, calculate the percentage of A recovered. *Ans.* Approximately 99.98 per cent.

3.6 Based on data in Example 3.4, including the mass transfer coefficients given in Figure 3.29, calculate the oxygen content in ppm to be expected in fresh water from the tower in Figure 3.26 when operated without scavenging gas, with entering water saturated with oxygen, and nitrogen in contact with atmospheric air at pressure 1.01 bar, water flow rate 0.048 m^3/s, and:

(a) water temperature 21 °C and absolute pressure 25 mm mercury;

(b) water temperature 25 °C and pressure 25 mm mercury;

(c) water temperature 21 °C, the two 2.44 m high sections separated by a liquid seal, the top section operated at 25 mm mercury, and the bottom section at 20 mm mercury. *Ans.* (a) 0.37 ppm, (b) 0.07 ppm, (c) 0.13 ppm.

3.7 Calculate (graphical integration) the necessary height of packing if 95 per cent of the CO_2 in a gas containing 60 per cent by volume CO_2 and 40 per cent inert gases is to be recovered in an aqueous solution of monoethanolamin in an absorption tower where the following is given:

CO_2 in liquid to the tower	0.1 mol CO_2/mol amin
CO_2 in liquid from the tower	0.6 mol CO_2/mol amin
Temperature of gas and liquid	40 °C
Amin concentration	0.153 kg amin/kg amin + water
Density of the aqueous solution	1000 kg/m^3
Gas side mass transfer coefficient	4×10^{-7} kg/(m^2 s Pa)
Liquid side mass transfer coefficient	5.6×10^{-5} m/s
Interfacial area per unit volume of packed section	$a = 126$ m^2/m^3

Equilibrium data as given in Table a.

Table a Partial pressure of CO_2 in equilibrium with 15.3 per cent by weight monoethanolamin at 40 °C

x	$\dfrac{\text{mol } CO_2}{\text{mol amin}}$	0.383	0.438	0.471	0.518	0.542	0.576
p_{CO_2}	kPa	0.134	0.667	1.332	4.00	6.67	13.32

Table a (continued)

x	$\dfrac{\text{mol } CO_2}{\text{mol amin}}$	0.641	0.639	0.657	0.672	0.686	0.705
p_{CO_2}	kPa	26.6	40.0	53.3	68.1	80.0	101.3

146

Symbols

A	Component A
A	Absorption factor, $A = \dfrac{P}{H}\dfrac{L'}{G'}$ (p. 107)

a	kmole of component A, kmol
a	Area of interface per unit volume, m^2/m^3
B	Component B
C	Component C
C_g	Gas film factor
c	Concentration, $kmol/m^3$
D	Diffusivity, m^2/s
d	Packing diameter (nominal), m
d_h	Hydraulic diameter, m
G	Superficial molar gas flow rate, $kmol/(m^2\,s)$ or mass flow rate, $kg/(m^2\,s)$
G	Inert gas B, kmol/s
G'	Inert gas B and solute A in gas, kmol/s
H	Henry's law constant, Pa/mol fraction
h	Height of packing, m
K	Overall mass transfer coefficient, $kmol/(m^2\,s\ \text{unit driving force})$
k	Individual mass transfer coefficient, $kmol/(m^2\,s\ \text{unit driving force})$
L	Liquid C, kmol/s or superficial mass flow rate, $kg/(m^2\,s)$
L'	Liquid C and solute A in liquid, kmol/s
m	Slope of equilibrium line, dy/dx
N	Molar flux relative to a fixed point, $kmol/(m^2\,s)$
N	Number of theoretical stages
n	Stage number
n	Number of kmoles
P	Total pressure, $N/m^2 = Pa$
p	Partial pressure, $N/m^2 = Pa$
R	Gas constant, $R = 8314.3\ J/(K\,kmol)$
S	Cross-sectional area, m^2
V	Velocity, m/s
v	Specific volume, $m^3/kmol$
x	Mole fraction in liquid, kmol/kmol liquid
x^*	Mole fraction in liquid in equilibrium with gas, kmol/kmol liquid
y	Mole fraction in gas, kmol/kmol gas
y^*	Mole fraction in gas in equilibrium with liquid, kmol/kmol gas
z	Height, m
ε	Voidage
γ	Surface tension, N/m
γ	Activity coefficient
μ	Dynamic viscosity, $N\,s/m^2 = kg/(s\,m)$
ρ	Density, kg/m^3

Subscripts

0	Inlet to first theoretical stage
1	First theoretical stage
c	Concentration as driving force
e	Equimolar counterdiffusion
g	Partial pressure as driving force
l	Liquid phase

lm	Logarithmic mean, equations (3.16) and (3.19)
N	Last theoretical stage
$N + 1$	Inlet to last theoretical stage
n	n-th theoretical stage
$n + 1$	Stage $n + 1$
$n - 1$	Stage $n - 1$
p	Partial pressure as driving force
s	Stefan diffusion, i.e. diffusion through stagnant inert
s	Surface
t	Total concentration (all components)
x	Liquid phase
y	Gas phase

References

1. *International Critical Tables*, Vol. 3, p. 302, McGraw-Hill, New York, 1928.
2. Kremser, A., *Natl. Petrol. News*, **22** (21), (21 May), 42 (1930) and *Proc. Calif. Natural Gasoline Ass.*, **5** (No. 2) (1930).
3. Souders, M., and G. G. Brown, Fundamental design of absorbing and stripping columns for complex vapors, *Ind. Eng. Chem.*, **24**, 519–522 (1932).
4. Khoury, F. M., Simulate absorbers by successive iteration, *Chem. Eng.*, **87** (29 Dec), 51–53 (1980).
5. Khoury, F. M., Calculate the right density, *Hydroc. Proc.*, **57** (Dec.), 155–167 (1978).
6. Charpentier, J. C., A review of the data on mass transfer parameters in most of gas–liquid reactors. *Two-phase Flows and Heat Transfer*, Kakac, S. and T. N. Veziroglu (Eds), Vol. II, Hemisphere, Washington, 1977.
7. Norton, *High Performance Metal Tower Packing HY-PAK*, Norton Co., Akron, Ohio, 1978.
8. Norman, W. S., *Distillation, Absorption and Cooling Towers*, Longmans, London, 1961.
9. Morris, M. A., and J. Jackson, *Absorption Towers*, Butterworths, London, 1953.
10. Norton, *Intalox Metal Tower Packing*, Norton Co., Akron, Ohio, 1977.
11. Norton, *Design Information for Packed Towers*, Bulletin DC-11, Norton Co., Akron, Ohio, 1977.
12. Laurent, A., and J. C. Charpentier, Aires interfaciales et coefficients de transfert de màtiere dans les divers types d'absorbeurs et de réacteurs gaz–liquide, *Chem. Eng. J.*, **8**, 85–101 (1974).
13. Booton, R., D. Cosserat, and J. C. Charpentier, Mass transfer in bubble columns operating at high gas throughputs, *Chem. Eng. J.*, **20**, 87–94 (1980).
14. Sridhar, T., and O. E. Potter, Gas holdup and bubble diameters in pressurized gas–liquid stirred vessel, *Ind. Eng. Chem. Fundamentals*, **19**, 21–26 (1980).
15. Joosten, G. E. H., and P. V. Danckwerts, Chemical reaction and effective interfacial areas in gas absorption, *Chem. Eng. Sci.*, **28**, 453–461 (1973).
16. Perry, R. H., and C. H. Chilton, *Chemical Engineer's Handbook*, 5th edn, McGraw-Hill, New York, 1973.
17. Teller, A. J., Absorption with chemical reaction, *Chem. Eng.*, **67** (11 July), 111–124 (1960).
18. van Krevelen, D. W., and P. J. Hoftijzer, Kinetics of gas–liquid reactions, *Rec. Trav. Pays-Bas*, **67**, 563–586 (1948).
19. van Krevelen, D. W., and P. J. Hoftijzer, Graphical design of gas–liquid reactors, *Chem. Eng. Sci.*, **2**, 145–156 (1953).
20. Brian, P. L. T., Gas absorption accompanied by an irreversible reaction of general order, *A.I.Ch.E.J.*, **10**, 5–10 (1964).

148

21. Danckwerts, P. V. and E. Alper, Design of gas absorbers. Part III. Laboratory 'point' model of a packed column absorber, *Trans. Instn Chem. Engrs*, **53**, 34–69 (1975).
22. Charpentier, J. C., What's new in absorption with chemical reaction, *Trans. Instn Chem. Engrs*, **60**, 131–156 (1982).
23. Midoux, N., and J. C. Charpentier, Les reacteurs gaz–liquide á curve agitée mécaniquement. Partie 1: Hydrodynamique, *Entropie*, **15** (88), 5–38 (1979).
24. Laurent, A., and J. C. Charpentier, Anwendung experimenteller Labormodelle bei der Voraussage der Leistung von Gas/Flüssigkeits-Reaktoren, *Chem.-Ing.-Techn.*, **53**, 244–251 (1981).
25. Satterfield, C. N., Trickle-bed reactors, *A.I.Ch.E.J.*, **21**, 209–228 (1975).
26. Leva, M., *Tower Packings and Packed Tower Design*, U.S. Stoneware Co., Akron, Ohio, 1953.
27. Sharma, M. M., and P. V. Danckwerts, Chemical methods of measuring interfacial area and mass transfer coefficients in two-fluid systems, *Brit. Chem. Eng.*, **15**, 522–528 (1970).
28. Calvert, S., Venturi and other atomizing scrubbers efficiency and pressure drop, *A.I.Ch.E.J.*, **16**, 392–396 (1970).
29. O'Connell, H. E., Plate efficiency of fractionating columns and absorbers, *Trans. A.I.Ch.E.*, **39**, 741–755 (1946).
30. Sharma, M. M., R. A. Mashelkar, and V. D. Mehta, Mass transfer in plate columns, *Brit. Chem. Eng.*, **14** (No. 1), 70–76 (1969).
31. Hagen, R. I., and H. Kolderup, *Flue Gas Desulfurization Process Study* Phase 1. Survey of Major Installations; NATO Committee on the Challenges of Modern Society, Brussels, 1979 (PB-295 005/3BE).
32. Bromley, L. A., *Inst. J. Sulfur Chem.*, Part B, **7** (No. 1) (1972).
33. Bøckman, O. K., P. E. Gramme, S. G. Terjesen, E. Thurmann-Nielsen, and A. Tokerud, Process for removal of sulphur dioxide from flue-gases by absorption in seawater, *International Scandinavian Congress on Chemical Engineering*, Kem-Tek 3, Copenhagen, 28 Jan, 1974.
34. Campbell, J. M., *Gas Conditioning and Processing*, Vol. II, Campbell Petroleum Series, Norman, Oklahoma, 1978.
35. Freshour, K. D., Determination of dissolved oxygen content of waters at subatmospheric pressures, M.Sc. Thesis, University of Oklahoma, Norman, Oklahoma, 1973.
36. Shelton, J. W., The effect of pressure on gas-stripping of dissolved gases from seawater, M.Sc. Thesis, University of Oklahoma, Norman, Oklahoma, 1973.
37. Frank, W. J., Efficient removal of oxygen in a waterflood by vacuum deaeration, *47th Annual Technical Conf. and Exhibition, Soc. Petroleum Engineers of AIME in San Antonio, Texas*, Oct. 8–11, 1972. SPE paper no. 4064.
38. Lydersen, A. L., *Fluid Flow and Heat Transfer*, John Wiley & Sons, Chichester, 1979.

CHAPTER 4

Liquid–liquid extraction and leaching

Leaching of sugar-beets was carried out early in the 1800s, while large scale liquid–liquid extraction dates only from the early 1930s. It then answered the need for a method for removing aromatic hydrocarbons from the kerosene fraction in oil-refining. The pioneering work in liquid–liquid extraction for separation of metals, often called *solvent extraction*, came from the nuclear materials industry in the 1940s, and large-scale copper extraction with chelating extractants came into operation in the 1970s.

Liquid–liquid extraction

In liquid–liquid extraction a homogeneous liquid solution containing a component A is brought in contact with another liquid which is insoluble or only

Table 4.1 Examples of commercial extraction processes (reference 1 and other sources)

Feed	Solute	Solvent	Notes
Pertoleum industry			
Petroleum fractions	S-compounds or mercaptans	Caustic soda	Petroleum refining
Petroleum fractions	Gum formers	Furfural, nitrobenzene	Remove gum formers
Diesel oil	S-containing and cyclic compounds	Furfural	Improves cetane number
Wax containing heavy crude residium	Wax and asphaltic materials	Propane	
Gasoline and kerosene	Aromatics	Sulpholane	Gives high-purity aromatic hydrocarbons
Catalytic reformates	Aromatics	Dimethyl sulphoxide (DMSO)	Negligible corrosion of carbon steel
Coal tar industry			
Coke-oven oil	Aromatics	Diethylene glycol–water	Udex process
Crude tar distillate	Tar acids	Aqueous methanol and hexane	Fractional extraction
Commercial tar acid fraction	2–4 and 2–5 xylenol	Aqueous NaOH and toluene	Dissociation extraction

Table 4.1 (continued)

Feed	Solute	Solvent	Notes
Oils and fats			
Vegetable oils	Unsaturated glycerides	Furfural	Production of drying oils
Vegetable oil and animal fats	Unsaturated glycerides and vitamins	Propane	Solexol process
Tallow		Propane	Decolorization
Pharmaceuticals			
Fermentation broth	Penicillin	Methyl–iso-butyl ketone	Multiple rapid extractions
Soya-bean meal fermented beer	Bacitracin	Butanol	
Miscellaneous organics			
Dilute acetic acid	Acetic acid	Ethyl acetate or diethyl ether	Recovery from cellulose acetate manufacture
Pulpmill black liquor	Acetic acid and formic acid	Methyl ethyl ketone (MEK)	MEK salted out from the raffinate
Oxidized sulphite liquor	Vanillin	Butanol, benzene	Recovery
Metallurgical industry			
Copper ore leach solution	Copper	LIX64N Kelex or Acorga in organic diluent	Copper recovery
Low-quality sulphuric acid ore leach liquor	Uranium salt	Secondary and tertiary amines in organic diluent	Amex process
Uranium concentrate solution in nitric acid	Uranium nitrate	20% Tri-butylphos-phate (TBP) in kerosene	For final purification
Mixed fluorides of niobium and tantalum	Tantalum	Hexone or TBP	
Miscellaneous inorganics			
Phosphoric acid solution from phosphate rock digestion	Phosphoric acid	C_4–C_5 alcohols	Israel Mining Ind. process. Extraction with reflux
Bromide salt in brine	Bromine	Tetra-bromo-ethane	IMI process

partly soluble in the one containing A, and A diffuse into the other liquid phase. After this extraction of *solute* A, the liquid that has given off A is called the *raffinate*, and that receiving A the *extract* phase. Usually the two liquids are one aqueous and one organic liquid.

The purpose of liquid–liquid extraction is either to purify the raffinate or to purify the solute or to obtain a more concentrated solution of the solute. Table 4.1 gives examples of extraction processes.

Insoluble liquids

In a two-phase liquid system where a solute A is distributed between two liquids that are practically insoluble in each other, equilibrium data can be given as the *distribution coefficient*,

$$K = Y/X \qquad (4.1)$$

where Y = mass fraction of A in the extract phase at equilibrium, kg A/kg solvent S, and X = mass fraction of A in the raffinate B, kg A/kg B.

Another representation of equilibrium data is a plot giving mass solute per unit volume in the two phases. This is shown in Figure 4.1, which is an extraction that also utilizes the formation of a metal complex in the extract phase.[2] In this type of extraction, one may even utilize the kinetics of the system in order to separate different metal ions. As an example, Figure 4.2 gives the percentage extracted as a function of stirring time in a stirred tank.[3] Here, selective extraction of copper is obtained if the contact time for the two phases is kept to 2 min or less.

Phase diagrams

The compositions of partly soluble three-component systems are often given in triangular equilateral diagrams where the three apices represent the pure

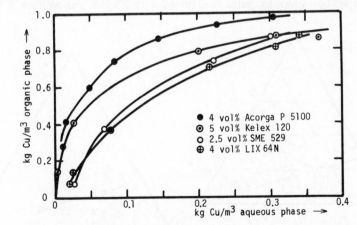

Figure 4.1 Extraction isotherms for various copper extraction commercial systems. Reproduced by permission of *Journal of Metals*

152

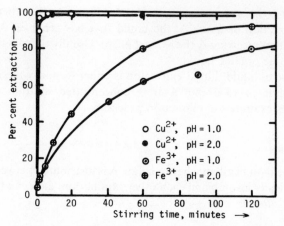

Figure 4.2   Extractions of Cu^{2+} and Fe^{3+} by Kelex 100 with aqueous phase containing 2 kg metal ions per m^3 and organic phase containing 10 vol per cent Kelex 100 in Shellsol R/iso-decanol. Reproduced by permission of the Institution of Mining and the Metallurgy

components, (Figure 4.3). In point M the mass fractions of each of the components are shown as X_A, X_B, and X_S.

If pure component B is added to the mixture at point M, the resulting mixture will be located on the straight line M–B. The lever-arm rule is valid, i.e. the new mixture will be located at point N where the ratio of the length MN to the length NB in the diagram equals the ratio of the amount of B to the amount of M added. If component B is stripped off completely, the final mixture is represented by point D on the side that represents binary mixtures of A and S.

The lever-arm rule is easily established by reference to Figure 4.4. Here R kg of the composition given by point R is mixed with E kg of composition E to give

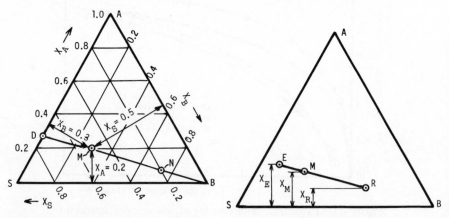

Figure 4.3   Coordinates in an equilateral triangular diagram

Figure 4.4   The lever-arm rule

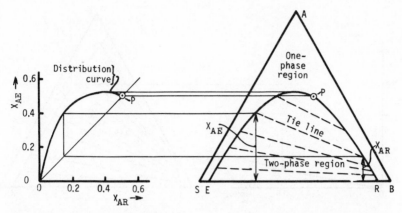

Figure 4.5 Type I system. Ethylene glycol (A), water (B), and furfural (S) at 25 °C.[4] P is the plait point. The solid curve E–P represents the extract phase with mass fraction X_{AE} of A, and P–R the raffinate phase, mass fraction X_{AR} of A

$(R + E) = M$ kg of composition M. A balance for component A gives

$$RX_R + EX_E = (R + E)X_M$$

where X = mass fraction of component A. This equation is rearranged to give

$$\frac{R}{E} = \frac{X_E - X_M}{X_M - X_R'}$$

Comparing similar triangles in Figure 4.4, the right-hand side of the last equation can be substituted by the ratio $\overline{EM}/\overline{MR}$, giving the lever-arm rule

$$R/E = \overline{EM}/\overline{MR}$$

The right-hand side of Figure 4.5 is a phase diagram for the usual case where the solute A is distributed between two liquids with some mutual solubility. Points inside the solid dome E–P–R are in the two-phase region and points outside the dome in the one-phase region. The solid curve E–P represents the extract and P–R the raffinate phase. The end points of a *tie-line* represent compositions in equilibrium. The length of the tie-lines decrease as the content of A increases, and they disappear at the *plait point* P. Figure 4.5 shows a type I system, i.e. only one of the binary pairs has a two-phase region.

The left-hand side of Figure 4.5 shows the construction of the distribution curve, which is also useful for interpolation between the tie-lines in the triangular diagram.

Figure 4.6 represents a type II system where two of the binary pairs have a two-phase region. Some type I systems may change into type II systems by change in temperature.

Two other types of graphs for the three-component systems are shown in Figures 4.7 and 4.8. The right-angled triangle in Figure 4.7 is plotted on ordinary plotting paper where the scales can be set individually at will. Concentrations of

154

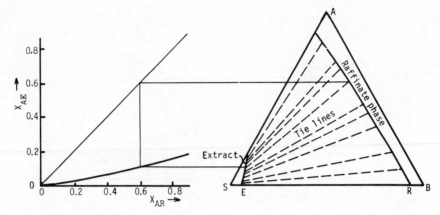

Figure 4.6 Type II system. Methylcyclohexane (A), n-heptane (B), and aniline (S) at 25 °C.[5] Distribution curve to the left

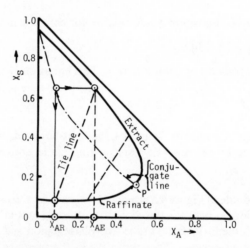

Figure 4.7 Right-angled diagram for the ternary mixture ethylene glycol (A), water (B), and furfural (S) at 29 °C

two of the components are plotted along the axes, and the third component calculated by difference.

The location of all tie-lines can be determined from a single curve, the *conjugate line* shown in Figure 4.7. This line is the locus of the points of intersection of horizontal and vertical lines drawn through the end points of the tie-lines (the *conjugate phases*) as shown for one of the tie-lines.

In Figures 4.8 and 4.9 the same information is translated into a type of diagram first proposed by Janecke,[6] with two of the components as the base mixture. Depending on the type of system, either A + B or S + B may be chosen as the base mixture. In the following, Z' is used for the mass fraction of solvent with A + B as

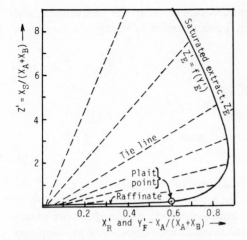

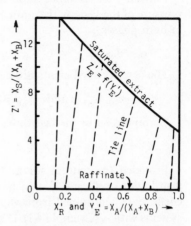

Figure 4.8 Janecke plot for the mixture in Figure 4.5. $X'_R = X_A/(X_A + X_B)$ in the raffinate and $Y'_E = X_A/(X_A + X_B)$ in the extract phase

Figure 4.9 Janecke plot for the mixture in Figure 4.6. The curve for saturated raffinate is practically at the abscissa. $X'_R = X_A/(X_A + X_B)$ and $Y'_E = X_A/(X_A + X_B)$ in the extract phase

base mixture,

$$Z'_E = X_S/(X_A + X_B) \tag{4.2}$$

in the extract phase, and

$$Z'_R = X_S/(X_A + X_B) \tag{4.3}$$

in the raffinate phase.

The mass fractions of A on a solvent-free basis are given as

$$X'_R = X_A/(X_A + X_B) \tag{4.4}$$

in the raffinate phase, and

$$Y'_E = X_A/(X_A + X_B) \tag{4.5}$$

in the extract phase. The primes indicate a solvent-free basis.

Thermodynamic considerations

Equilibrium data may be expressed in terms of the *distribution coefficient,*

$$K = X_{AE}/X_{AR} \tag{4.6}$$

where X_{AE} = mass fraction of A in the extract phase, and X_{AR} = mass fraction of A in the raffinate phase, or

$$K' = x_{AE}/x_{AR} \tag{4.7}$$

where x_{AE} and x_{AR} are mole fractions of A in the two phases.

Choosing as reference state the pure components at the temperature and pressure of the system, the equilibrium is characterized by the same activity of A in both phases,

$$a_{AE} = a_{AR} \tag{4.8}$$

The activity coefficients are defined as

$$\gamma_{AE} = a_{AE}/x_{AE} \quad \text{and} \quad \gamma_{AR} = a_{AR}/x_{AR} \tag{4.9}$$

Combination of the last three equations gives

$$K' = x_{AE}/x_{AR} = \gamma_{AR}/\gamma_{AE} \tag{4.10}$$

Provided the mutual solubility of the two solvents may be neglected, the distribution coefficient K' may be estimated from equation (4.10) with the activity coefficients determined from vapour–liquid equilibria obtained for the binary systems with A (Example 4.1). Distribution coefficients obtained in this way may give the order of magnitude, but should never be used as a basis for the final design.

For computer-implemented interpolation and estimation methods, see the work by Prausnitz et al.[7]

If the aim of the extraction includes the separation of two solutes, consideration has to be given to the selectivity of the solvent for the two solutes A and C. This is measured by the *separation factor*, α,

$$\alpha_{AC} = K'_A/K'_C = \gamma_{AR}\gamma_{CE}/\gamma_{AE}\gamma_{CR} \tag{4.11}$$

Activities different from unity may arise from interactions such as polarity differences or hydrogen bonding, or from chemical interaction. In the latter case, the separation factor can vary significantly with the concentration, as there will be competition between the solutes for the available solvent if this becomes limited.

In some cases the acidity can have a strong influence on the distribution coefficient. As an example, a low pH favours the extraction of penicillin from an aqueous solution to an organic phase. Hence, pH in the filtered fermentation broth is reduced to between 2 and 2.5 before the extraction of penicillin with *n*-butyl acetate, while the following extraction back into an aqueous phase is carried out at a pH of about 6, as shown schematically in Figure 4.10. Because penicillin degrades rapidly at low values of pH, the first extraction has to be completed within a short time after the acidification.

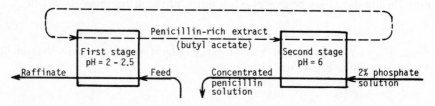

Figure 4.10   Recovery of penicillin

Solvent selection

The choice of solvent is usually a compromise between conflicting properties, and is critical in determining whether liquid–liquid extraction will be economically competitive for a given separation. The solvent must be recovered, often by distillation, and the cost of the recovery may be the major operating cost of the whole process.

Some of the properties to be considered are as follows:

1. *Selectivity.* A high degree of selectivity is important if a desired solute has to be separated from impurities.
2. *Recoverability.* The solvent must be recovered from both extract and raffinate, usually by distillation or by reversing the extraction process. In distillation a favourable relative volatility between the solvent and non-solvent components is desirable, and a low heat of vaporization of the volatile solvent.
3. *Distribution coefficient.* A large distribution coefficient of the extracted component makes it possible to use smaller amounts of solvent and obtain a higher concentration in the extract phase. Robbins[8] gives a qualitative guide for the selection of solvent based on the interaction between solute and solvent, and Treybal[9] gives distribution coefficients and literature references for 288 ternary systems.
4. *Solvent solubility.* A low solubility of the solvent in the raffinate phase reduces or eliminates the cost of recovery.
5. *Density.* It is essential to have a reasonable difference in density between the equilibrium phases.
6. *Interfacial tension.* High interfacial tension is an advantage for rapid coalescence of dispersed liquid drops. Experimental data are scarce, but Antonoff's rule seems to give fairly good estimates,[10,11]

$$\sigma_i = \sigma_1 - \sigma_2 \tag{4.12}$$

where σ_i = interfacial tension, N/m, and σ_1, σ_2 = surface tension of the mutual saturated phases as measured against a common gas or vapour, N/m. More elaborate estimation methods are given by Shain and Prausnitz.[12]
7. *Others.* Other desirable solvent properties are low viscosity for better mass transfer rates and to promote settling of dispersions, low cost, low corrosiveness, non-flammability, and low toxicity. Government regulations must be checked.

Single-stage equilibrium extraction

Figure 4.11 shows a single-stage extraction with solvent recovery from both the raffinate and the extract phase.

If F kg feed with weight fraction X_{AF} of component A is mixed with S kg pure solvent, the lever-arm rule gives the composition of the mixture M as shown in

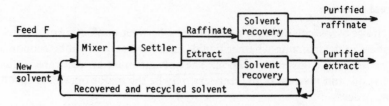

Figure 4.11 Single-stage extraction with mixer and settler and solvent recovery from both phases

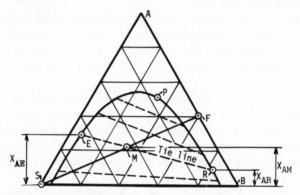

Figure 4.12 Determination of equilibrium compositions of extract E raffinate R

Figure 4.12. It is on the straight line between F and S, and

$$\overline{SM}/\overline{MF} = F/S$$

where $\overline{SM}$ = length of the line between S and M, and $\overline{MF}$ = length of the line between M and F.

The end points E and R of the tie-line through point M give the equilibrium compositions of the extract and the raffinate phase.

Continuous multistage countercurrent extraction

Immiscible solvents. Figure 4.13 is a schematic drawing of an extraction process with countercurrent flow. Each theoretical stage may be visualized as a mixer where equilibrium is obtained, and a settler giving complete separation of the two liquid phases before they are passed on to the neighbouring stages. The symbols are as follows: F = feed, kg/s, S_1 = solvent feed, kg/s, R = raffinate, kg/s, E = extract, kg/s, X = mass fraction in raffinate, and Y = mass fraction in extract.

A material balance around the n-th stage gives

$$E'Y_{n+1} + R'X_{n-1} = E'Y_n + R'X_n \tag{4.13}$$

where the primes indicate a solute-free basis, i.e. E'_n = kg S/s and R'_n = kg B/s, and

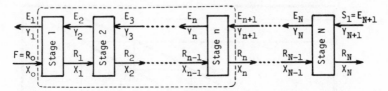

Figure 4.13 Continuous countercurrent extraction with N theoretical stages

Y is kg A per kg S and X kg A per kg B. If on this basis the distribution coefficient equation (4.1) is constant, X in equation (4.13) may be substituted by Y/K. Introducing this and the *extraction factor*,

$$\varepsilon = \frac{R'}{E'K} \tag{4.14}$$

and rearranging equation (4.13), gives

$$Y_n = \frac{Y_{n+1} + \varepsilon Y_{n-1}}{1 + \varepsilon} \tag{4.15}$$

This equation is identical to equation (3.7) with A substituted by ε, giving the Kremser–Souders–Brown equation,

$$\frac{Y_{N+1} - Y_1}{Y_{N+1} - Y_0} = \frac{\varepsilon^{N+1} - \varepsilon}{\varepsilon^{N+1} - 1} \tag{4.16}$$

or the number of stages required to obtain Y_1 in the extract,

$$N = \frac{\ln\left[\frac{Y_{N+1} - Y_0}{Y_1 - Y_0}\left(1 - \frac{1}{\varepsilon} \right) + \frac{1}{\varepsilon} \right]}{\ln \varepsilon} \tag{4.17}$$

where $Y_0 = $ mass fraction in extract in equilibrium with entering feed (Example 4.2). When the extraction factor $\varepsilon = 1$, the right-hand side of equation (4.16) reduces to $N/(N+1)$.

The assumption of unchanged liquid flow rates ($E'_n = E'_{n+1}, R'_n = R'_{n+1}$) is sometimes sufficient for all stages except the first one. In that case, the first stage can be calculated separately, and the rest by equation (4.17).

A material balance around the first n stages gives

$$RX_0 + EY_{n+1} = RX_n + EY_1$$

or the equation for the *operating line*,

$$Y_{n+1} = \frac{R}{E}X_n + \left(Y_1 - \frac{R}{E}X_0 \right) \tag{4.18}$$

This is an equation of a straight line, Y_{n+1} versus X_n, of slope R/E, and the number of theoretical stages required is stepped off between this line and the equilibrium curve (Figure 4.14 and Example 4.3).

160

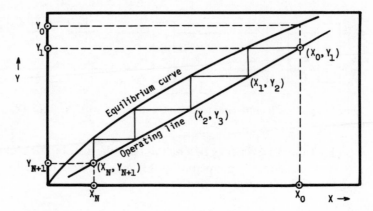

Figure 4.14 Determination of number of stages. Countercurrent flow of immiscible solvents

Partly miscible solvents. A mass balance around all stages in Figure 4.13 gives

$$F + S_1 = E_1 + R_N = M \tag{4.19}$$

where M is the mixture obtained if the two incoming streams were mixed. The composition of this mixture can be located on the line FS_1 in Figure 4.15 through a material balance for substance A,

$$FX_0 + S_1 Y_{N+1} = (F + S_1)X_M$$

or

$$X_M = \frac{FX_0 + S_1 Y_{N+1}}{F + S_1} \tag{4.20}$$

The point M is also located on the line $E_1 R_N$. Hence, if R_N is known, point E_1 is

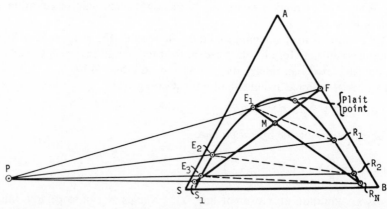

Figure 4.15 Countercurrent multistage extraction of the ternary system Figure 4.5. The dotted lines are tie-lines

located as the intersection of a straight line through R_N and M with the extract curve.

The net flow outwards from stage N (Figure 4.13) equals the net flow inwards at stage 1 and also at any intermediate stage n,

$$R_N - S_1 = F - E_1 = R_{n-1} - E_n = P \tag{4.21}$$

According to equation (4.21), the extended lines $R_N S_1$, FE_1, and $R_{n-1} E_n$ in Figure 4.15, must all intersect at point P, the *operating point*, giving the graphical solution as outlined in the following:

1. Plot the straight line $S_1 F$ between the two feed points and locate point M by the lever-arm rule, $S_1/F = \overline{FM}/\overline{MS_1}$, or by equation (4.20)
2. Locate point E_1 where the extension of the straight line through R_N and M intersects the curve for the extract phase.
3. Locate point P as the intersection of the extensions of the lines $R_N S_1$ and FE_1. Depending on the location of the feed F, point P can be either on the left- or on the right-hand side of the diagram.
4. The number of theoretical stages is determined as shown in Figure 4.15. The raffinate from stage 1 (R_1) is located by the tie-line from E_1, the extract from the second stage (E_2) on the line between P and R_1, the raffinate from the second stage (R_2) on the tie-line from E_2, and so on until stage N is reached.

Minimum solvent/feed ratio, $(S_1/F)_{min}$. As the solvent/feed ratio is decreased, the points M and E_1 in Figure 4.15 move to the right, and the number of stages increases. The minimum S_1/F ratio is reached when E_1 coincides with the plait point, giving an infinite number of stages.

Figure 4.16 shows the same equilibrium diagram with point P on the right-hand side of the diagram. Here, a decreased S_1/F ratio moves P to the left, giving an infinite number of stages when a line from P coincides with a tie-lie.

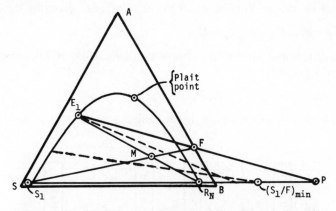

Figure 4.16 Location of minimum solvent/feed ratio. The dotted lines are tie-lines extended to intersection with the straight line through S_1 and R_N. With P on the right-hand side of the triangle, the intersection closest to S_1 represents the minimum solvent/feed ratio, $(S_1/F)_{min}$

162

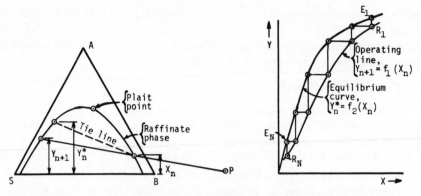

Figure 4.17 Transfer of coordinates to a distribution diagram. X represents the mass fraction of A in the raffinate and Y in the extract phase. The number of theoretical stages is stepped off, as shown in the diagram

Operating line and equilibrium curve. Figure 4.17 shows an $X–Y$ plot that may be more convenient when the number of stages is large. A few lines are drawn at random from point P (only one such line is shown), and the intersections with the two branches of the solubility curve give coordinates of the operating line, $Y_{n+1} = f_1(X_n)$. The coordinates of the equilibrium curve are read on both ends of the tie-lines, $Y_n^* = f_2(X_n)$, and the number of theoretical stages stepped off as shown in the diagram.

Janecke plot

The Janecke plot with mass fractions given on a solvent-free basis is especially suitable for type II systems (Figure 4.6), and for extractions with reflux. Referring to Figure 4.18, the net flow of $A + B$ to the right, is L', given by the balance

$$R'_{n-1} - E'_n = R'_N - E'_{N+1} = L' \tag{4.22}$$

where the primes indicate a solvent-free basis, i.e. mass $A + B$. The corresponding net flow of solvent is

$$R'_{n-1} Z'_{R,n-1} - E'_n Z'_{E,n} = R'_N Z'_{R,N} - E'_{N+1} Z'_{E,N+1} = L' Z'_L \tag{4.23}$$

where Z'_R = mass fraction solvent in the raffinate phase on a solvent-free basis,

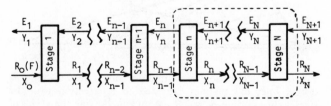

Figure 4.18 Material balance including the theoretical stages n and N

$Z'_R = S/(A + B)$, Z'_E = mass fraction of solvent in the extract phase on a solvent-free basis, $Z'_E = S/(A + B)$, and Z'_L = the mass fraction of solvent on a solvent-free basis, $Z'_L = S/(A + B)$, that gives the net flow of solvent S as $L'Z'_L$. Combining equations (4.22) and (4.23) give the ratio

$$\frac{R'_{n-1}}{E'_n} = \frac{Z'_{E,n} - Z'_L}{Z'_{R,n-1} - Z'_L} \tag{4.24}$$

The net flow to the right of component A (Figure 4.18), is

$$R'_{n-1}X'_{n-1} - E'_n Y'_n = R'_N X'_N - E'_{N+1} Y'_{N+1} = L'X'_L \tag{4.25}$$

where X'_L = the mass fraction of A on a solvent-free basis, $X'_L = A/(A + B)$, that gives the net flow of A as $L'X'_L$. If the fresh solvent feed is pure, $X'_L = X'_N$.

Equations (4.22) and (4.25) give

$$\frac{R'_{n-1}}{E'_n} = \frac{Y'_n - X'_L}{X'_{n-1} - X'_L} \tag{4.26}$$

or with equation (4.24),

$$\frac{R'_{n-1}}{E'_n} = \frac{Z'_{E,n} - Z'_L}{Z'_{R,n-1} - Z'_L} = \frac{Y'_n - X'_L}{X'_{n-1} - X'_L} \tag{4.27}$$

This equation corresponds to a straight line through the point $L'(X'_L, Z'_L)$ in the Janecke plot (Figure 4.19).

The coordinates X'_L and Z'_L for L' are determined as follows.

With *pure solvent S* as feed E_{N+1}, the net flow to the right of $A + B$ (Figure 4.18) is $L' = R'_N$, and of A, $L'X'_L = R'_N X'_N$, i.e.

$$X'_L = X'_N \tag{4.28}$$

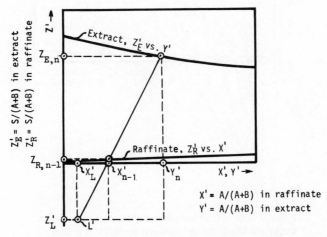

Figure 4.19   Janecke plot (solvent-free basis)

The net flow of solvent S to the right is

$$L'Z'_L = R'_N Z'_{R,N} - E_{N+1}$$

or

$$Z'_L = Z'_{R,N} - E_{N+1}/R'_N \qquad (4.29)$$

If the solvent feed E_{N+1} contains some A and/or B, combining the two last terms of equations (4.22) and (4.25) gives the abscissa,

$$X'_L = \frac{R'_N X'_N - E'_{N+1} Y'_{N+1}}{R'_N - E'_{N+1}} \qquad (4.30)$$

Combining the two last terms of equations (4.22) and (4.23) gives the ordinate,

$$Z'_L = \frac{R'_N Z'_{R,N} - E_{N+1}}{R'_N - E_{N+1}} \qquad (4.31)$$

The number of ideal stages can be determined by the Ponchon–Savarit method or by the McCabe–Thiele method, or in some cases by the Kremser equations.

The *Ponchon–Savarit* method is shown in Figure 4.20. The *operating point* L' is located by equations (4.28) and (4.29), or equations (4.30) and (4.31). The composition of the extract leaving the first theoretical stage, E'₁, is located as the

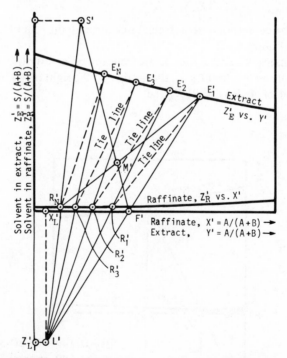

Figure 4.20 Determination of the number of ideal stages by the Ponchon–Savarit method in a Janecke plot

intersection of the straight line through L' and the feed point R'_0 with the extract curve. The raffinate leaving the first ideal stage, R'_1, is located at the end of the tie-line through E'_1, and the extract from the second stage, E'_2, is on the straight line through L' and R'_1. This procedure is continued until the raffinate R'_N is reached on the straight line L'–S' where S' has the coordinates of the entering solvent, Y'_{N+1} and Z'_{N+1}.

If the entering solvent E_{N+1} is pure S, $Z'_{N+1} = \infty$, and the lines L'–R'_N and R'_0–M' are vertical, i.e. M' is located vertically above the feed point R'_0 (Example 4.4).

The *McCabe–Thiele* diagram with coordinates on a solvent-free basis is shown in the lower part of Figure 4.21. In the upper diagram a few lines are drawn at random from point L' (only one such line is shown), and the intersections with the extract and the raffinate curves give the coordinates X'_{n-1} and Y'_n on the operating line. The end points of the tie-lines give the coordinates of the

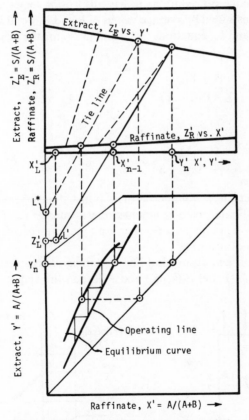

Figure 4.21 Graphical construction of the operating line and the equilibrium curve. The operating line is determined by lines drawn at random through L' with coordinates X'_L and Z'_L (equations (4.30) and (4.31))

equilibrium curve. The number of theoretical stages is stepped off between the operating line and the equilibrium curve.

If the ratio solvent-to-feed is decreased, the point L' in Figures 4.20 and 4.21 moves upwards. When it coincides with an extended tie-line (L* in Figure 4.21), it is a pinch corresponding to an infinite number of theoretical stages.

Extraction with continuous contercurrent contact

Continuous extraction, as incindicated in Figure 4.22, may be carried out in spray towers, packed towers, and mechanically agitated columns, in a similar manner to distillation and gas absorption. Applying the two-film theory, the mass fraction gradients for transfer of solute from the raffinate to the extract phase is shown in Figure 4.23, where the mass fractions at the interface, X_i and Y_i, are assumed to be in equilibrium.

Applying the same procedure as in gas absorption (equations (3.15)–(3.27)) with G' and L' substituted by average values of E and R, y and x by Y and X, and k_{ys} and k_{xs} by k_E and k_R, equations (3.25) and (3.27) for the height of the contact section become

$$z = \frac{E}{k_E a S_t} \int_{Y=Y_2}^{Y_1} \frac{(1-Y)_{lm}}{(Y-Y_i)(1-Y)} dY \qquad (4.32)$$

and

$$z = \frac{R}{k_R a S_t} \int_{X=X_1}^{X_2} \frac{(1-X)_{lm}}{(X_i-X)(1-X)} dX \qquad (4.33)$$

where k_E = extract phase volumetric mass transfer coefficient, kgA/(m² s mass fraction), k_R = raffinate phase volumetric mass transfer coefficient, kgA/(m² s mass fraction), a = interfacial area per unit volume of tower, m²/m³, and S_t = cross-section of tower, m².

As in distillation and absorption, the two integrals represent the number of transfer units (NTU), and $E/(k_E a S_t)$ and $R/(k_R a S_t)$ the height per transfer unit, HTU.

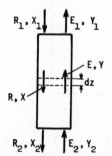

Figure 4.22 Extraction with continuous counter-current contact

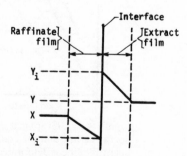

Figure 4.23 Mass fractions near the interface

Expressions for the overall mass transfer coefficients were derived in Chapter 1, equations (1.52) and (1.53), or with the symbols in this chapter,

$$\frac{1}{K_E} = \frac{1}{k_E} + \frac{K}{k_R} \tag{4.34}$$

and

$$\frac{1}{K_R} = \frac{1}{Kk_E} + \frac{1}{k_R} \tag{4.35}$$

where K = distribution coefficient, equation (4.1), $K = Y/X$.

Corresponding to equations (3.39) and (3.40) for gas absorption and with average values of E and R, this gives the contact height in terms of overall volumetric mass transfer coefficients $K_E a$ and $K_R a$,

$$z = \frac{E}{K_E a S_t} \int\limits_{Y=Y_2}^{Y_1} \frac{(1-Y)^*_{lm}}{(Y-Y^*)(1-Y)} dY \tag{4.36}$$

and

$$z = \frac{R}{K_R a S_t} \int\limits_{X=X_2}^{X_1} \frac{(1-X)^*_{lm}}{(X^*-X)(1-X)} dX \tag{4.37}$$

where

$$(1-Y)^*_{lm} = \frac{(1-Y^*) - (1-Y)}{\ln[(1-Y^*)/(1-Y)]} \tag{4.38}$$

$$(1-X)^*_{lm} = \frac{(1-X) - (1-X^*)}{\ln[(1-X)/(1-X^*)]} \tag{4.39}$$

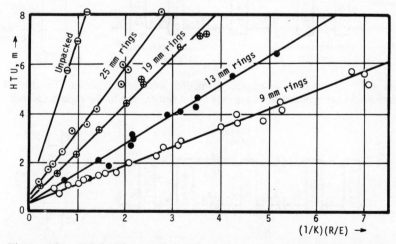

Figure 4.24 Height of overall transfer unit as a function of $(1/K)(R/E)$ for various packings in a 0.15 m diameter column where diethylamine is transferred from continuous water to dispersed toluene. Reproduced by permission of American Institute of Chemical Engineers

168

where Y^* = mass fraction of A in extract in equilibrium with raffinate with mass fraction X, X^* = mass fraction of A in raffinate in equilibrium with extract with mass fraction Y.

The integrals in equations (4.36) and (4.37) are the numbers of transfer units, NTU, and $E/(K_E a S_t)$ and $R/(K_R a S_t)$ the heights of the transfer units, HTU. Figure 4.24 is an example of a correlation of height of an overall transfer unit for packed and empty columns.[13]

Caution must be exercised in drawing conclusions from this kind of plot. The individual mass transfer coefficients are functions of the flow rates, and not only the ratio of the flow rates, R/E. Furthermore, impurities and especially axial mixing can have a considerable influence on the mass transfer, as shown for instance by Laddha and Degaleesan.[1] They also give methods for estimation of HTU.

Countercurrent extraction with reflux

In ordinary continuous countercurrent extraction, as shown in Figure 4.13, the richest possible product is only in equilibrium with the feed solution. But the use of reflux, as shown in Figure 4.25, can give an extract with a concentration of solvent that exceeds the concentration in equilibrium with the feed. The operation is similar to fractional distillation. Another possibility is the use of two solvents, as shown in Figure 4.26.

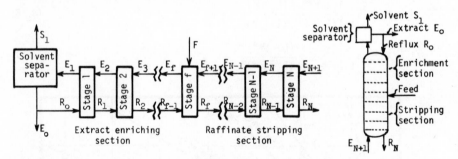

Figure 4.25 Continuous multistage extraction with solvent reflux. To the left an arrangement where each stage is a mixer and a settler, and to the right a column

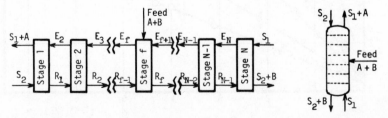

Figure 4.26 Continuous multistage extraction with two solvents S_1 and S_2. To the left an arrangement where each stage is a mixer and a settler, and to the right a column

The simplest calculation procedure depends on what quantities are given. For multistage extraction with reflux, the procedure outlined here can be used if the following data are fixed in advance:

1. The quantity and composition of the feed F.
2. The composition of the extract product E_0.
3. The composition of the raffinate product R_N.
4. The quantity and the composition of the entering solvent E_{N+1}. (The solvent S_1 removed from the solvent separator is returned to stage N as E_{N+1} after ΔS_1 fresh solvent is added to compensate for solvent losses in E_0 and R_N).
5. The reflux ratio R_0/E_0.
6. In addition, solvent lost in the product streams E_0 and R_N must be added to the stream E_{N+1}.

In the following, index m is used for the raffinate stripping section and n for the extract enriching section. A single prime indicates quantities and compositions on a solvent-free basis in the raffinate stripping section, and a double prime in the extract enriching section.

Material balances around the whole system in Figure 4.25 give

$$F = E_0 + R_N \tag{4.40}$$

and

$$FX_F = E_0 Y_0 + R_N X_N \tag{4.41}$$

where X_F, Y_0, and X_N are mass fractions of component A in feed, extract product, and raffinate product. The two equations are solved to give E_0 and R_N, and the reflux R_0 is determined from the reflux ratio R_0/E_0.

The next step is determination of the quantity and composition of the stream E_1. Assuming S_1 to be pure solvent, the composition of E_1 is given by the point on the extract phase that gives the known ratio A/B in the extract product E_0. The corresponding quantity E_1 is given by a mass balance for component A around the solvent separator,

$$E_1 Y_{A1} = (E_0 + R_0)X_{A0} \tag{4.42}$$

where $Y_{A1} =$ mass fraction of A in E_1, and $X_{A0} =$ mass fraction of A in E_0 and R_0.

If the stream S_1 contains mass fraction Y_{AS} of component A and Y_{BS} of component B, mass balances around the solvent separator give

$$E_1 = S_1 + (E_0 + R_0) \tag{4.43}$$

$$E_1 Y_{A1} = S_1 Y_{AS} + (E_0 + R_0)X_{A0} \tag{4.44}$$

$$E_1 Y_{B1} = S_1 Y_{BS} + (E_0 + R_0)X_{B0} \tag{4.45}$$

where $Y_{B1} =$ mass fraction of B in E_1, and $Y_{B0} =$ mass fraction of B in E_0 and R_0.

Eliminating E_1 and S_1 in the three equations gives,

$$\frac{Y_{A1} - Y_{AS}}{Y_{B1} - Y_{BS}} = \frac{X_{A0} - Y_{AS}}{X_{B0} - Y_{BS}} \tag{4.46}$$

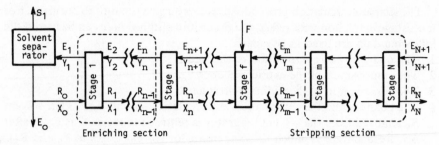

Figure 4.27 Material balances around portions in the extract enriching and the raffinate stripping section

where Y_{A1} and Y_{B1} are determined as the values read on the curve for the extract phase that satisfies equation (4.46). With Y_{A1} known, equations (4.43) and (4.44) give E_1 and S_1.

The material balance shown to the right in Figure 4.27 is identical to the balance shown in Figure 4.18. Substituting index n by m in equations (4.22)–(4.27) gives equation (4.27) as

$$\frac{R'_{m-1}}{E'_m} = \frac{Z'_{E,m} - Z'_L}{Z'_{E,m-1} - Z'_L} = \frac{Y'_m - X'_L}{X'_{m-1} - X'_L} \tag{4.47}$$

This equation corresponds to straight lines through the operating point L' in the plot (Figure 4.28) with coordinates for L' given by equations (4.28) and (4.29).

The same procedure applied to the extract enriching section in Figure 4.27, and using double primes for this section, yields

$$\frac{R''_{n-1}}{E''_n} = \frac{Z''_L - Z''_{E,n}}{Z''_L - Z''_{E,n-1}} = \frac{X''_L - Y''_n}{X''_L - X''_{n-1}} \tag{4.48}$$

corresponding to straight lines through a point L'' in the plot (Figure 4.28) with coordinates derived by equations for the net flow to the left in the extract enrichment section in Figure 4.27. On a solvent-free basis, this net flow is

$$E''_1 - R''_0 = L'' \tag{4.49}$$

$$E''_1 Y''_1 - R''_0 X''_0 = L'' X''_L \tag{4.50}$$

$$E''_1 Z''_{E,1} - R''_0 Z''_{R,0} = L'' Z''_L \tag{4.51}$$

or

$$X''_L = (E''_1 Y''_1 - R''_0 X''_0)/L'' \tag{4.52}$$

and

$$Z''_L = (E''_1 Z''_{E,1} - R''_0 Z''_{R,0})/L'' \tag{4.53}$$

Points on the equilibrium curve and the operating lines are located as shown in the lower part of Figure 4.28. The number of theoretical stages is stepped off as shown in the diagram.

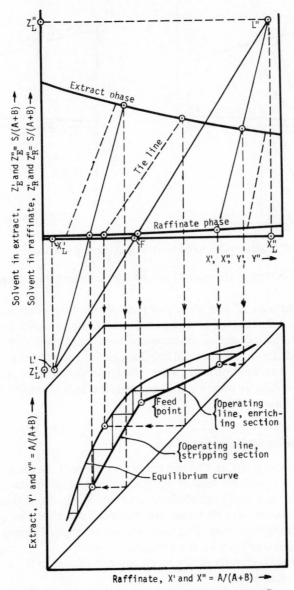

Figure 4.28 Countercurrent extraction with reflux.
Determination of the equilibrium curve and the operating lines. Primes refer to the raffinate stripping section
and double primes to the extract enriching section

Total reflux. With total reflux (infinite reflux ratio) $Z_L'' = \infty$ and $Z_L' = -\infty$, and the operating lines coincide with the diagonal. This gives the minimum number of theoretical stages.

Minimum reflux. An infinite number of stages is required if a line through L′ for

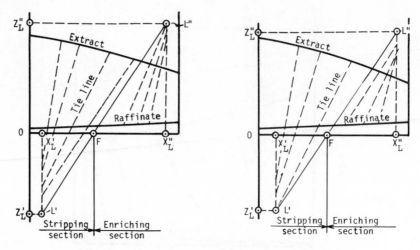

Figure 4.29 Countercurrent extraction with reflux

the stripping section or through L″ for the enriching section coincides with a tie-line.

The line between L′ and L″ passes through the feed point F (Figure 4.29). Reducing the reflux moves L″ downwards and L′ upwards. In Figure 4.29, tie-lines in the enriching section are extended to X_L'' and tie-lines in the stripping section extended to X_L'. If a vertical line through F is rotated clockwise around F, it corresponds to minimum reflux when it first intersects with an extended tie-line at X_L'' or X_L', whichever occurs first. Frequently the tie-line passing through F will establish minimum reflux.

Equipment for liquid–liquid extraction

The key to carrying out an effective extraction lies with the choice of a suitable solvent and of a suitable type of equipment. Special problems are caused by materials that are easily emulsified, have small density differences, or are corrosive. It should also be noted that the most effective contact is obtained during the initial dispersion. This accounts for the high efficiency of units with disengaging sections between the contacts.

The rate of mass transfer of solute A from the raffinate to the extract phase is given by the equation

$$\frac{dw}{dt} = ka\Delta c \tag{4.54}$$

where dw/dt = mass transfer per unit time, kg/s, k = mass transfer coefficient per unit volume, $kg/[s(m^2/m^3)(kg/m^3)]$, a = interfacial area per unit volume, m^2/m^3, and Δc = deviation of the actual solute concentration in the two phases from the equilibrium concentration difference, kg/m^3.

In liquid–liquid extraction one phase is *dispersed* as drops in a *continuous phase*, and the interfacial area per unit volume, *a*, depends on the hold up and the drop size of the dispersed phase.

The transport process involves both molecular and eddy diffusion. The eddy diffusion may be several orders of magnitude greater than the molecular diffusion. Movements within drops moving through the continuous phase are caused by the drag at the interface, and for larger drops also by oscillations in shape. The smallest drops, however, behave as solid spheres with molecular diffusion as the only mechanism of mass transport within the drop.

Increased agitation decreases the drop size and thus increases the interfacial area *a*. It also increases the turbulence in the continuous phase. But the reduction in drop size reduces the mass transport within the drops, and beyond a certain point, increased agitation may reduce the product of the mass transfer coefficient and the interfacial area. In addition, the succeeding separation of the two phases will be more difficult.

Adequate agitation is desirable, while excessive agitation produces undesirable effects.

Maximum driving force Δc in equation (4.54) is obtained if the two phases have perfect countercurrent flow. With differential contactors such as columns, back-mixing and forward mixing may give appreciable deviations from perfect countercurrent flow. Such undesirable mixing is due to back-mixing in the continuous phase, entrainment of drops, entrainment behind the drops, and different flow rates due to the drop-size distribution.

Stable systems are usually obtained with the minority phase dispersed. But

Table 4.2 Liquid capacity of the combined streams, stage efficiency, and applications of commercial extraction equipment. *HETS* = height equivalent of one theoretical stage (references 1, 2, and other sources)

Equipment	Fig. no.	Capacity, $(m^3/(m^2 h))$	Stage efficiency	Typical applications
Mixer–settler	4.31 4.32 4.33	—	75–95%	Duo-Sol lubricating oil process, nuclear fuel processing
Spray column	4.34	15–75	$HETS = 1–2$ m, 1–2 stages	NaCl from NaOH, using $(NH_3)_{aq}$
Packed column		6–45		Phenol recovery
Perforated-plate	4.34	3–60	30% for 100–200 mm plate spacing	Furfural lubricating oil process
Rotating disc-contractor (RDC)	4.34	3–60	See reference 10	Gasoline desulphurization, phenol from waste water
Oldshue–Rushton	4.34	3–60	0.8–1 stage stage per m	Organic chemicals
Podbielniak centrifugal extractor	4.36	4–95 m^3/h	3.4–12.5 stages	Pharmaceutical manufacture, petroleum processing

over a limited range of flow ratios it is possible to disperse either phase, depending on the start-up conditions. To promote settling, it is usually best to disperse the phase with the higher viscosity.

Table 4.2 gives liquid capacity, stage efficiency, and typical applications of some commercial extractors. Figure 4.30 is a selection guide[14,15] which can be used as a rule-of-thumb only. Most extractors react differently to the various liquid properties, influenced also by trace impurities and the direction of mass transfer (Marangoni effects). For pilot plant developments it is advisable to operate on as large a scale as possible with real process liquids.[16]

Mixer–settlers Mixer–settlers were used in some of the earliest commercial liquid–extraction plants. They are still widely used because of features such as high stage efficiency, handling of wide solvent ratios and of liquids with high viscosity, reliable scale-up, and high capacity (up to $0.4\,\mathrm{m}^3/\mathrm{s}$ in the mining industry). Their main disadvantages are their size and the inventory of material held up in the equipment.

Figure 4.31 shows a box configuration of a battery of mixer–settlers, Figure 4.32 a mixer–settler designed for the extraction of copper, and Figure 4.33 a vertical stack of stages of a type used extensively for the separation of aromatic and aliphatic hydrocarbons.

The rate of extraction is a function of power input, and mixers can be scaled up by geometric similarity and constant power input per unit volume.

In large industrial mixer–settlers the volume of the settler is at least three times

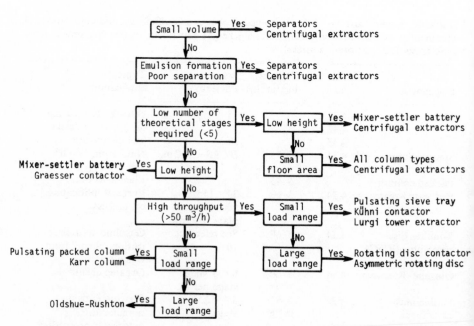

Figure 4.30   Criteria for preliminary selection of liquid–liquid extractors[14,15]

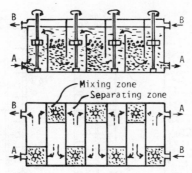

Figure 4.31 Battery of mixer–settlers with heavy phase A and light phase B

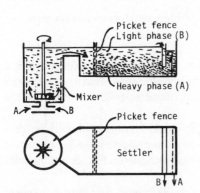

Figure 4.32 General Mills mixer–settler. The mixer is provided with baffles and the settler with a picket fence

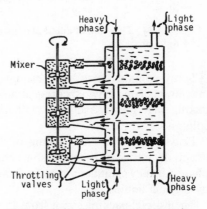

Figure 4.33 Lurgi extractor consisting of stacked mixer–settler units. Capacity for aromatic extraction up to 1600 t/h with column diameter 3 m

the volume of the mixer, and the processes taking place are quite complicated. A review is given by Hanson and co-workers,[17] and a detailed description by Jeffreys and Davies.[18]

Different methods being used to promote sedimentation and coalescence in the settler are installation of horizontal or inclined plates, installation of woven-mesh packing wetted by the dispersed phase or by both phases, application of an electric field, and the use of centrifugal force. Valves for withdrawal of accumulated matter in the dispersion band and of sludge in the bottom may also be included.

Unagitated columns The *spray column* as shown to the left in Figure 4.34 is the simplest of all column extractors. It consists of an empty shell and provisions for introducing and removing the liquids. The droplets of the dispersed phase are generated by spray nozzles.

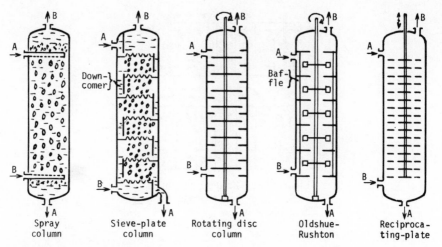

Figure 4.34   Diagram of different continuous column extractors with heavy phase A
and light phase B

Spray columns have high throughputs per unit cross-sectional area, but a
maximum of one to two theoretical stages. The low efficiency is partly due to axial
mixing that greatly increases with increased diameter-to-length ratio of the
column.

Packed columns. The packing of a simple packed column should be preferentially wetted by the continuous phase to prevent coalescence of the dispersed
phase. Usually ceramics are preferentially wetted by aqueous liquids and plastics
by organic liquids.

The normal packing size is from 15 to 25 mm. Reissinger and Schröter report
2.5 theoretical stages per metre with 15 mm packing in a 72 mm diameter
column.[15] An increase in the column diameter increases the height per theoretical
stage, due to increased axial mixing.

Perforated-plate columns. Figure 4.34 shows a perforated-plate column with
the light phase dispersed. The heavy phase passes over the plates and through the
downcomers to the tray below, while the light phase is dispersed again each time
it passes through the perforations. Columns with the heavy phase dispersed are
turned upside-down, so the downcomers become 'upcomers'. Treybal[19] gives
details of tray hydraulics, hold-up, and plate efficiencies.

Agitated columns   Pulsed perforated-plate columns are fitted with horizontal
perforated plates or sieve plates which occupy the entire cross-section of the
column. The free area of the plates is 20–25 per cent. Pulsations are superimposed
on the main flow by means of a reciprocating piston or bellows. The frequencies of
the pulsations are 1.5–4 Hz, and the amplitudes 6–25 mm. The main use is in the
nuclear industry.

Logsdail and Thornton[20] give data for scale-up from laboratory data, based
on tests with column diameters from 75 to 300 mm.

Rotating disc columns (Figure 4.34), use the shearing action of rapidly rotating discs to interdisperse the phases. It is widely used in the petroleum industry for furfural and sulphur dioxide extraction, propane de-asphalting, sulpholane extraction of aromatic from aliphatic hydrocarbons, and caprolactam purification. Columns up to 4.3 m in diameter are in service.[21]

Marr and co-workers[22] give semi-empirical data for determination of column diameter, Reman and Olney[23,24] for the number of theoretical stages per unit height, and Strand and co-workers[25] for mass transfer.

Asymmetric rotating disc columns (ARD) have rotating discs mounted on a shaft that is off-centre in the column, and three different types of stationary baffles in each compartment.

Oldshue–Rushton columns (Figure 4.34) consist essentially of compartments separated by horizontal stator-ring baffles, each fitted with vertical baffles and a flat-blade disc turbine impeller mounted on a central shaft.

Residence time, throughput, and efficiency can be controlled by the aperture size of the stator-ring baffles. With narrow apertures, 1–3 theoretical stages per metre can be achieved, and with greater apertures 0.8–1.0 stage.[15] Columns up to 2.7 m in diameter are in service.[21]

Kühni columns have stator discs made of perforated plates and shrouded turbine impellers. Data for droplet size, hold-up, and back-mixing are available for pilot-scale columns.[26,27]

In separation of aromatics from aliphatic hydrocarbons, a column 2.56 m in diameter containing 52 actual stages each 0.6 m high, had a stage efficiency of 29 per cent at a total throughput of 30 m^3/(m^2 h).[28]

Reciprocating-plate columns, as shown schematically in Figure 4.34, consist of an outer shell with perforated baffles and a stack of reciprocating perforated plates. Reciprocating-plate columns are built with a diameter of up to 0.9 m with a throughput of 18 m^3/(m^2 h). Lo[29] gives a good review of published data and scale-up procedures.

The *Graesser raining bucket contactor* was developed for handling the difficult settling systems in the coal-tar industry. The basic features are shown in Figure 4.35. It is a horizontal design where the phases are interdispersed by a 'water-

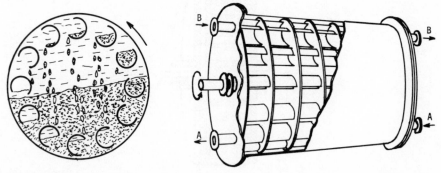

Figure 4.35 Graesser raining bucket contactor where A is heavy and B light liquid

178

wheel' with buckets or dippers resembling rain gutters. The phases pass from one chamber to the next through the annular space between the rotating discs and the cylindrical shell. Units have been built up to 1.8 m in diameter and with a throughput of 8.6 m³/(m² h). Data on axial mixing have been reported for a test extractor 150 mm in diameter.[30]

Separators and centrifugal extractors   A separator is a mixer–settler where the phase separation is accomplished by means of centrifugal force.

The centrifugal extractor operates as a sieve-plate column with radial flow of the two phases. Capital costs and probably also maintenance costs are higher than for other types of extractor, but they have other advantages such as the ability to handle liquid pairs having density differences less than 20 kg/m³. Other important advantages are small space requirements, low product inventory, and a short contact time. This is particularly important for chemically unstable systems.

Podbielniak extractors were the first centrifugal units to gain commercial acceptance in the early 1950s. Figure 4.36 shows the basic features schematically. Concentric sieve trays are located around a horizontal axis. The liquids enter through the shaft with inlet pressures of 4–7 bar and pass through the sieve trays in countercurrent flow. The extractor is especially useful for applications where a very short residence time is essential (e.g. extraction of penicillin). It is not suitable for processing emulsions and for liquid pairs with density differences less than 50 kg/m³. Three to five theoretical stages have been achieved at throughputs up to 130 m³/h.[15]

The *Alfa-Laval extractor* is a vertical centrifuge with a number of concentric, perforated cylindrical baffles. Every second baffle has perforations at the top and the others at the bottom. The liquids are introduced through the shaft and pass up and down in spirals between the baffles in countercurrent flow, and with intimate mixing when the two phases pass through the perforations in opposite directions. The two standard units have maximum capacities of 5.4 and 21 m³/h.

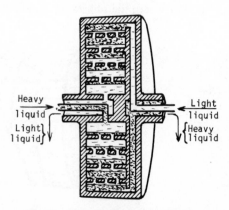

Figure 4.36   Podbielniak extractor

Applications are extraction of penicillin, erythromycin, tetracycline, alkaloids, etc.

For other types such as the Quadronic, the Westfalia, and the Robatel centrifugal extractors, see reference 8.

Leaching

Leaching is the process of extraction of a soluble constituent from a solid by means of a solvent. It is a technique used either to extract a valuable constituent from the solid, such as sugar from suger-beet slices, or to free a solid from a soluble material with which it is contaminated, e.g. purification of a pigment. Table 4.3 lists some commercial liquid–solid extraction processes.

Leaching is sometimes distinguished from washing and diffusional extraction,[31] where leaching includes a chemical reaction with one or more substances in the solid matrix, so as to render them soluble. The leaching of copper sulphate is a typical example.

Washing of natural materials is carried out after crushing to break the cell walls, thus permitting the soluble material to be washed from the solid matrix. Sugar recovery from sugar-cane is an example.

Table 4.3 Some commercial leaching processes[31,32]

Solute	Raw material	Solvent and remarks
Copper sulphate	Copper ore	Sulphuric acid–ferric sulphate
Nickel	Ores	Aqueous ammonia or acids
Phosphoric acid	Phosphate rock	Sulphuric acid, nitric acid, or hydrochloric acid
Rare earths	Rare earth concentrate	Hot sulphuric acid
Sodium aluminate	Bauxite	Caustic solution (Bayer process)
Sodium dichromate	Chromite ore	Hot sulphuric acid
Titanyl sulphate	Ilmentite	Sulphuric acid
Zinc	Zinc ore	Sulphuric acid
Apple juice	Apple chunks	Water (75–85 min extraction time)
Coffee	Coffee beans	Hot water
Fish oils	Fish, fish meal	Alcohols, hexane, etc. (15–60 min extraction time)
Flavours, odours, etc.	Natural products	Hexane, benzene, petroleum ether
Gelatin	Bones and skins	Aqueous solution, pH = 3–4
Hop-flavoured brewing	Hops	Brewing wort (1.5–2 h extraction time)
Sugar	Sugar-beet, sugar-cane	Aqueous solution containing lime
Turpentine and wood resin	Stump wood	Naphtha
Vanilla	Vanilla beans	35% ethanol–65% water
Vegetable oils	Oilseeds	Furfural, propane, or hexane

In diffusional extraction the soluble product diffuses across the denaturated, unbroken cell walls, as in the recovery of beet sugar.

Rate of extraction

The rate of extraction in solid–liquid systems is affected by temperature, concentration of solvent, particle size, porosity and pore size distribution, and agitation.

Increased temperature increases the rate of extraction due to decreased liquid viscosity and increased diffusivity. Maximum temperature is limited by the boiling-point of the solvent, the stability of the product or solvent, or by economic factors.

Solvent concentration is important in the case of aqueous solutions with chemical reaction. It has less influence in cases where diffusion in the solid matrix is rate-controlling, as in oilseed extraction.

Decreased particle size increases the total surface area available for reactions or diffusion. Table 4.4 gives some typical leaching times.

According to the simple film theory, the mass flux of solute from the solid surface in kmol/s is

$$N = k(y_i - y) \qquad (4.55)$$

where k = mass transfer coefficient, $kmol/(m^2s$ mole fraction), y_i = mole fraction solute at the interface, and y = mole fraction solute in the bulk of the liquid.

Elaborate expressions for the mass flux have been developed based on the film penetration theory combined with equations for first-order irreversible reactions,[33] based on equations for capillary flow,[34] and based on unsteady-state diffusion in the solid matrix.[32] However, based on a careful study of liquid–solid extraction in the beet-sugar industry, Brüniche-Olsen[35] concluded that in some cases of countercurrent processes, calculations may, with reasonable accuracy, be based on relatively simple equations with constant k. For stepwise countercurrent extraction with complete mixing in the individual cells, he developed the expression

$$\frac{Z_n - Y_{n-1}}{Z_{n-1} - Y_{n-1}} = \frac{Q}{1 + k_1 t_m} \qquad (4.56)$$

and for systematic movement of the material through the cells, but complete

Table 4.4 Typical leaching times[31]

Technique	Solid matrix	Size	Time
Leaching in place	Copper ore	150 mm	3–4 years
Leaching in place	Copper ore	6 mm	5 days
Leaching by agitation	Copper ore	20 mesh	4–8 h
Leaching by agitation	Phosphate rock	45 mesh	1–5 h

mixing in the extract phase,

$$\frac{Z_n - Y_{n-1}}{Z_{n-1} - Y_{n-1}} = Q \, e^{-kt_m} \tag{4.57}$$

where Z = mass fraction solute in the solid, Y = mass fraction solute in the overflow, Q and k = constants, and t_m = average residence time in the cell, s. The indices indicate the cell from which the stream originates.

Ideal and equilibrium stages

Leaching calculations are based on the concept of an *ideal stage*, defined as a stage from which the resultant solution is of the same composition as the solution adhering to the solids leaving the stage after settling, filtration, or drainage.

A stage with infinite residence time followed by perfect separation of liquid and solid is an *equilibrium stage*. In practice, however, the solute may be incompletely dissolved due to inadequate contact time, and also because the solids leaving the stage will retain some liquid with its associated dissolved solute. This is taken into account by either using a practical overall stage efficiency for the leaching operation in question, or using 'practical equilibrium data' which take the stage efficiencies directly into account.

In two special cases of multistage leaching the number of stages can be calculated analytically. It is in the case of constant underflow, and in the case of constant solvent-to-inert ratio in the underflow. In all other cases the number of stages must be determined by stage-by-stage calculation or by graphical solution. The phase diagrams used are of the same type as in liquid–liquid extraction.

Analytical solutions

As the underflow to the first stage, L_0 kg/s, is different from the underflow L_n between stages, it is convenient to calculate the first stage by material balances.

Constant underflow. In the case of constant underflow after stage 1, McCabe and Smith[36] derived an equation that can be written

$$N - 1 = \frac{\ln\left[(Y_{N+1} - X_N)/(Y_2 - X_1)\right]}{\ln\left[(Y_{N+1} - Y_2)/(X_N - X_1)\right]} \tag{4.58}$$

where N = number of ideal stages, Y = mass fraction of solute in the overflow, inert-free basis, and X = mass fraction of solute in the underflow, inert-free basis. The subscript refers to the stage where the stream originates (Figure 4.37).

Constant solvent-to-inerts ratio. With constant solvent-to-inerts ratio, equation (4.58) becomes

$$N - 1 = \frac{\ln\left[(Y'_{N+1} - X'_N)/(Y'_2 - X'_1)\right]}{\ln\left[(Y'_{N+1} - Y'_2)/(X'_N - X'_1)\right]} \tag{4.59}$$

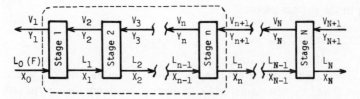

Figure 4.37 Countercurrent multistage leaching with N ideal stages and mass fraction solute Y in the overflow V and X in the underflow L

where $X' =$ mass ratio of solute to solvent in the underflow, and $Y' =$ mass ratio of solute to solvent in the overflow.

The use of equation (4.58) is shown in Example 4.5.

Stage-by-stage calculation

Material balances as indicated by the dotted cage in Figure 4.37, yield

$$V_{n+1} + L_0 = V_1 + L_n \tag{4.60}$$

$$V_{n+1}Y_{n+1} + L_0X_0 = V_1Y_1 + L_nX_n \tag{4.61}$$

Eliminating V_{n+1} gives the equation for the operating line

$$Y_{n+1} = \frac{1}{1 + (V_1 - L_0)/L_n}X_n + \frac{V_1Y_1 - L_0X_0}{L_n + V_n - L_0} \tag{4.62}$$

Stage-by-stage calculation of the number of ideal stages may be carried out by the following steps:

1. Calculate quantities and compositions of all terminal streams, using a convenient quantity of the terminal streams as a basis.
2. Using material balances and equilibrium data, calculate the stream compositions for ideal stage number 1.
3. Continue the calculation for each successive ideal stage, using equation (4.62) for the material balance combined with the equilibrium data, until the result corresponds to the desired terminal conditions.

This method is illustrated in Example 4.5.

Graphical solution

The graphical solution is similar to the solution shown in Figure 4.21 for liquid–liquid extraction, carrying out the following steps shown in Figure 4.38:

1. Plot the underflow curve.
2. Plot the solvent-free feed L_0 at $X = 1.0$ and draw a straight line through L_0 and the overflow discharge point V_1.
3. Plot the underflow discharge point L_N and the overflow feed point V_{N+1}, and draw a straight line through the two points until intersection with the line through L_0 and V_1. The point of intersection is the difference point Δ.

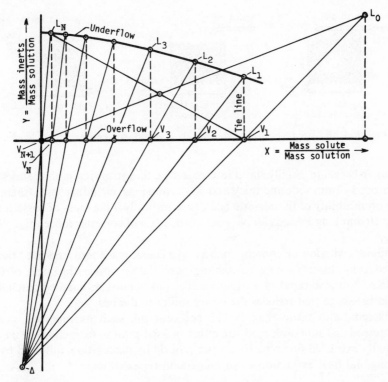

Figure 4.38 Janecke plot for determination of the number of ideal stages in leaching

4. Starting in V_1, draw the tie-line to intersection with the underflow curve. The tie-line is vertical by solid-free overflow. The point of intersection is L_1.
5. The straight line between L_1 and Δ intersects the overflow curve at V_2. By solid-free overflow is the intersection at the abscissa.
6. Determine ideal stage 2, starting at point V_2 and repeating the procedures under 4 and 5. Continue until the point L_N is reached. The last step may be a fraction of a step.

Methods of operation

Figure 4.39 shows the three principal methods used in leaching. To the left is percolation, as used in the extraction of soya beans on a conveyor or in washing crystals in a pusher centrifuge. Figure 4.39b shows intermittent drainage that can be carried out as indicated in the figure, or hoisting out the solids placed in a basket, as done in the leaching of seaweeds. Figure 4.39c shows fully immersed solids where the liquid continuously passes on to the next stage.

Depending partly on the method of operation, the solids may have to be modified prior to extraction. Ores are crushed to suitable particle size, and organic materials can have different kinds of pretreatment.

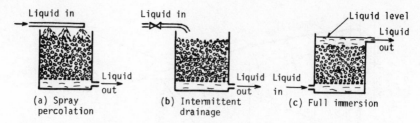

Figure 4.39 The three principal methods of liquid–solid contacting in leaching

Sugar-beets are usually sliced into cossettes, thin strips resembling shoestring potatoes 3–7 mm wide and thick and 50–80 mm long, including some fragments. The premeability of the cossette bed drops radically if the mass of cossettes less than 10 mm long exceeds 10 per cent of the mass of cossettes more than 50 mm long.[37]

Oilseeds with low oil content, such as soya beans, grape seed, and rice bran, are flaked prior to extraction, resembling corn flakes with a thickness of 0.25–0.4 mm,[38] to give rapid oil extraction. The flakes must be handled carefully to avoid breakage that reduces the permeability of the bed.

Oilseeds with more than 20–25 per cent oil, such as rapeseed, peanut, cottonseed and sunflower seed, are either pressed prior to extraction, or they are partially extracted down to 10–15 per cent oil by percolation, followed by wet flaking and final extraction in an immersion-type extractor.[39]

Equipment for leaching

The choice of equipment and solvent for leaching depends on the physical properties of the solid, the quantity to be leached, and the recovery of solvent.

The solid will often need a pretreatment, such as grinding, cutting in pieces, or reforming in shapes such as flakes, in order to increase the surface area. But fine particles may prevent the free flow of liquid, or require difficult clarification or filtration. In the case of cellular materials, rupture of the cell walls may also increase the extraction of undesired compounds. Increased temperature increases the leaching rate. But in the case of natural materials, such as coffee, tea, and sugar-beets, too high a temperature results in excessive extraction of undesired material.[8]

Coarse solids that form an open, permeable mass throughout the leaching operation, may be leached by percolation in fixed bed or moving bed equipment.

Table 4.5[32] lists continuous extractors in common use in the food industries in Europe and in the U.S.A. The extractor (Figure 4.40) consists of a series of sector-shaped compartments that rotate around a central axis. As a compartment passes the feed point, it receives a load of feed, usually introduced as a slurry with recycled miscella. Then, as the filled compartments move around the cycle, extract is percolated through the beds of flakes in countercurrent fashion, by pumping the extract drained from progressively more exhausted flakes to the top

Table 4.5 Continuous extractors currently used in food industries. Reproduced by permission of American Institute of Chemical Engineers.

Type	Fig. no.	Capacity, solids (t/day)	Size† (m)	Products and materials processed††	Efficiency (%) L mass of liquid F mass of solids	Solid retention time (min)	No. of units in use
Rotary, deep bed sector-shaped compartments	4.40	Soy 0.2–3 000 Sugar-cane 1200	Rotocel: D = 3.4–11.3 H = 6.4–7.3 B = 1.3–3.0 Carroucel: D = 1–8 B = 0.5–2.5	BD, CO, FM, FCS, FOS, GY, H, POS, RI, SC, TE, TW	Soy 97–98% $L/F = 0.8$–0.85 SC 92–97% $L/F = 0.99$	Soy 18 FCS 85	365
Stationary cells, rotating feed and discharge		Soy 61–2700 FCS to 1350 PCS to 3400	D = 2.4–10.4 H = 9–15 B = 1.8–3.6	Fish meal flaked oilseeds, prepressed oilseeds	Soy 97–98% $L/F = 1.1$ FCS 96.5% $L/F = 1.3$	Soy 45 FCS 60	
Drag chain, horizontal loop		Oilseeds 1–1500 BA 1000–6000	L = 3–24 W = 0.13–3.7 B = 0.1–0.76 L = 16–28 B to 1.8	Bagasse, flaked oilseeds, prepressed oilseeds, sugar-cane	Soy 96.3–98.5% $L/F = 0.8$–1.0 BA, SC 95.5–97.6% $L/F = 1.0$–1.12	Soy 40 Bagasse 40	155
Double screw conveyor	4.41 4.42	0.7–4500 (mini-units down to 0.2)	D = 2–3.7 L = 21–27 H = 8–11 W = 5.5–8.8	AP, CA, CH, CO, FM, GR, GE, GL, PE, RB, SB, SC, SP, ST, TE, YP	AP 92–95%, $L/F = 0.75$ BA 96.1%, $L/F = 1.03$ CH 99%, $L/F = 2.15$ SB 97%, $L/F = 1.09$	AP 75–80 BA 33 SB 60–70	435

Type		Capacity	Dimensions	Material	Efficiency		
Vertical screw tower		800–6000	L = 3–7.9, H = 10.2–14.1	Flaked oilseeds, sugar-beet	98.3%, L/F = 1.2–1.25	70–90	> 260
Horizontal helix		600–10000	D = 4.7–7, L = 34–56, H = 7–10.5, B = 1.2–2.7	Sugar-beet	98.5%, L/F = 1.1–1.2	60–90	240
Vertical basket conveyor	4.43	180–900	L = 9.6, H = 14, W = 1.5–2.0, B = 0.5–0.7	Flaked oilseeds	Soy 97.3%, L/F = 0.96		95
Horizontal basket conveyor		3–4000	B = 0.5–0.7	COA, HO, FM, RI	Soy 97–98%, HO 75–80%		145
Continuous belt	4.44	FOS 10–3000 SB, SC 700–7200	L = 7–37, W = 0.5–9.5, B = 0.8–2.6	AP, BA, FOS, POS, SB, SC, TW	SC 97–98%, L/F = 0.95–1.1	SC 40–60	> 350

[†] D = diameter, L = length, H = height, W = width, B = bed depth.

[††] AP = apples, BA = bagbasse (prepressed sugar-cane), BD = botanical drugs, CA = carrots, CH = cherries, CO = coffee, COA = cocoa shells, pulp, or residue, FCS = flaked cottonseed, FM = fish meal, FOS = flaked oilseeds, GR = grapes, GE = gelatin, GL = glue, GY = glycosides, H = hops, HO = high-oil-content oilseeds, PE = pears, POS = prepressed oilseeds, RB = red beet, betanines, RI = rich bran, SB = sugar-beet, SC = sugar-cane and flaked soya beans, SP = soya-bean protein, ST = straw, TE = tea, TW = tallow, WG = wheat germ, YP = yeast protein.

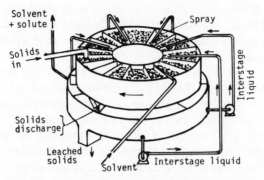

Figure 4.40 Schematic arrangement of rotary extractors with sector-shaped compartments.[19] In the Rotocel (Rosedowns, England) the spent solids are discharged through the hinged screen bottom in each cell, while in Carroucel (Extraktionstechnik, West Germany) the cell partitions rotate over a stationary screen with a sector-shaped opening for solid discharge

of cells containing progressively richer flakes. At the end of the cycle, the perforated floor of the compartment opens and drops the extracted solids into a conveyor, which transfers the solids with 25–30 per cent solvent to the toaster–desolventizer.

In the Carroucel extractor the cell partitions rotate over a stationary screen of concentric wedge-bar rings. The spent solids drop through a sector-shaped unscreened opening, and the miscella sumps and pumps are arranged to give countercurrent flow.

The French stationary basket extractor uses the same principle of operation as the Rotocel, but the cells do not move. Instead the solids feed spout, the solids discharge hopper, and the cell bottom unlatching mechanism rotate.

In drag chain extractors the solids are deposited between slatted partitions

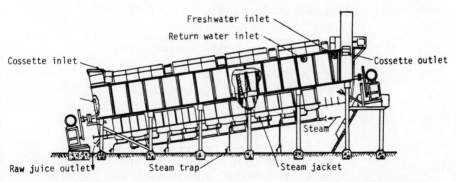

Figure 4.41 D.D.S. double screw extractor. Reproduced by permission of H. Brüniche–Olsen, Copenhagen

pulled by a drag chain over an enclosed loop. The horizontal drag chain extractor has one horizontal and one slightly tilted nearly horizontal leg, while the vertical ones have two or more vertical loops. All of them have countercurrent extraction in part of the loops.

The double screw conveyor extractor (Figure 4.41) was developed by the Danish Sugar Corporation (D.D.S.) for sugar-beet extraction,[35] and has turned out to be one of the most widely used and versatile extractors available for food industries. The solids are propelled through the extractor by the two screws oppositely pitched and turning in opposite directions in the two parallel cylindrical troughs (Figure 4.42). The troughs are sloped to permit the gravity flow of water and to extract countercurrent from the solids that are fed in through a hopper in the lower end and lifted out by means of a scoop-wheel with perforated scoops at the upper end. In Europe the slope for sugar-beet cossette extraction is usually 7–8°, and in the U.S. approximately 11° in order to compensate for greater resistance to flow of liquid through the finer cossettes. The heat economy is especially good because the solids are preheated by heat exchange with the hot extract in the first 3–4 m of passage in the troughs, and the cold water entering the top section is preheated by passage through the hot cossettes.

In D.D.S. extractors used for sugar-cane and bagasse extraction, the speed of each screw is varied slightly and periodically out-of-phase, thus providing a squeezing action.[40]

Niro Atomizers produce smaller units for products other than beet and cane sugar, including pilot-scale extractors with a diameter down to 0.15 m and length 2 m.

Compared with pressing, the D.D.S. extractor used for fruit-juice extraction increases the yield for sliced apples by 15–25 per cent.[32,41] This extraction is not used in the U.S. because the standards of identity for apple juice forbid any dilution.

Vertical screw tower extractors are mostly used for sugar-beet extraction. Screws carry the cossettes upward through the tower, countercurrent to the descending extract. Prior to extraction, the cossettes are scalded by heating them with extract from the bottom of the tower.

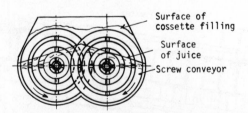

Figure 4.42 Section through screws and troughs in the D.D.S. extractor.[35] The screw flights are formed of metal ribbons with slight gaps between them to permit flow of extract through the screw surface

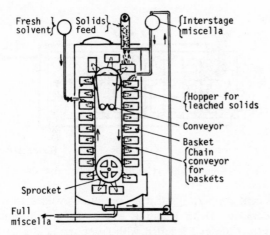

Figure 4.43   Vertical basket conveyor

The rotary helix or rotary Tirlemont extractor has a horizontal cylinder rotating on trunnions with 0.3–0.45 rpm. A perforated baffle lifts the cossettes that are transferred to the next compartment. A scroll in the centre acts as an Archimedes screw that conveys the liquid in the opposite direction to the solids. The extractor is a highly flexible, trouble-free unit which has a good turn-down ratio and can handle a wide variety of cossettes, in terms of thickness and quality. Because of a complicated interior construction, it is expensive.[32]

The vertical basket conveyor (Figure 4.43) consists of a series of large wire-mesh-bottomed baskets that move vertically inside a vapour-tight steel tank. Fresh flakes are loaded into the descending baskets and leached in parallel flow by a dilute solution of solvent and oil (half miscella) from the ascending leg with countercurrent leaching. After a short drainage, the baskets are turned around at the top and the flakes discharged in a hopper. Advantages are low power consumption, minimum breakage of thin flakes, and good clarity of final miscella. The major disadvantage is the possibility of channelling.

The vertical basket conveyor extractor has been largely superseded by the horizontal design that permits one-floor operation. The liquid flow in each stage is controlled by a combination of recirculation and advancement of the miscella. Horizontal chain and basket extractors are currently produced by H.L.S. Ltd., Israel, Lurgi Apparate-Technik, West Germany, and Gianazza of Legnano and Bernardini, both Italy.[42]

Perforated belt extractors are extensively used in cane-sugar, beet-sugar, and oilseed extraction. The solids are deposited on a perforated, moving belt; the liquid percolates through and drips into catch basins, and is pumped to the top of subsequent sections of the bed, as shown in Figure 4.44.

The metallurgical industries probably handle the largest tonnages for leaching. The operations range from in-place (in situ) leaching and solution mining, to leaching in pressurized autoclaves.

In-place leaching has been applied even to low-grade copper ores containing as

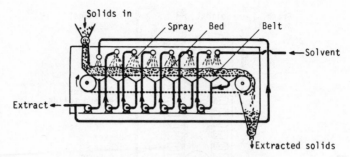

Figure 4.44 Continuous belt conveyor

little as 0.2 per cent copper,[43] and it is continuously used to recover salt from underground deposits. Heap leaching where the ore, as mined, is piled in prepared drainage pads and leached with sulphuric acid solutions, is another example of recovery of minerals which cannot otherwise be recovered economically. The leaching process may take months.

On leaching in tanks or vats, the raw material is prepared by crushing or grinding to optimal size before being bedded into the large tanks where the ore is leached by percolation (Figure 4.39a) or by intermittent drainage (Figure 4.39b). Examples are leaching of non-porous copper, gold and silver ores, and small tonnages of uranium and vanadium ores.[44]

Leaching vessels may also be equipped with stirrers (Figure 4.45), rotate on trunnions (Figure 4.46), or be agitated by compressed air (Figure 4.47). Classifiers and Dorr thickeners are other examples of equipment used in leaching of minerals.[45]

Multistage equipment for countercurrent leaching or washing is often arranged in a cascade. The solids may be transferred from one vessel to the next directly as a slurry from a thickener or a hydrocyclone, or it can pass a filter to reduce the amount of liquid in the underflow.

Bautista[44] gives an excellent review of leaching used in preprocessing of metals, and O'Kane[45] describes different types of pressure leaching autoclaves and gives an example of how laboratory and pilot plant data are translated to a commercial scale.

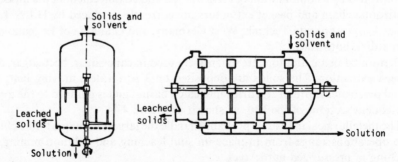

Figure 4.45 Agitated vessels for batch leaching

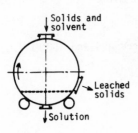

Figure 4.46 Leaching vessel rotating on trunnions

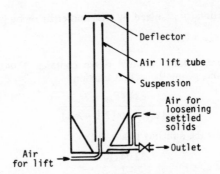

Figure 4.47 Pachuca tank, agitation by compressed air

Example 4.1 ESTIMATION OF DISTRIBUTION COEFFICIENTS

Hydrate formation in pipelines with natural gas and some condensate may be avoided by heating or by injection of methanol at the wellhead. Methanol dissolved in the condensate can be recovered by liquid–liquid extraction with water.[46]

Lacking experimental data for the distribution of methanol between the two phases, estimate for mole fractions of methanol in the oil up to $x_{AR} = 0.04$ the order of magnitude of the distribution coefficient based on vapour–liquid equilibrium data (Table a), and assuming the condensate to behave as n-heptane.

Table a Constants in Van Laar's equations (equations (2.11) and (2.12)) for mixtures of methanol–heptane and of methanol–water, calculated from vapour–liquid equilibrium data at atmospheric pressure[47]

	Methanol (1)–n-heptane (2)	Methanol (1)–water (2)
A_{12}	0.9775	0.3655
A_{21}	1.0702	0.2137

Solution

Equation (4.10) is rearranged to give the equilibrium condition as

$$\log x_{AE} + \log \gamma_{AE} = \log x_{AR} + \log \gamma_{AR} \tag{a}$$

The right-hand side of equation (a) is calculated for selected values of x_{AR} with $\log \gamma_{AR}$ calculated by equation (2.11) as shown below for $x_{AR} = 0.02$:

$$\log 0.02 + \frac{0.9775 \times 0.98^2}{(0.02 \times 0.9775/1.0702 + 0.98)^2} = -0.7352$$

This value inserted in equation (a) and combined with equation (2.11) gives the equilibrium mole fraction x_{AE} in the extract,

$$\log x_{AE} + \frac{0.3655(1 - x_{AE})^2}{[0.3655 x_{AE}/0.2137 + (1 - x_{AE})]^2} = -0.7352 \tag{b}$$

Equation (b) is solved by trial and error or by iterations, giving

$$x_{AE} = 0.0217$$

Here, 1 kmol organic phase contains $32.04x_{AR}$ kg methanol and $100.2(1 - x_{AR})$ kg heptane, giving the weight fraction

$$X_{AR} = \frac{32.04x_{AR}}{32.04x_{AR} + 100.2(1 - x_{AR})} \tag{c}$$

and for the aqueous phase,

$$X_{AE} = \frac{32.04x_{AE}}{32.04x_{AE} + 18.02(1 - x_{AE})} \tag{d}$$

The four equations give the results in Table b.

Table b   Equilibrium data for methanol in the two phases

x_{AR}	0.005	0.01	0.02	0.03	0.04	Notes
$\log \gamma_{AR}$	0.9901	0.9818	0.9637	0.9458	0.9280	eq. (2.11)
$\log (\gamma_{AR} x_{AR})$	-1.3101	-1.0182	-0.7352	-0.5771	-0.4699	
x_{AE}	0.0051	0.0104	0.0217	0.0342	0.0482	eq. (a)
X_{AR}	0.0016	0.0032	0.0065	0.0098	0.013	eq. (c)
X_{AE}	0.0090	0.018	0.038	0.059	0.083	eq. (d)
K	5.6	5.7	5.9	6.1	6.3	eq. (4.6)

Example 4.2 COUNTERCURRENT EXTRACTION WITH CONSTANT DISTRIBUTION COEFFICIENT

Claffey et al.[48] measured the distribution coefficient by extraction with kerosene of the nicotine in aqueous solutions from tobacco trade wastes. Water and kerosene are essentially insoluble, and the distribution coefficient for nicotine, $K = Y/X$, was practically constant for nicotine concentrations below 30 kg/m³ aqueous solution, but strongly depending on temperature:

Temperature °C	5	20	35	50	65	80
$K = \dfrac{\text{kg nicotine/m}^3 \text{ organic phase}}{\text{kg nicotine/m}^3 \text{ aqueous phase}}$	0.29	0.66	1.36	2.34	3.31	3.43

(a) Calculate the minimum ratio of organic to aqueous phase, $(E/R)_{\min}$, if the aqueous phase is to be stripped from 25 to 0.25 kg nicotine/m³ with kerosene at a temperature of 50 °C.

(b) Calculate the number of theoretical stages needed if the stripping is carried out with $(E/R) = 1.4 (E/R)_{\min}$.

Solution

(a) With the symbols in Figure 4.13, the E_1 m³ organic phase contains $25 - 0.25 = 24.75$ kg nicotine per m³ aqueous phase R_0. The minimum organic phase corresponds to equilibrium at stage 1, i.e.

$$\frac{24.75/E}{25/R} = K = 2.34$$

or

$$(E/R)_{min} = 0.423 \, m^3 \text{ organic phase/m}^3 \text{ aqueous phase}$$

(b) Equation (4.14) gives the extraction factor,

$$\varepsilon = \frac{1}{1.4 \times 0.423} = 1.69$$

Equation (4.17) with $Y_{N+1} = 0$, $Y_1 = 24.75/(1.4 \times 0.423) = 41.8$, and $Y_0 = 2.34 \times 25 = 58.5$, gives number of theoretical stages,

$$N = \frac{\ln\left[\frac{0-58.5}{41.8-58.5}\left(1 - \frac{1}{1.69}\right) + \frac{1}{1.69}\right]}{\ln 1.69} = 1.34$$

Example 4.3 COUNTERCURRENT EXTRACTION OF COPPER

Table a gives some of the equilibrium data reported by DeMent and Merigold[49] for copper in an aqueous solution of dump leach liquors in contact with kerosene with 15 vol per cent LIX 64N.

Table a   Equilibrium data for Cu in aqueous solutions, pH = 2.0, in contact with kerosene with 15 vol per cent LIX 64N at 23 °C.

X kg Cu/m³ aqueous solution	0.08	0.25	0.67	1.80
Y kg Cu/m³ organic solution	0.31	0.76	1.84	3.26

Using these data, calculate the number of theoretical stages required to reduce the copper content of the aqueous phase from 3 kg/m³ to 0.1 kg/m³ by extraction with 0.98 m³ kerosene solution per m³ aqueous solution.

Solution

The equilibrium curve from Table a is plotted in Figure 4.48 together with the operating line given by equation (4.18),

$$Y_{n+1} = \frac{R}{E}X_n + \left(Y_1 - \frac{R}{E}X_0\right)$$

where $R/E = 1/0.98 \, m^3$ aqueous phase/m³ organic phase, $X_0 = 3 \, kg/m^3$ aqueous phase, and $Y_1 = (3 - 0.1)/0.98 = 2.96$, giving

$$Y_{n+1} = X_n/0.98 + (2.96 - 3/0.98) = 1.02X_n - 0.10$$

Figure 4.48 gives three theoretical stages to reduce the copper content in the aqueous phase from 3 to 0.1 kg/m³ solution, with $Y_1 = 2.96$ kg Cu per m³ organic solution leaving stage 1.

194

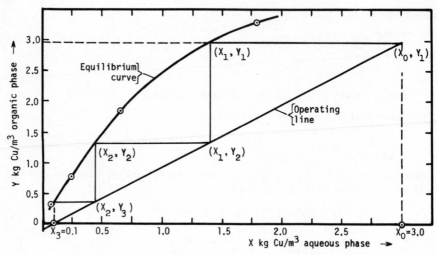

Figure 4.48 Determination of number of theoretical stages

Example 4.4 LIQUID–LIQUID EXTRACTION

Here, 300 kg/h of a mixture with mass fraction $X_A = 0.20$ acetic acid and $X_B = 0.80$ diethylether is to be extracted with pure water (S) to give an organic phase R with mass fraction acetic acid $X_A = 0.01$. The prime indicates a solvent-free basis, i.e. $X'_A = kgA/(kgA + kgB)$.

Wittenberger and Fritz[50] give the equilibrium data in Table a for this system where A, B, and S are mass fractions.

Table a Liquid–liquid equilibrium data for diethylether–water

Aqueous A_E	0	0.051	0.088	0.138	0.184	0.231	0.279
(extract) S_E	0.933	0.88	0.84	0.782	0.721	0.650	0.557
phase							
Organic A_R	0	0.038	0.073	0.125	0.181	0.236	0.287
(raffinate) S_R	0.023	0.036	0.050	0.072	0.104	0.151	0.236
phase							

(a) Plot the equilibrium curves in a Janecke plot with $Z' = kgS/(kgA + kgB)$ as ordinate and X' and $Y' = kgA/(kgA + kgB)$ as abscissa, X' for the organic raffinate and Y' for the aqueous extract phase. Plot the conjugate curve instead of the tie-lines.

(b) Determine the minimum flow rate of water S_{min} in kg/h to obtain the required extraction in countercurrent flow.

(c) Determine the location of the mixing point M', the composition of the leaving extract E_1, and the operating point L' for $S = 1.3 S_{min}$.

(d) Transfer the equilibrium data to a McCabe–Thiele diagram (Y' versus X') and determine the number of theoretical stages.

Solution

(a) The data are recalculated to a solvent-free (S-free) basis by the equations:

$$X' = \frac{kgA}{kgA + kgB} = \frac{A_R}{A_R + [1 - (A_R + S_R)]} = \frac{A_R}{1 - S_R} \tag{a}$$

$$Z'_R = \frac{kgS}{kgA + kgB} = \frac{S_R}{A_R + [1 - (A_R + S_R)]} = \frac{1}{1/S_R - 1} \tag{b}$$

$$Y' = \frac{A_E}{A_E + [1 - (A_E + S_E)]} = \frac{A_E}{1 - S_E} \tag{c}$$

$$Z'_E = \frac{S_E}{A_E + [1 - (A_E + S_E)]} = \frac{1}{1/S_E - 1} \tag{d}$$

These equations with the data in Table a give the results in Table b.

Table b Data from Table a recalculated to a solvent-free basis

									Notes
Organic X' (raffinate)	0	0.039	0.077	0.135	0.202	0.278	0.367		eq. (a)
phase Z'_R	0.024	0.037	0.053	0.078	0.116	0.178	0.309		eq. (b)
Aqueous Y' (extract)	0	0.425	0.550	0.633	0.659	0.660	0.630		eq. (c)
phase Z'_E	13.93	7.33	5.25	3.59	2.58	1.86	1.26		eq. (d)

Points in Table b are plotted in Figure 4.49, and the conjugate line is determined as shown for one point on the line.

(b) The intersection of the extract curve and the straight line through the feed point $R'_0 (X'_0 = 0.2)$ that coincides with a tie-line gives $E'_{1, min}$ which is the extract from stage 1 when minimum S is used. The corresponding mixing point for the feeds, M'_{min}, is determined as shown in Figure 4.49. It is the intersection between a line through R'_N and $E'_{1, min}$, and the vertical line through R'_0 (with pure solvent, $Z'_S = \infty$), i.e. $Z'_{M, min} = 0.68$. This gives

$$S_{min}/(A + B) = 0.68, \qquad S_{min} = 0.68 \times 300 = 204 \text{ kg/h}$$

(c) Water $S = 1.3 \times 204 = 265$ kg/h and $Z'_M = 265/300 = 0.883$ which gives M' in Figure 4.49. The straight line through R'_N and M' intersects the extract curve at E'_1. The operating point L' is located on a straight line through E'_1 and the feed point R'_0, and a line with abscissa $X'_L = X'_N$ (equation (4.28) for the case with pure solvent feed).

Point E'_1 in Figure 4.49 is located at

$$Y' = A/(A + B) \tag{e}$$

and

$$Z' = S/(A + B) \tag{f}$$

Combining the two equations with the sum

$$A + B + S = 1 \tag{g}$$

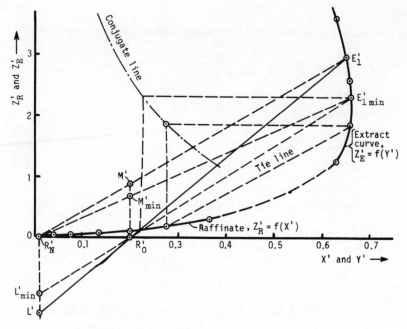

Figure 4.49   Janecke plot of the data in Table a

gives the following mass fractions:

Acetic acid	(A)	0.167
Diethylether	(B)	0.089
Water	(S)	0.744

(d) Points on the equilibrium curve are taken directly from Table a. Additional points are obtained from the two end points of tie-lines as shown for point P in Figure 4.50.

The coordinates of points on the operating line are determined by the intersections of the raffinate and extract curves with straight lines through L', as shown for point Q in Figure 4.50. Stepping off the stages gives approximately six theoretical stages.

Example 4.5 LEACHING

One of the steps in the production of caustic soda by the lime–soda process gives a slurry containing 9 kg water and 0.8 kg sodium hydroxide per kg calcium carbonate particles.[51] The slurry is washed countercurrently with water in four stages (four thickeners). The solids discharged from each stage contains 3 kg water per kg calcium carbonate.

Calculate the amount of wash-water needed when the discharged calcium carbonate after drying contains a maximum of 0.0075 kg sodium hydroxide per kg calcium carbonate.

Solution

Except for the feed to the first stage, the solvent (water) underflow is constant, and equation (4.58) can be used. The mass fractions in this equation are determined by mass

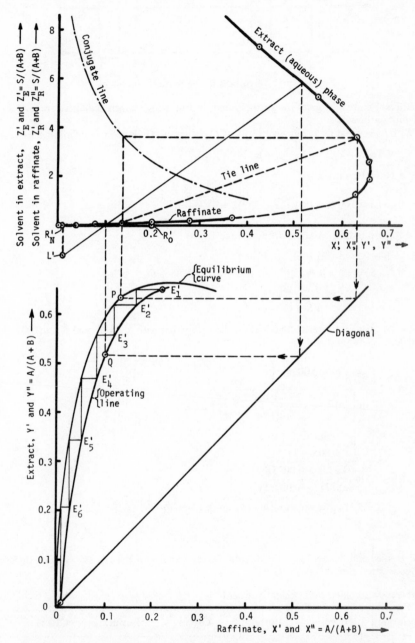

Figure 4.50 Operating line and equilibrium curve determined from the Janecke plot in the upper diagram. The number of theoretical stages is stepped off between the two lines as shown in the lower diagram

198

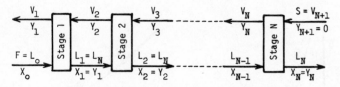

Figure 4.51 Symbols in the material balances

balances. Figure 4.51 gives the symbols, and the following material balances are for 1 kg calcium carbonate.

With $L_0 = 9$ and $L_N = 3\,\text{kg water/kg CaCO}_3$, a material balance around all stages gives:

$$S + 9 = 3 + V_1, \qquad V_1 = S + 6$$

and a NaOH balance,

$$0 + 0.8 = 0.0075 + (S + 6)Y_1, \qquad Y_1 = \frac{0.7925}{S + 6}$$

The corresponding balances around stage 1 give

$$9 + V_2 = 3 + (S + 6), \qquad V_2 = S$$
$$0.8 + SY_2 = 3 \times 0.7925/(S + 6) + (S + 6)0.7925/(S + 6)$$
$$Y_2 = \frac{2.3325 - 0.0075S}{S(S + 6)}$$

These numbers are inserted in equation (4.58), together with $X_1 = Y_1$ and $X_N = 0.0075/3 = 0.0025$, giving

$$4 - 1 = \frac{\ln\left[(0 - 0.0025)\Big/\left(\dfrac{2.3325 - 0.0075S}{S(S + 6)} - \dfrac{0.7925}{S + 6}\right)\right]}{\ln\left[\left(0 - \dfrac{2.3325 - 0.000\,75S}{S(S + 6)}\right)\Big/\left(0.0025 - \dfrac{0.7925}{S + 6}\right)\right]}$$

$$3 = \frac{\ln\dfrac{0.000\,25S(S + 6)}{0.805 - 3325}}{\ln\dfrac{2.3325 - 0.0075S}{S(0.7775 - 0.0025S)}}$$

$S = 7.31$ kg wash-water/kg calcium carbonate.

Problems

4.1 Henley and Seader[4] give the data in Tables a and b for the ternary system ethylene glycol–water–furfural at 25 °C.

Table a Miscibility data, weight per cent

Furfural	Glycol	Furfural	Glycol	Furfural	Glycol
94.8	0.0	40.6	47.5	10.2	32.2
84.4	11.4	33.8	50.1	9.2	28.1
63.1	29.7	23.2	52.9	7.9	0.0
49.4	41.6	20.1	50.6		

Table b Tie-line data, weight per cent glycol

Water layer	49.1	32.1	11.5	7.7	6.1	4.8	2.3
Furfural layer	49.1	48.8	41.8	28.9	21.9	14.3	7.3

(a) What are the compositions and the quantities of the two phases if 1 kg containing 30 per cent by weight glycol and 70 per cent water is mixed with 0.5 kg pure furfural?

(b) Plot a Janecke diagram with furfural as solvent.

4.2 Acetone (A) in an aqueous solution containing 0.20 kg acetone per kg water (B) is to be purified by extraction of 90 per cent of the acetone with kerosene (S) and extraction back into pure water. The equilibrium relationship is[1],

$$Y = 0.155X$$

where Y = kg acetone/kg kerosene, and X = kg acetone/kg water. Water and kerosene are practically insoluble.

(a) Calculate the number of stages needed in the first extraction, using 6.45 kg kerosene per kg water in countercurrent flow. The entering kerosene contains 0.002 kg acetone per kg kerosene.

(b) Using the same number of theoretical stages in countercurrent flow in the second extraction and 0.17 kg pure water per kg kerosene, calculate the acetone content in the final aqueous solution.

Ans. $Y_1 = 0.152$.

4.3 Determine by stage-by-stage calculation the amount of wash-water needed in Example 4.5.

4.4 Seeds containing 22 per cent by weight of oil are extracted countercurrently with oil-free hexane as solvent. Calculate the number of theoretical stages needed if 90 per cent of the oil is recovered in extract with 40 per cent by weight of oil, and the amount of liquid (solvent and oil) in the underflow from each stage is 0.6 kg per kg insoluble matter. *Ans.* $N = 4$.

Symbols

A	Component A, kg or kg/s
A	Van Laar constants
a	Activity
a	Interfacial area per unit volume, m^2/m^3
B	Component B, kg or kg/s
E	Extract, kg or kg/s
F	Feed, kg or kg/s
K	Distribution coefficient based on mass fractions
K'	Distribution coefficient based on mole fractions
k	Volumetric mass transfer coefficient, kg $A/(m^3$ s mass fraction)
k	Mass transfer coefficient, $kmol/(m^2$ s mole fraction)
L	Underflow in leaching
L	Net flow, equation (4.22)
L'	Operating point, coordinates, equations (4.29) and (4.30)
L''	Operating point, coordinates, equations (4.52) and (4.53)
P	Difference point, equation (4.21)
R	Raffinate, kg or kg/s

S	Solvent, kg or kg/s
S_t	Cross-section of tower, m^2
t	Time, s
V	Overflow in leaching
X	Mass fraction solute, kg/kg solvent or kg/kg solvent + solute
x	Mole fraction
Y	Mass fraction solute in extract phase, kg/kg solvent or kg/kg solvent + solute
Y_0	Mass fraction in extract in equilibrium with entering feed
y	Mole fraction
Z	Mass fraction solvent
z	Height of contact section, m
α	Separation factor, equation (4.11)
α	Activity coefficient
ε	Extraction factor, equation (4.14)

Subscripts

A	Component A
B	Component B
E	Extract phase
F	feed
i	At interface
L	Composition of net flow, equation (4.25)
lm	Logarithmic mean, equations (3.16), (3.19), (4.38), (4.39)
N	Number of theoretical stages
n	Stage number
0	Entering feed
R	Raffinate phase
S	Solvent

Superscripts

′	Solvent-free basis
″	Solvent-free basis, extract enriching section
*	Equilibrium composition

References

1. Laddha, G. S., and T. E. Degaleesan, *Transport Phenomena in Liquid Extraction*, Tata-McGraw-Hill, New Delhi, 1976.
2. Barthel, G., Solvent extraction recovery of copper from mine and smelter waters, *J. Metals*, **30** (July), 7–12 (1978).
3. Flett, D. S., Solvent extraction in copper hydrometallurgy, *Inst. Min. Metall., Trans., Sect. C*, **83**, C30–38 (1974).
4. Henley, E. J., and J. D. Seader, *Equilibrium-Stage Separation Operations in Chemical Engineering*, John Wiley & Sons, New York, 1981.
5. Varteressian, K. A., and M. R. Fenske, The system methylcyclohexane–aniline–*n*-heptane, *Ind. Eng. Chem.*, **29**, 270–277 (1937).
6. Janecke, E., Über eine neue Darstellungsform der wässerigen Lösungen zweier und dreier gleichioniger Salze, resiproker Salzpaare und der van't Hoffschen Untersuchungen über ozeanische Salzablagerungen, *Z. Anorg. Chemie*, **51**, 132–157 (1906).
7. Prausnitz, J., T. Anderson, E. Grens, C. Eckert, R. Hsieh, and J. O'Connell, *Computer Calculations for Multicomponent Vapor–Liquid and Liquid–Liquid Equilibria*, Prentice-Hall, Englewood Cliffs, N. J., 1980.

8. *Handbook of Separation Techniques for Chemical Engineers*, P. A. Schweitzer (Ed.), McGraw-Hill, New York, 1979.
9. Perry, R. H., and C. H. Chilton, *Chemical Engineers' Handbook*, 5th edn., McGraw-Hill, New York, 1973.
10. Gambill, W. R., Surface and interfacial tensions, *Chem. Eng.*, **65** (5 May), 143–146 (1958).
11. Treybal, R. E., *Liquid Extraction*, 2nd edn., McGraw-Hill, New York, 1963.
12. Shain, S. A., and J. M. Prausnitz, Thermodynamics and interfacial tension of multicomponent liquid–liquid interfaces, *A.I.Ch.E.J.*, **10**, 766–773 (1964).
13. Leibson, I., and R. B. Beckmann, The effect of packing size and column diameter on mass transfer in liquid–liquid extraction, *Chem. Eng. Progr.*, **49**, 405–416 (1953).
14. Hanson, C., *Neuere Fortschritte der Flüssig–Flüssig Extraktion*, Saarländer, Aarau, Switzerland, 1974.
15. Reissinger, K.-H., and J. Schröter, Selection criteria for liquid–liquid extractors, *Chem. Eng.*, **85**, (6 Nov.), 109–118 (1978).
16. Bailes, P. J., J. C. Godfrey, and M. J. Slater, Designing liquid/liquid extraction equipment, *Chem. Eng. (London)* (No. 370), 331–333 (1981).
17. Bailes, P. J., C. Hanson, and M. A. Hughes, Liquid–liquid extraction: the process, the equipment, *Chem. Eng.*, **83** (19 Jan.), 86–100 (1976).
18. *Recent Advances in Liquid–Liquid Extraction*, C. Hanson (Ed.), Pergamon Press, Oxford, 1971.
19. Treybal, R. E., *Mass-Transfer Operations*, 3rd Edn., McGraw-Hill, Tokyo, 1980.
20. Logsdail, D. H., and J. D. Thornton, The effect of column diameter upon performance and throughput of pulsed plate columns, *Trans. Inst. Chem. Engrs*, **35**, 331–342 (1957).
21. Kirk-Othmer, *Encyclopedia of Chemical Technology*, Vol. 9, John Wiley & Sons, New York, 1980.
22. Marr, R., G. Husung, and F. Moser, Die Auslegung von Drehscheibenextraktoren, *Verfahrenstechnik*, **12**, 139–144 (1978).
23. Reman, G. H., and R. B. Olney, The rotating-disc contactor, *Chem. Eng. Progr.*, **51**, 141–146 (1955).
24. Reman, G. H., Extraction equipment outside the U. S., *Chem. Eng. Progr.*, **62**, Sept., 56–61 (1966).
25. Strand, C. P., R. B. Olney, and G. H. Ackerman, Fundamental aspects of rotating disk contactor performance, *A.I.Ch.E.J.*, **8**, 252–262 (1962).
26. Fischer, A., Die Tropfengrösse in Flüssig–Flüssig Rührextraktionskolonnen, *Verfahrenstechnik*, **5**, 360–365 (1971).
27. Ingham, J., J. R. Bourne, and A. Mögli, Backmixing in a Kühni liquid–liquid extraction column, *Proc. Intn. Solvent Extraction Conference* 1974, **2**, 1299–1317, Soc. of Chem. Industry, London, 1974.
28. Lo, T. C., Recent development in commercial extractor, paper at Engineering Foundation Conference on Mixing Research, Rindge, N. H., The Eng. Foundation, New York, Aug. 1975.
29. Lo, T. C., Commercial liquid–liquid extraction equipment, Section 1.10 in *Handbook of Separation Techniques for Chemical Engineers*, P. A. Schweizer (Ed.), McGraw-Hill, New York, 1979.
30. Sheikh, A. R., J. Ingham, and C. Hanson, Axial mixing in a Graesser raining bucket liquid–liquid contactor, *Trans. Instn Chem. Engrs*, **50**, 199–207 (1972).
31. Rickles, R. H., Liquid–solid extraction, *Chem. Eng.*, **72** (15 march), 157–172 (1965).
32. Schwartzberg, H. G., Continuous counter-current extraction in the food industry, *Chem. Eng. Progr.* **76**, 67–85 (1980).
33. Huang, C., and C. Kuv, General mathematical model for mass transfer accompanied by chemical reaction, *A.I.Ch.E.J.*, **9**, 161–167 (1963).
34. Othmer, D. F., and W. A. Jaatmen, Extraction of soybeans. Mechanism with various solvents, *Ind. Eng. Chem.*, **51**, 543–546 (1959).

202

35. Brüniche-Olsen, H., *Solid–Liquid Extraction*, Nyt Nordisk Forlag, Arnold Busck, Copenhagen, 1962.
36. McCabe, W. L., and J. C. Smith, *Unit Operations of Chemical Engineering*, 3rd edn McGraw-Hill, New York, 1976.
37. McGinnis, R. A., *Beet Sugar Technology*, Beet Sugar Dev. Foundation, Ft. Collins, Colorads, 1971.
38. Hutchins, R. P., Continuous solvent extraction of soybeans and cottonseed, *J. Am. Oil Chem. Soc.*, **53**, 279–282 (1976).
39. Bernardini, R., Batch and continuous solvent extraction, *J. Am. Oil Chem. Soc.*, **53**, 275–278 (1976).
40. Hugot, E., *Handbook of Sugar Cane Engineering*, 2nd edn, Elsevier, Amsterdam, 1972.
41. Binkley, C. R., and R. C. Wiley, Continuous diffusion–extraction method to produce apple juice, *J. Food Sci.*, **43**, 1019–1023 (1978).
42. Milligan, E. D., Survey of current extraction equipment, *J. Am. Oil Chem. Soc.*, **53**, 286–290 (1976).
43. Rampacek, C., The impact of R & D on the utilization of low-grade resources, *Chem. Eng. Progr.*, **73** (No. 2), 57–68 (1977).
44. Bautista, R. G., Hydrometallurgy, T. B. Drew and J. W. Hopes (Eds), *Advances in Chemical Engineering*, Vol. 9, Academic Press, 1974.
45. O'Kane, P. T., Pressure leach autoclave design, Prepr. Pap. 13th Chem. Eng. Conf., Montreal, 19–23, Oct. 1963.
46. McClintock, W. A., *et al.*, Reduction of hydrate formation in natural gas stream by contacting with anti-freeze agent, U. S. Pat. No. 3,886,757, 1975.
47. Hála, E., L. Wichterle, J. Polák, and T. Boublik, *Vapour–liquid Equilibrium Data at Normal Pressures*, Pergamon Press, Oxford, 1968.
48. Claffey, J. B., C. O. Badgett, J. J. Skalamera, and G. W. M. Phillips Nicotine extraction from water with kerosene, *Ind. Eng. Chem.*, **42**, 166–171 (1950).
49. DeMent, E. R., and C. R. Merigold, LIX 64N–a process report on the liquid ion exchange of copper, Paper at 99th Annual Meeting AIME, Denver, Colorado, 15–19 Feb. 1970.
50. Wittenberger, W., and W. Fritz, *Rechnen in der Verfahrenstechnik und chemischen Reaktionstechnik*, Springer-Verlag, Vienna, 1981.
51. Coulson, J. M., and J. F. Richardson, *Chemical Engineering*, 3rd edn, Pergamon Press, Oxford, 1978.

CHAPTER 5
Humidification

In this chapter, the word *humidification* is used for the vaporization of water into a gas that is practically insoluble in water, and *dehumidification* is used for transfer of water vapour from an inert gas to liquid water. The processes are simpler than absorption, as there is no concentration gradient in the liquid phase. On the other hand, both mass transfer and heat transfer are important and influence one another.

Most of the following derivations are made for gas mixtures of air and water vapour at a total pressure $P = 1$ bar $= 10^5$ Pa $= 0.987$ atm, which is closer to the average atmospheric pressure over land areas than 1 atm (760 mm mercury at $0\,°C$). Also, the gas is assumed to be an ideal gas mixture of water (A) with molecular weight 18.0, and of air with molecular weight 29.0.

Definitions

The *absolute humidity* H is defined as mass of vapour per unit mass of dry air. Partial pressure of water vapour p and total pressure P gives $p/(P - p)$ kmol water vapour/kmol dry air, or

$$H = \frac{18p}{29(P - p)} \tag{5.1}$$

Saturation humidity H'' is the humidity of air in equilibrium with water at the temperature of the air,

$$H'' = \frac{18p_s}{29(P - p_s)} \frac{\text{kg H}_2\text{O}}{\text{kg dry air}} \tag{5.2}$$

where p_s = vapour pressure of water at the temperature of the air–water vapour mixture.

Percentage absolute humidity is the ratio of absolute humidity to saturation humidity,

$$H_P = 100 \frac{H}{H''} \tag{5.3}$$

or expressed in terms of partial pressure,

$$H_P = 100 \frac{p/(P - p)}{p_s/(P - p_s)} \tag{5.4}$$

Percentage relative humidity is the partial vapour pressure of water vapour as a percentage of the saturation pressure,

$$H_R = 100\frac{p}{p_s} \tag{5.5}$$

Humid volume v_H is the volume in m^3 of 1 kg dry air and the water vapour it contains. The ideal gas law combined with Amagat's law, gives the relationship

$$v_H = \frac{R}{29}\frac{T}{P} + \frac{R}{18}\frac{T}{P}H = \frac{RT}{P}\left(\frac{1}{29} + \frac{H}{18}\right) \tag{5.6}$$

or with $R = 8314$ J/(kmol K) and $P = 10^5$ N/m^2,

$$v_H = 0.08314T(1/29 + H/18) \tag{5.6a}$$

Humid heat of a mixture of air and water vapour, c_H, is the heat capacity of the mixture per kg dry air. The heat capacity of dry air in the temperature range from 0 to 100 °C is approximately 1.00 kJ/(kg K), and of water vapour 1.88 kJ/(kg K),

$$c_H = 1.00 + 1.88H \quad \text{kJ/(kg dry air K)} \tag{5.7}$$

Total enthalpy of the air–vapour mixture, h_y, is the enthalpy of 1 kg dry air plus the enthalpy of the water vapour it contains. With air and water at temperature T_0 as reference state (enthalpy = 0), the total enthalpy in kJ/(kg dry air) is

$$h_y = c_H(T - T_0) + H(\Delta h_v)_{T_0}$$
$$= (1.00 + 1.88H)(T - T_0) + H(\Delta h_v)_{T_0} \tag{5.8}$$

where $(\Delta h_v)_{T_0} = $ enthalpy of vaporization of water at the reference temperature T_0, K.

$$(\Delta h_v)_\theta = 2501.6 - 2.275\theta - 0.00018\theta^2 \quad \text{kJ/kg} \tag{5.9}$$

where $\theta = $ temperature, °C.

For temperature $0 < \theta < 124$ °C, equation (5.9) gives Δh_v within ± 0.04 per cent, and to 160 °C within 0.5 per cent.

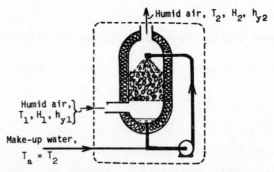

Figure 5.1 Adiabatic saturator. At equilibrium $T_2 = T_s$, $H_2 = H''$ and $h_{y2} = h_y''$

Adiabatic saturation temperature. Figure 5.1 shows an insulated tower where an air–water vapour mixture is contacted with recycled liquid water.

Unless the entering air is saturated with water vapour, it will be cooled as some of the water vaporizes. Assuming ideal insulation, the enthalpy of the exit gas equals the enthalpy of the entering gas,

$$h_{y2} = h_{y1} \tag{5.10}$$

If equilibrium is obtained, the exit air temperature will be the *adiabatic saturation temperature*, T_s, and the air will have the saturation humidity H''. With T_s as reference temperature and the same enthalpy of entering and leaving humid air, equation (5.8) gives

$$(1.00 + 1.88H_1)(T_1 - T_s) + H_1(\Delta h_v)_{T_s} = (1.00 + 1.88H'')(T_s - T_s) + H''(\Delta h_v)_{T_s}$$

or

$$\frac{H_1 - H''}{T_1 - T_s} = -\frac{1.00 + 1.88H_1}{(\Delta h_v)_{T_s}} \tag{5.11}$$

Equations (5.10) and (5.11) are based on the assumption that the make-up water is at the temperature of the exit gas. If T_a in Figure 5.1 differs from T_2, an enthalpy balance, as shown with the broken line, gives

$$h_{y2} = h_{y1} + 4.19(H_2 - H_1)(T_a - T_2) \tag{5.12}$$

where 4.19 = specific heat capacity of liquid water, kJ/(kg K).

Wet bulb temperature. The adiabatic saturation temperature is attained at steady state when a large amount of water is contacted with the gas under adiabatic conditions. The wet bulb temperature is attained at the liquid when a small liquid surface comes into contact with a large amount of gas under adiabatic, steady-state conditions.

Figure 5.2 shows a dry and a wet bulb thermometer located in an insulated air duct with turbulent flow. The continuous air stream is large, and the temperature

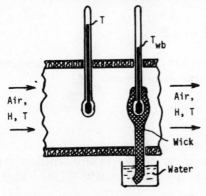

Figure 5.2   Measurement of dry and wet bulb temperature

T and the humidity H is practically unchanged. But the liquid evaporating from the wick of the wet bulb thermometer cools the wick. At steady state the latent heat of evaporation is exactly balanced by the heat transfer by convection from the air to the wick,

$$M_A N_A A(\Delta h_v)_{Twb} = hA(T - T_{wb}) \tag{5.13}$$

where M_A = molecular weight of water, 18 kg/kmol, N_A = molar flux of water vapour, kmol/(m²s), A = surface area of the wick, m², $(\Delta h_v)_{Twb}$ = enthalpy of vaporization at temperature T_{wb}, J/kg, h = heat transfer coefficient, W/(m²K), T = air temperature, K, and T_{wb} = wet bulb temperature, K.

Equation (3.18) gives the molar flux of A (water vapour) through a stagnant B (air).

$$N_A = -k_{ys}\frac{y - y_i}{(1 - y)_{lm}} \tag{3.18}$$

where k_{ys} = mass transfer coefficient by diffusion through a stagnant inert, kmol/(m²s mol frac.), $(1 - y)_{lm}$ = logarithmic mean mole fraction of air, and y_i = mole fraction water vapour at the surface, assumed to be equal to the mole fraction by saturation at the wet bulb temperature, y_{wb}. In our case $(1 - y)_{lm}$ is almost unity. Also, the flux is from the liquid to the gas, and equation (3.18) may be approximated by

$$N_A \approx k_{ys}(y_{wb} - y) \tag{5.14}$$

where y_{wb} = mole fraction of water vapour in the gas at the surface, and y = mole fraction of water vapour in the bulk of the gas. Expressed in terms of absolute humidity,

$$y = \frac{H/M_A}{1/M_B + H/M_A} \tag{5.15}$$

where M_A and M_B are the molecular weights of water and air, 18 and 29. Since H is small, equation (5.15) may be approximated by

$$y \approx \frac{HM_B}{M_A} \quad \text{and} \quad y_{wb} \approx \frac{H_{wb}M_B}{M_A} \tag{5.16}$$

Substituting equation (5.16) into (5.14), and then substituting the resultant into equation (5.13) and rearranging, gives

$$\frac{H - H_{wb}}{T - T_{wb}} = -\frac{h/M_B k_{ys}}{(\Delta h_v)_{Twb}} \tag{5.17}$$

The numerator on the right-hand side, $h/M_B k_{ys}$, is called the *psychrometric ratio*.[†] For water vapour–air mixtures in turbulent flow, experimental data on the psychrometric ratio is approximately $0.96 - 1.005^1$ kJ/(kg K), which is almost

[†] This ratio is also called the Lewis relation or in German literature 'Lewische Kennzahl after W. K. Lewis.[2]

identical to the specific heat capacity of air, and to the numerator on the right-hand side of equation (5.11), i.e.

$$\frac{H_1 - H''}{T_1 - T_s} \approx \frac{H - H_{wb}}{T - T_{wb}} \qquad (5.18)$$

The left-hand side of equation (5.18) is the slope of a line that connects points of constant enthalpy in a plot of H versus T. The right-hand side of the same equation is the slope of a psychrometric line, i.e. a line that connects the actual state with the state at the wet surface. The important conclusion is that in a H–T chart for water vapour–air mixtures, *psychrometric lines* and *constant enthalpy lines* are close to parallel. Hence, constant enthalpy lines can also be used for wet bulb lines with reasonable accuracy.

For organic vapours in air, the psychrometric ratio is larger than for water vapour in air, and the psychrometric lines are steeper than the constant enthalpy lines.

The *dew-point temperature* is the temperature at which a given mixture of air and water vapour becomes saturated as it is cooled at constant pressure and humidity. On further cooling, some water vapour would condense.

Humidity charts

Humidity charts are useful tools for calculations involving humid air. Different types of humidity charts are in use. The Mollier chart is common in German literature. It has the enthalpy plotted versus the moisture content. The chart with absolute humidity plotted versus the dry bulb temperature was introduced by

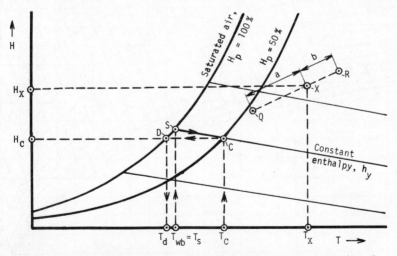

Figure 5.3 Simplified humidity chart showing how to determine the humidity H_C and the dew-point T_d from the dry bulb temperature T_C and the wet bulb temperature T_{wb}, and how to determine the temperature T_X and the humidity H_X of a mixture of humid air Q and humid air R

208

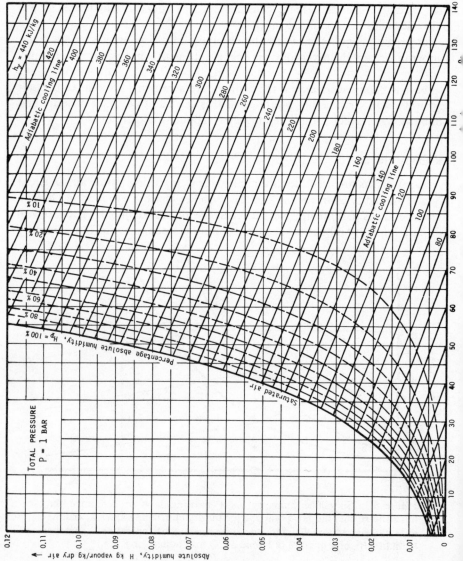

Figure 5.4 Humidity chart for air–water vapour at total pressure 1 bar (10^5 Pa)

Grosvenor.[3] It is the type of chart commonly used in English texts. Its use is illustrated in Figure 5.3, and Figure 5.4 is a detailed diagram for total pressure $P = 1$ bar.

Following the arrows in Figure 5.3, a wet bulb temperature T_{wb} and a dry bulb temperature T_C give point C where the absolute humidity is read on the ordinate. The percentage humidity is read on the H_p-line and the enthalpy on the constant enthalpy line. Interpolation between the lines may be necessary. The dew-point T_d is the abscissa for the point D where the constant humidity line intersects the curve for saturated air.

If q kg dry air with humidity and temperature corresponding to point Q in the chart is mixed with r kg dry air with humidity and temperature corresponding to point R, the resulting mixture is located at point X on the straight line between Q and R. The distance from Q is given by the lever-arm rule, $a/b = r/q$.

Measurement of humidity

The *psychrometric method* is based on measurements of the dry bulb and the wet bulb temperature. The humidity is read from a psychrometric chart as shown in Figure 5.3.

Dew-point methods. By one of the methods, a polished metal disc is inserted in the gas stream and cooled gradually. The temperature at which mist just forms on the surface is the dew-point.

Another method is based on compression of samples of the gas and adiabatic expansion through a converging–diverging nozzle into a chamber with radioactive tape to give nuclei for fog formation. The temperature after the expansion is given by the equation

$$T_1/T_2 = (P_1/P_2)^{(\kappa - 1)/\kappa} \tag{5.19}$$

where $T_1/T_2 = $ ratio of absolute temperature before expansion to absolute temperature after expansion, K/K, $P_1/P_2 = $ pressure ratio, Pa/Pa, and $\kappa = $ specific heat capacity ratio for the gas, c_p/c_v, for air 1.4.

The dew-point is determined from equation (5.19) with the lowest pressure ratio that gives fog formation.

Electric conductivity method. The water content of certain salts, e.g. lithium chloride, and as a result also the electric conductivity is a function of the temperature and the humidity of the surrounding air. This is utilized in a humidity-sensing instrument where the salt is deposited in a thin layer on the outside of a rod.

Direct methods. Direct methods consist of weighing of the vapour absorbed in a suitable liquid or absorbed on an appropriate adsorption medium from a known volume of gas.

Air conditioning

In this section the term 'air conditioning' is used for processes to obtain air with a desired humidity and temperature.

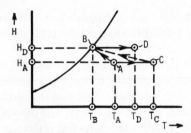

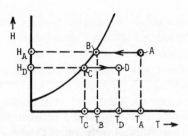

Figure 5.5 Air conditioning with humidification. Path 1, $A \to B \to D$, Path 2, $A \to C \to B \to D$

Figure 5.6 Dehumidification of air by cooling to temperature T_C below the dew-point T_B, followed by heating to temperature T_D

Humidification may be obtained by two different processes shown in the humidity chart (Figure 5.5). Air with temperature T_A and absolute humidity H_A is heated and humidified to temperature T_D and humidity H_D. Path 1 consists of humidifying with hot water to the dew-point B, followed by heating at constant absolute humidity H_D to temperature T_D. Path 2 consists of heating at constant humidity H_A to the temperature T_C at the constant enthalpy line through point B, humidification at adiabatic conditions, followed by heating to temperature T_D.

Dehumidification may be obtained by cooling below the dew-point in a heat exchanger or by contact with cold water. Figure 5.6 shows this process, including heating of the dehumidified air to temperature T_D.

Drying of gas or air to very low humidity is usually carried out by absorption, e.g. in glycol solutions, or by adsorption on media such as silica gel, alumina, and molecular sieves.

Evaporative cooling of water

Evaporative cooling of water is used in order to conserve cooling water for reuse. It is also used for cooling of water before it is discarded in a canal, river, lake, or

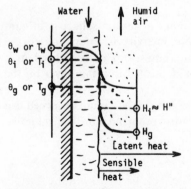

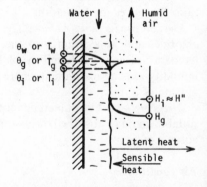

Figure 5.7 Conditions at top of cooling tower

Figure 5.8 Conditions at bottom of cooling tower

estuary where too high a temperature may give an objectionable growth of algae.

The water is cooled in spray ponds or more commonly by cascading down in cooling towers, countercurrent to a natural or forced draught of air. At the top of a cooling tower, the water is cooled as latent heat of vaporization is taken from the remaining water, and also by convection if the water temperature exceeds the temperature of the air (Figure 5.7). At the bottom, however, the water may be colder than the air (Figure 5.8). Here, the vaporization cools the water while sensible heat from the air is transferred to the water by convection. Hence, the water cannot be cooled quite as much as the adiabatic saturation temperature of the incoming air.

Calculation procedure. The procedure to be outlined here was developed by Merkel.[4] It is based on enthalpy balances.

The enthalpy balance around the bottom part of the tower in Figure 5.9 gives the equation for the *air operating line,*

$$Gh_y + Lc_w\theta_{w1} = Gh_{y1} + Lc_w\theta_w$$

or

$$h_y = \frac{L}{G}c_w\theta_w + \left(h_{y1} - \frac{L}{G}c_w\theta_{w1}\right) \tag{5.20}$$

where c_w = specific heat capacity of water, $4.19\,\text{kJ/(kg\,°C)}$, h_y = enthalpy of humid air, kJ/kg dry air, L = water flow rate, kg/s, G = dry air flow rate, kg/s, and θ_w = temperature of water, °C. In equation (5.20), the reference state for the enthalpy of water and air is at $0\,°C$.

In the next step, an equation is derived for the mass transfer within the height dz of the tower (Figure 5.10), based on the following approximations.[†]

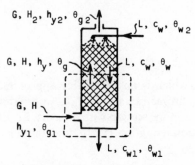

Figure 5.9   Enthalpy balance

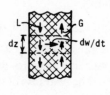

Figure 5.10 Section of tower with mass transfer from the water surface with area dA to the gas G

[†] For calculations with less approximations, see the specialized literature.[5]

1. The temperature gradient in the liquid is neglected, i.e. $\theta_w \approx \theta_i$.
2. The air at the interface is saturated at the temperature of the liquid.
3. The liquid flow rate L is constant.
4. In the equation for diffusion of water vapour (A) through stagnant air (B), the logarithmic mean mole fraction of air in the humid air is approximated by 1.0, giving equation (5.21),

$$N_A \approx k_{ys}(y_s - y) \tag{5.21}$$

where y_s = mole fraction of water vapour at saturation.
5. The mole fraction given by equation (5.15) is approximated by equations (5.16),

$$y_s \approx \frac{M_B}{M_A} H'' \quad \text{and} \quad y \approx \frac{M_B}{M_A} H \tag{5.16}$$

Equation (5.16) substituted in equation (5.21) give the flux of A in kmol/(m² s),

$$N_A \approx k_{ys} \frac{M_B}{M_A}(H'' - H) \tag{5.22}$$

This gives the rate of mass transfer $N_A M_A \, dA = dw/dt$ kg/s in a small section inside the tower with interfacial area dA,

$$dw/dt = M_B k_{ys}(H'' - H)\, dA \tag{5.23}$$

The corresponding latent heat of vaporization in kJ/s = kW, is

$$dQ_1 = (\Delta h_v)_{\theta_w} dw/dt = M_B k_{ys}(H'' - H)(\Delta h_v)_{\theta_w}\, dA \tag{5.24}$$

where $(\Delta h_v)_{\theta_w}$ = enthalpy of vaporization at temperature θ_w, equation (5.9).
  The sensible heat transferred across the small area dA, is

$$dQ_s = h(\theta_w - \theta_g)\, dA \tag{5.25}$$

or in terms of k_{ys},

$$dQ_s = M_B k_{ys} c_H(\theta_w - \theta_g)\, dA \tag{5.26}$$

where h is substituted by $M_B k_{ys} c_H$, which is an acceptable approximation for water vapour–air mixtures with psychrometric ratio $h/M_B k_{ys}$ approximately 1.0 (see p. 206). The heat loss of the water in the section with area dA, $Lc_w d\theta_w$, is the sum of the latent heat of vaporization (equation (5.24)), and the sensible heat transferred to the gas (equation (5.26)),

$$Lc_w d\theta_w = M_B k_{ys}[(H'' - H)(\Delta h_v)_{\theta_w} + c_H(\theta_w - \theta_g)]\, dA \tag{5.27}$$

The expression in brackets should be exchanged with a function of enthalpies. Air and water at 0 °C is chosen as the reference state. With the water evaporated at the temperature θ_w, the enthalpy h_y'' of saturated air at this temperature is

$$h_y'' = c_p \theta_w + H''[c_w \theta_w + (\Delta h_v)_{\theta_w}] \tag{5.28}$$

where c_p = specific heat capacity of dry air, approximately 1.00 kJ/(kg °C), and

c_w = specific heat capacity of water, 4.19 kJ/(kg °C)

The enthalpy of humid air at temperature θ_g and absolute humidity H, is

$$h_y = c_p \theta_g + H[c_w \theta_w + (\Delta h_v)_{\theta_w} + c_{pv}(\theta_g - \theta_w)] \tag{5.29}$$

where c_{pv} = specific heat capacity of water vapour, 1.88 kJ/(kg °C). Equation (5.29) subtracted from equation (5.28) yields

$$h_y'' - h_y = (H'' - H)(\Delta h_v)_{\theta_w} + c_p(\theta_w - \theta_g) + \{H'' c_w \theta_w - H[c_w \theta_w + c_{pv}(\theta_w - \theta_g)]\} \tag{5.30}$$

At moderate temperatures the expression in braces in equation (5.30) is small (Figure 5.11) and may be neglected, i.e.

$$h_y'' - h_y \approx (H'' - H)(\Delta h_v)_{\theta_w} + c_p(\theta_w - \theta_g) \tag{5.31}$$

Substituting equation (5.31) into equation (5.27), with $c_H \approx c_p$, rearranging and integrating, gives the necessary interfacial area in m²,

$$A = \frac{L c_w}{M_B k_{ys}} \int_{\theta_{w1}}^{\theta_{w2}} \frac{d\theta_w}{h_y'' - h_y} \tag{5.32}$$

The product $c_w d\theta_w$ is the differential of the enthalpy of the water, dh_w. Hence, equation (5.32) may also be written in the form

$$\frac{A M_B k_{ys}}{L} = \int_{h_{w1}}^{h_{w2}} \frac{dh_w}{h_y'' - h_y} \tag{5.33}$$

The integral at the right-hand side of this equation is called the number of transfer units, NTU.

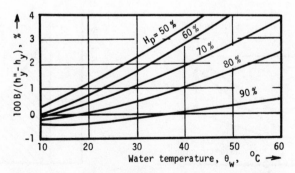

Figure 5.11 The expression in braces in equation (5.30), $B = \{H'' c_w \theta_w - H[c_w \theta_w + c_{pv}(\theta_w - \theta_g)]\}$, as a percentage of the enthalpy difference $h_y'' - h_y$ as a function of the water temperature θ_w for the temperature difference $\theta_w - \theta_g = 10 °C$, and the percentage absolute humidity H_p from 50 to 90 per cent

214

In order to solve the integral in equations (5.32) or (5.33), the enthalpy difference $h_y'' - h_y$ must be known as a function of the water temperature θ_w. Equation (5.20) gives h_y as a function of θ_w. The enthalpy h_y'' may be calculated as follows:

The saturation temperature of water in $N/m^2 = Pa$ can be reproduced within ± 0.04 per cent with the simple equations

$$p_s = \exp[23.7093 - 4111/237.7 + \theta)] \tag{5.34}$$

for $0 < \theta < 57°C$, and

$$p_s = \exp[23.1863 - 3809.4/(226.7 + \theta)] \tag{5.35}$$

for $57 < \theta < 135°C$.

With total pressure $P = 1$ bar $= 10^5$ Pa, equation (5.2) becomes

$$H'' = \frac{18p_s}{29(10^5 - p_s)} \tag{5.36}$$

With $0°C$ as reference state, equations (5.8) and (5.9) give

$$h_y'' = (1.00 + 1.88H'')\theta_w + 2501.6H'' \tag{5.37}$$

where $2501.6 =$ enthalpy of vaporization of water at $0°C$ [kJ/kg]. Figure 5.12 illustrates the water and air relationships and the driving potential in terms of enthalpy difference in a cooling tower with countercurrent flow, and Figure 5.13 shows the graphical solution of the integral in equation (5.32).

Table 5.1 gives h_y'' calculated by equations (5.34), (5.36) and (5.37). Figure 5.14 is a flow sheet for calculation computation of the integral in equation (5.32), and the program is given in the Appendix at the end of this chapter.

The same flow sheet and program is valid also for dehumidification, except that the sign of the integral is changed and the conditional $1/(h_y'' - h_y)_{\theta_{w1} + n} < 0$ is changed to > 0.

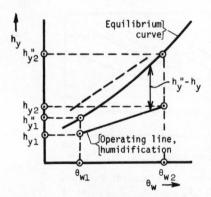

Figure 5.12 Temperature enthalpy diagram with equilibrium curve and operating lines for dehumidification and humidification (cooling tower)

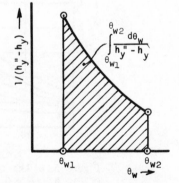

Figure 5.13 Graphical integration of the integral in equation (5.32)

Table 5.1 Enthalpy of saturated air, h_y'' kJ/kg dry air, with air and water at $0\,°C$ as reference state and total pressure $P = 1$ bar

θ_w (°C)	h_y'' (kJ/kg)	θ_w (°C)	h_y'' (kJ/kg)	θ_w (°C)	h_y'' (kJ/kg)	θ_w (°C)	h_y'' (kJ/kg)	θ_w (°C)	h_y'' (kJ/kg)
0	9.5	22	64.8	32	111.4	42	185	55	357
5	18.7	24	72.6	34	123.4	44	205	60	464
10	29.4	26	81.0	36	136.7	46	226	65	609
15	42.2	28	90.3	38	151.3	48	250	70	812
20	57.7	30	100.3	40	167.3	50	277	75	1108

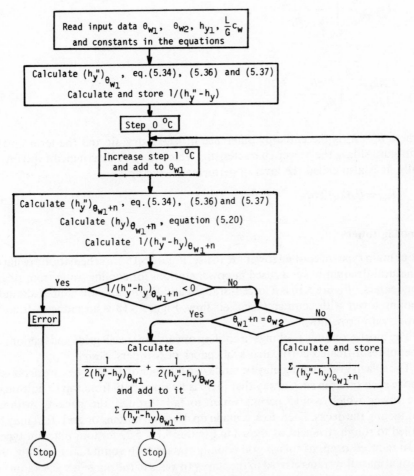

Figure 5.14 Flow sheet for calculation of $\int d\theta_w/(h_y'' - h_y)$ in equations (5.32) and (5.39). Program in the Appendix at the end of this chapter

In humidification and dehumidification the mass transfer is often expressed in terms of a mass transfer coefficient k_G, giving kg water transferred as water vapour per unit area, per unit time, and per unit partial pressure difference, in SI units kg water/(m^2 s Pa), while k_{ys} is kmol/(m^2 s mol fraction). With the definition $p_A \equiv y_A P$, or

$$y_A = p_A/P \tag{5.38}$$

$$k_{ys} = Pk_G/M_A$$

where M_A = molecular weight of water, 18 kg/kmol.

Substituting into equation (5.32) with $M_B = 29$ kg/kmol and $c_w = 4.19$ kJ/(kg °C), gives

$$A = \frac{2.6L}{Pk_G} \int_{\theta_{w1}}^{\theta_{w2}} \frac{d\theta_w}{h_y'' - h_y} \tag{5.39}$$

or

$$\frac{PAk_G}{2.6L} = \int_{\theta_{w1}}^{\theta_{w2}} \frac{d\theta_w}{h_y'' - h_y} \tag{5.40}$$

where $h_y'' - h_y$ is the enthalpy difference in kJ/kg dry air and the term on the left-hand side is the 'tower characteristic' which may be determined experimentally. It is also called the 'overall performance factor',[6]

$$K_m = PAk_G/2.6L \tag{5.41}$$

Cooling towers

Four main types of cooling tower are in use. Figure 5.15 is a schematic drawing of a natural-draught tower as used in power stations for cooling water from steam condensers.[7] Figure 5.16 is a forced-draught and Figure 5.17 an induced-draught cooling tower with countercurrent air flow. Figure 5.18 is an induced-draught tower with cross-flow.

The most economic packings in cooling towers are splash grids and sections of phenol-resin-treated corrugated kraft paper of Munters' design.

The splash grid packing may be made of a number of decks. They are arranged above each other in such a way that there is no open path from top to bottom of the tower which would permit water to fall through the packing without contacting the decks. Each deck is made up of rough battens or grids that may be nailed to rough stringers, as shown in Figures 5.19–5.22 for four different types.

In induced-draught splash grid cooling towers with countercurrent flow, the normal mass flow rate referred to the empty tower is in the range 1.5–2.4 kg/(m^2 s) for air and 2.5–5 kg/(m^2 s) for water. For the splash grids shown in Figures 5.19–5.22, Kelly and Swenson[8] reported the following empirical correlation for the

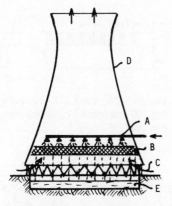

Figure 5.15 Hyperbolic natural-draught cooling tower. A is the water distributor, B the packing, C the air inlet, D the stack giving natural draught due to buoyancy of warm, humid air, E is the cooled water

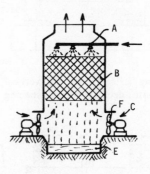

Figure 5.16   Forced-draught cooling tower with countercurrent air flow. Symbols as in Figure 5.15. F is the fan

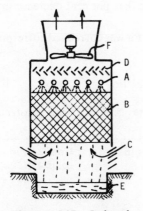

Figure 5.17 Induced-draught cooling tower with countercurrent air flow. A, water distributor; B, packing; C, air inlet (louvres); D, mist eliminator; E, cooled water; F, fan

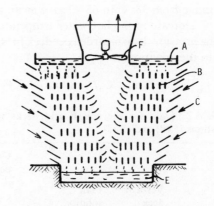

Figure 5.18 Induced-draught cooling tower with cross-flow of air. Symbols as in Figure 5.17

218

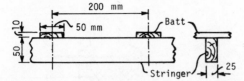

Figure 5.19 Deck I: vertical spacing 230 mm; deck II: vertical spacing 300 mm

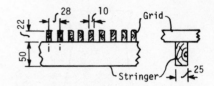

Figure 5.20 Deck III: vertical spacing 280 mm; deck IV: vertical spacing 610 mm

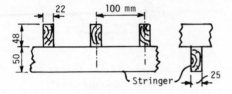

Figure 5.21 Deck V: vertical spacing 610 mm

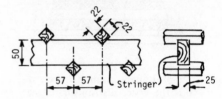

Figure 5.22 Deck VI: vertical spacing 610 mm

overall performance factor:

$$K_m = 0.07 + C_1 N(L/G)^{-n} \tag{5.42}$$

where 0.07 = factor for end effects, kg °C/kJ, N = number of decks, L/G = ratio of mass flow of water to mass flow of dry air, and C_1 and n are empirical constant (Table 5.2).

Equation (5.42) was established for a hot-water temperature of approximately 50 °C, and a wet bulb temperature of 24 °C. Tests with hot-water temperatures in the range from 38 °C to 65 °C gave as an average 0.35 per cent decrease in K_m per °C increase in the hot-water temperature.

The pressure drop recorded per deck in N/m² = Pa was correlated corresponding to the equation

$$\frac{\Delta p}{N} = C_2 G^2/\rho_G + C_3 L \sqrt{Sp/r}(G + 3.37 \sqrt{Sp/r})^2/\rho_G \tag{5.43}$$

Table 5.2 Data for the splash grid decks in Figures 5.19–5.22, and constants in equations (5.42) and (5.43)

Deck no.	Fig. no.	Vertical deck spacing, Sp (m)	Plan solidity fraction, r	Sp/r (m)	Constants in eq. (5.42) C_1 $\left(\dfrac{kJ}{kg\,°C\,deck}\right)$	n	Constants in eq. (5.43) C_2	C_3 $\left(\dfrac{m^{1,s}\,s}{kg}\right)$
I	5.19	0.23	0.250	0.92	0.060	0.62	0.50	0.021
II	5.19	0.30	0.250	1.20	0.070	0.62	0.50	0.021
III	5.20	0.38	0.333	1.14	0.092	0.60	0.59	0.027
IV	5.20	0.61	0.333	1.83	0.119	0.58	0.59	0.027
V	5.21	0.61	0.219	2.79	0.100	0.51	0.38	0.014
VI	5.22	0.61	0.550	1.11	0.127	0.47	1.10	0.050

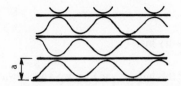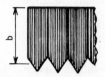

Figure 5.23 Packing of Munters' design, made of corrugated kraft paper treated with phenol resin. Seen from the
top (left), and seen from the side (right). One type has
distance $a = 3.7$ mm and heights $b = 191$, 292, 418, and
622 mm. The other type has distance $a = 7$ mm and height
$b = 565$ mm

where $G =$ air mass velocity, kg dry air/(m² s), $L =$ water mass velocity, kg
water/(m² s), $Sp =$ deck spacing, m, $r =$ fraction of cross-section covered by the
battens, C_2 and C_3, empirical constants (Table 5.2), and $\rho_G =$ density of dry air,
kg dry air/m³. The last term in equation (5.43) takes into account the relative
velocity between air and falling water drops.

Existing splash grid towers can sometimes be upgraded economically by
increasing the air rate, improving the water distribution, and increasing
the heat transfer surfaces.[9]

The corrugated kraft-paper packings have dimensions as shown in Figure 5.23.
Experimental data are reported in the literature[10,11] for water flow rates of 1.75
and 3 kg/(m² s) and air velocities from 1.1 to 3 m/s for packing with $a = 3.7$ mm,
and water flow rates from 1.75 to 6.1 kg/(m² s) and air velocities from 1.6 to
3.4 m/s for packing with $a = 7$ mm.

Deposits may accumulate in the narrow ducts. In practice, this has been taken
care of by renewal of the packing after some years of operation.

Sizes and prices of cross-flow, induced-draught type cooling towers with a
variable-pitch axial-flow fan are given by Uchiyama.[12]

Spray ponds are an alternative to cooling towers. A spray pond has a number of
nozzles, usually located 2–2.5 m above the pond, spraying water into the air. The
pond may be enclosed with louvre fences to reduce drift loss. For details and
calculations, see the literature.[13]

EXAMPLE 5.1   DEW-POINT AND WET BULB TEMPERATURE

Find the vapour pressure, the absolute humidity, the dew-point, the wet bulb temperature,
and the humid volume of air of temperature 383 K (110 °C), percentage relative humidity
$H_R = 5$ per cent, and total pressure $P = 1.00$ bar $= 10^5$ Pa.

Solution

The saturation pressure given by equation (5.35) is

$$p_s = \exp\left[23.1863 - 3809.4/(226.7 + 110)\right] = 1.433 \times 10^5 \text{ Pa}$$

Vapour pressure (equation (5.5)),

$$p = H_R p_s/100 = 5 \times 1.433 \times 10^5/100 = 7163 \text{ Pa}$$

Absolute humidity (equation (5.1)),

$$H = \frac{18 \times 7163}{29(10^5 - 7163)} = 0.048 \text{ kg/kg dry air}$$

The corresponding dew-point is read at this humidity in Figure 5.4 as shown in Figure 5.3,

Dew-point 39 °C

The wet bulb temperature is obtained following an adiabatic cooling line in Figure 5.4 from air temperature 110 °C and humidity $H = 0.048$ kg/kg dry air to the saturation line, giving a wet bulb temperature of 47 °C.

Equation (5.6a) gives the humid volume,

$$v_H = 0.08314 \times 383(1/29 + 0.048/18) = 1.18 \text{ m}^3/\text{kg dry air}$$

Example 5.2 MIXING OF HUMID AIR

Air entering a drier is heated to 135°C after it is made up of 3000 m³/h air with temperature 25 °C and percentage humidity $H_p = 80$ per cent, and 2000 m³/h recycled air with temperature 90 °C and wet bulb temperature 45 °C. The process is carried out at a total pressure of approximately 1 bar.
(a) Calculate the humidity and the volume of the air entering the drier.
(b) Calculate the heat added to the mixture and the amount of water evaporated in the drier.

Solution
(a) Figure 5.4 gives humidity $H_1 = 0.017$ kg/kg dry air for air with temperature 25°C and percentage humidity $H_p = 80$ per cent, and $H_2 = 0.047$ kg/kg dry air at 90°C and wet bulb temperature 45°C.

Equation (5.6a) applied to the two streams gives

$$v_{H1} = 0.083\,14(273 + 25)(1/29 + 0.017/18) = 0.878 \text{ m}^3/\text{kg dry air}$$

and

$$v_{H2} = 0.083\,14(273 + 90)(1/29 + 0.047/18) = 1.119 \text{ m}^3/\text{kg dry air}$$

Total mass of dry air passing through the drier.

$$3000/0.878 + 2000/1.119 = 3417 + 1787 = 5204 \text{ kg dry air/h}$$

and of water vapour entering the drier,

$$0.017 \times 3417 + 0.047 \times 1787 = 142 \text{ kg/h}$$

Absolute humidity of mixture,

$$H_3 = 142/5204 = 0.0273 \text{ kg/kg dry air}$$

Humid volume (equation (5.6a)),

$$v_{H3} = 0.083\,14(273 + 135)(1/29 + 0.0273/18) = 1.221 \text{ m}^3/\text{kg dry air}$$

(b)

	Air (135 °C) $H = 0.0273$	Cold air, (25 °C) $H_p = 80\%$	Recycled air (90 °C) $H = 0.047$	Notes
h_y (kJ/kg dry air)	210	68	217	Figure 5.4
Gh_y (kJ/h)	717 570	232 356	387 779	(mass dry air/h) $\times h_y$

Heat added to the mixture

$$717\,570 - (232\,356 + 387\,779) = 97\,430 \text{ kJ/h}$$

or

$$97\,430/3600 = 27\,\text{kW}$$

Evaporated in the drier,

$$5204 \times 0.047 - 142 = 103 \text{ kg/h}$$

Example 5.3 SPLASH-GRID COOLING TOWER

Approximately 22 kg/s cooling water with temperature 48 °C is cooled to 24 °C in a splash-grid cooling tower with induced draught and decks of type I in Figure 5.19. The temperature of the ambient air is 25 °C, the wet bulb temperature 20 °C, and the pressure 1 bar = 10^5 N/m^2.
(a) Calculate the enthalpy and the density of the entering air.
(b) Calculate the minimum air flow (infinite tower height).
(c) Estimate the number of decks needed by an air flow 1.5 times the minimum air flow.
(d) Select a tower cross-section, preferably within the normal operating range.
(e) Estimate the amount of evaporated water as a percentage of the circulating water.
(f) Estimate the pressure drop through the packing, in Pa and in mm water gauge.

Solution

(a) Equation (5.34) with the wet bulb temperature 20 °C gives the vapour pressure,

$$p_s = \exp\,[23.7093 - 4111/(237.7 + 20)] = 2337 \text{ N/m}^2$$

Equation (5.36),

$$H'' = \frac{18 \times 2337}{29(10^5 - 2337)} = 0.0149 \; \frac{\text{kg water vapour}}{\text{kg dry air}}$$

The enthalpy of the ambient air equals the enthalpy of saturated air at the wet bulb temperature, equation (5.37),

$$h_{y1} = (1.00 + 1.88 \times 0.0149)20 + 2501.6 \times 0.0149 = 57.8 \text{ kJ/kg dry air}$$

(the same value can be read from the humidity chart, Figure 5.4).

(b) The equilibrium curve (Table 5.1) and the enthalpy of the entering air (point 1) are plotted in Figure 5.24 as a function of the water temperature θ_w. Line B corresponds to minimum air flow. It is a tangent to the equilibrium curve at $\theta_w \approx 35$ °C where equations (5.34), (5.36), and (5.37) give the enthalpy $h_y'' = 129.8$ kJ/kg. The corresponding slope is

$$\frac{129.8 - 57.7}{35 - 24} = 6.56 \; \frac{\text{kJ/kg dry air}}{°\text{C}}$$

or (equation (5.20))

$$Lc_w/G = 6.56, \qquad L/G = 6.56/4.19 = 1.57 \; \frac{\text{kg water}}{\text{kg dry air}}$$

$$G = 22/1.57 = 14 \text{ kg dry air/s}$$

(c) Here, 1.5 times minimum G gives

$$Lc_w/G = 6.56/1.5 = 4.37 \text{ kJ/(kg °C)}$$
$$L/G = 4.37/4.19 = 1.04$$
$$G = 22/1.04 = 21 \text{ kg dry air/s}$$

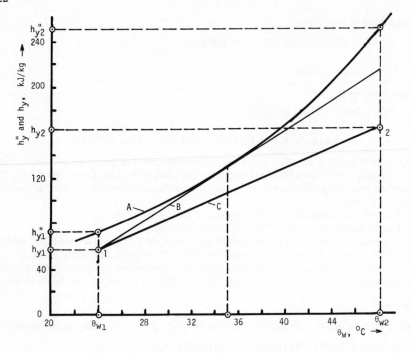

Figure 5.24 Equilibrium curve A and operating lines in Example 5.3. Line B corresponds to minimum air flow, line C is 1.5 times minimum air flow

Equation (5.20) gives the enthalpy h_{y2} at the temperature $\theta_{w2} = 48\,°C$,

$$h_{y2} = 4.37 \times 48 + (57.7 - 4.37 \times 24) = 162.6 \text{ kJ/kg}$$

The integral in equation (5.32) may be solved by the use of a programmable calculator or by manual graphical solution. Both methods are shown here:

(i) Solution by a pocket calculator. The numbers $\theta_{w1} = 24\,°C$, $\theta_{w2} = 48\,°C$, $h_{y1} = 57.8$ kJ/kg dry air, $Lc_w/G = 4.37$ kJ/(kg °C), and the constants in equations (5.34), (5.36), and (5.37), stored in the program in the Appendix, give

$$\int_{24}^{48\,C} \frac{d\theta}{h_y'' - h_y} = 0.96 \text{ kg °C/kJ}$$

(ii) Graphical integration. Table a gives h_y'' from Table 5.1 and h_y calculated by equation (5.20),

$$h_y = 4.37\theta_w + (57.7 - 4.37 \times 24) = 4.37\theta_w - 47.2 \tag{a}$$

Table a   The integrand $1/(h_y'' - h_y)$ at different temperatures

θ_w (°C)	24	28	32	36	40	44	48	Notes
h_y'' (kJ/kg)	72.6	90.3	111.4	136.7	167.3	205	250	Table 5.1
h_y (kJ/kg)	57.7	75.2	92.6	110.1	127.6	145.1	162.6	eq. (a)
$1/(h_y'' - h_y)$	0.0670	0.0661	0.0533	0.0376	0.0252	0.0167	0.0114	

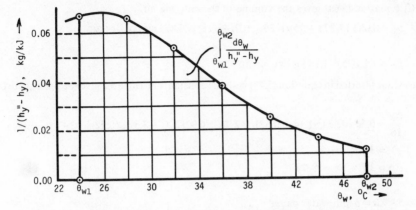

Figure 5.25 Diagram for determination of the integral in equation (5.32)

The numbers in the last line are plotted in Figure 5.25 as a function of the water temperature θ_w.

The area under the curve from $\theta_{w1} = 24\,°C$ to $\theta_{w2} = 48\,°C$ in Figure 5.25 contains approximately 47.8 squares. Each square represents $(0.01\text{ kg/kJ}) \times (2\,°C) = 0.02\text{ kg }°C/kJ$,

$$\int_{24}^{48} \frac{d\theta_w}{h_y'' - h_y} = 47.8 \times 0.02 = 0.96\text{ kg }°C/kJ$$

This performance factor inserted in equation (5.42) with the constants in Table 5.2 gives

$$0.96 = 0.07 + 0.06 \times 1.04^{-0.62}\,N$$

$N = 15.2$ decks, i.e. 16 decks of type I, Figure 5.19. Height of packing, $16 \times 0.23 = 3.68\text{ m}$.

(d) The normal ranges, given on p. 216, correspond to the following cross-sections:

For water, Between $22/5 = 4.4\text{ m}^2$ and $22/2.5 = 8.8\text{ m}^2$

For air, between $21/2.4 = 8.75\text{ m}^2$ and $21/1.5 = 14\text{ m}^2$

Selected cross-section $3 \times 3 = 9\text{ m}^2$, giving

$22/9 = 2.44\text{ kg water}/(\text{m}^2\text{ s})$

$21/9 = 2.33\text{ kg dry air}/(\text{m}^2\text{ s})$

(e) The exact number cannot be calculated as the temperature of the exit air is unknown. It is probably somewhere between $1\,°C$ and $5\,°C$ colder than the entering water.

Equation (5.8) with $0\,°C$ as the reference temperature and $(\Delta h_v)_{0\,°C} = 2501.6\text{ kJ/kg}$ gives the enthalpy of the exit air

$$162.6 = (1.00 + 1.88H)\theta_2 + 2501.6H$$

With

$\theta_2 = 48 - 1 = 47\,°C, \qquad H_2 = 0.0446\text{ kg water vapour/kg dry air}$

With

$\theta_2 = 48 - 5 = 43\,°C, \qquad H_2 = 0.0463\text{ kg water vapour/kg dry air}$

An average value of $H_2 = 0.0454$ gives the evaporated water $G(H_2 - H_1) = 21(0.0454 - 0.0149) = 0.641\text{ kg/s}$ or $100 \times 0.641/22 = 2.9$ per cent of the circulating water.

(f) Equation (5.6a) gives the volume of the entering air,

$$v_H = 0.083\,14(273 + 25)(1/29 + 0.0149/18) = 0.875\,\text{m}^3/\text{kg dry air}$$

or

$$\rho_G = 1/0.875 = 1.14\,\text{kg dry air/m}^3$$

This value inserted in equation (5.43) with the constants in Table 5.2 give the pressure drop per deck in $\text{N/m}^2 = \text{Pa}$,

$$\frac{\Delta p}{16} = 0.50 \times 2.33^2/1.14 + 0.021 \times 2.44\,\sqrt{0.92}(2.33 + 3.37\,\sqrt{0.92})^2/1.14$$

$$= 3.7\,\text{Pa/deck}$$

$$\Delta p = 3.7 \times 16 = 59\,\text{Pa}$$

or

$$59/9.8 = 6\,\text{mm water gauge}$$

Example 5.4 PERFORMANCE FACTOR OF COOLING TOWER

An induced-draught cooling tower with decks as shown in Figure 5.19 is designed for cooling of 86 m^3/h water from 47 °C to 24 °C with 65 000 m^3/h inlet air of temperature 25 °C, wet bulb temperature 20 °C, and pressure 1 bar.
(a) Calculate the performance factor.
(b) The acceptance test was carried out with 72 m^3/h water that was cooled from 39 °C to 21 °C. The inlet air temperature was 18 °C, the wet bulb temperature 17 °C, and the pressure 1 bar. Compare the test data with the design data.

Solution

(a) Humidity from Figure 5.4 with air temperature 25 °C and wet bulb temperature 20 °C,

$$H = 0.013\,\text{kg/kg dry air}$$

giving the humid volume (equation (5.6a)),

$$v_H = 0.083\,14(273 + 25)(1/29 + 0.013/18) = 0.872\,\text{m}^3/\text{kg dry air}$$

or

$$G = \frac{65\,000}{3600 \times 0.872} = 20.7\,\text{kg dry air/s}$$

$$L = 86 \times 1000/3600 = 23.9\,\text{kg/s}$$

$$\frac{L}{G}c_w = \frac{23.9}{20.7}4.19 = 4.84\,\text{kJ/(kg °C)}$$

The enthalpy of the ambient air equals the enthalpy of saturated air at the wet bulb temperature (equation (5.37)),

$$h_{y1} = (1 + 1.88 \times 0.013)20 + 2501.6 \times 0.013 = 53.0\,\text{kJ/kg dry air}$$

Equations (5.40) and (5.41) give the performance factor,

$$K_m = \frac{10^5\,Ak_G}{2.6 \times 23.9} = \int_{24}^{47} \frac{d\theta_w}{h''_y - h_y}$$

where the integral may be solved by the program in the Appendix, giving

$K_m = 0.883$

(b) The liquid flow rate is

$L = 72 \times 1000/3600 = 20.0 \, \text{m}^3/\text{s}$

corresponding to

$$\frac{L}{G} c_w = \frac{20.0}{20.7} 4.19 = 4.05 \, \text{kJ/(kg} \, ^\circ\text{C)}$$

Figure 5.4 gives the humidity of the ambient air,

$H = 0.0115 \, \text{kg/kg dry air}$

giving the enthalpy of the entering air (equation (5.37)),

$h_{y1} = (1 + 1.88 \times 0.0115)17 + 2501.6 \times 0.0115 = 46.1 \, \text{kJ/kg dry air}$

Performance factor,

$$K_m = \frac{10^5 A k_G}{2.6 \times 20.0} = \int_{21}^{39} \frac{d\theta_w}{h_y'' - h_y} = 0.950$$

where the integral is solved as under (a).

According to equation (5.42), the lower value of L/G in the acceptance test should give a somewhat higher value of the performance factor.

With $n = 0.62$ (Table 5.2), equation (5.42) gives

$C_1 N = (0.883 - 0.07)(23.9/20.7)^{0.62} = 0.889$

for the design data, corresponding to

$K_m = 0.07 + 0.889(20.0/20.7)^{-0.62} = 0.978$

for the conditions under the acceptance test. Hence, the measured data give a performance factor

$$100 \frac{0.978 - 0.950}{0.978} = 3 \text{ per cent}$$

below what could be expected from the design data.

Problems

5.1 Find the vapour pressure, the absolute humidity, the dew-point, the wet bulb temperature, and the humid volume of air of temperature 383 K (110 °C), percentage relative humidity $H_R = 5$ per cent, and total pressure $P = 2.00 \, \text{bar} = 2 \times 10^5 \, \text{Pa}$.

5.2 In the continuous drier Figure 5.26 250 kg/h water is evaporated. The total pressure is approximately 1 bar. The air temperature at point 1 is $\theta_1 = 10\,^\circ\text{C}$ and the wet bulb temperature $\theta_{wb1} = 75\,^\circ\text{C}$. Without recycling of air, the temperatures at point 2 are $\theta_2 = 85\,^\circ\text{C}$ and $\theta_{wb2} = 50\,^\circ\text{C}$. Neglecting heat loss to the surroundings and the sensible heat of the material passing through the drier, calculate:
(a) the air flow through the drier;
(b) the heat input in the air heater;
(c) the temperature of the air entering the drier, θ_3.

226

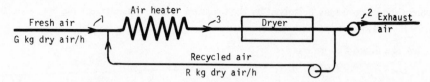

Fresh air
G kg dry air/h

Air heater

Dryer

2 Exhaust
air

Recycled air
R kg dry air/h

Figure 5.26 Arrangement of a continuous drier

(d) One-third of the air passing through the drier is recycled air, and the exhaust air temperature is $\theta_2 = 90\,°C$ and the wet bulb temperature $\theta_{wb2} = 55\,°C$. Calculate the air temperature θ_3 in front of the drier and the heat input to the air heater.

5.3 An induced-draught cooling tower with decks of type I (Figure 5.19) is designed to cool $L = 22$ kg/s water from 48 °C to 26 °C, with $G = 20$ kg dry air/s, with air temperature 25 °C, wet bulb temperature 20 °C, and total pressure $P = 1.00$ bar. Estimate the temperature of the water from the tower, θ_{w1}, if G is increased to 28 kg dry air/s.

5.4 Write a calculator program for calculation of:
(a) total enthalpy h_y of humid air with 0 °C as reference state and temperature and absolute humidity as input variables;
(b) total enthalpy h_y'' of saturated air with total pressure and temperature as input variables in the temperature range $0 < \theta < 57\,°C$;
(c) the wet bulb temperature with absolute humidity, total pressure, and dry bulb temperature as input variables in the same temperature range as under (b).

Symbols

A	Area, m²
a	Area per unit volume, m²/m³
c_H	Humid heat capacity, kJ/(kg dry air °C) or J/(kg dry air °C)
c_p	Specific heat capacity of dry air at constant pressure, 1.00 kJ/(kg °C)
c_{pv}	Specific heat capacity of water vapour at constant pressure, 1.88 kJ/(kg °C)
c_v	Specific heat capacity of air at constant volume, 0.714 kJ/(kg °C)
c_w	Specific heat capacity of water, 4.19 kJ/(kg °C)
G	Dry air flow rate, kg/s
H	Absolute humidity, kg H_2O/kg dry air
H''	Saturated humidity, kg H_2O/kg dry air
H_p	Percentage absolute humidity, per cent
H_R	Percentage relative humidity, per cent
h	Heat transfer coefficient, W/(m² K) = W/(m² °C)
h_y	Enthalpy of humid air, kJ/kg dry air or J/kg dry air
Δh_v	Enthalpy of vaporization of water, kJ/kg or J/kg
K_m	Performance factor, equation (5.42)
k_G	Mass transfer coefficient of water, kg water/(m² s Pa)
k_{ys}	Gas phase mass transfer coefficient by Stefan diffusion, kmol/(m² s mol fraction)
L	Water flow rate, kg/s
M_A	Molecular weight of water, 18 kg/kmol
M_B	Molecular weight of air, 29 kg/kmol
N	Number of decks
N_A	Molar flux of water vapour, kmol/(m² s)
P	Total pressure, Pa = N/m²
p	Partial pressure, Pa = N/m²
p_s	Vapour pressure of water at the temperature of the air–water vapour mixture, Pa = N/m²

Δp	Pressure drop, Pa $= \mathrm{N/m^2}$
Q	Heat flux, $\mathrm{kW/m^2}$ or $\mathrm{W/m^2}$
R	Gas constant, 8134 J/(kmol K)
T	Temperature, K
T_0	Reference temperature, K
t	Time, s
v_H	Humid volume, $\mathrm{m^3/kg}$ dry air
w	Mass of water, kg
y	Mole fraction
θ	Temperature, °C
κ	Specific heat capacity ratio, c_p/c_v, for air 1.40
ρ	Density, $\mathrm{kg/m^3}$

Subscripts

A	Component A (water)
B	Component B (air)
g	Gas
i	At the interface
p	Percentage
s	At saturation
T_0	At reference temperature T_0
wb	At wet bulb temperature
w	Water
θ	At temperature θ °C

Superscript

″ Saturated air

References

1. Geankoplis, C. J., *Transport Processes and Unit Operations*, Allyn and Bacon, Boston, 1978.
2. Lewis, W. K., The evaporation of a liquid into a gas, *Mech. Engn.*, **44**, 445–446 (1922).
3. Grosvenor, W. M., Calculations for dryer design, *Trans. A.I.Ch.E.*, **1**, 184–202 (1908).
4. Merkel, F., *Vendunstungskühlung*, VDI-Forschungsheft 275, VDI-Verlag, Berlin, 1925.
5. Poppe, M., *Wärm- und Stoffübertragung bei der Verdunstungskühlung im Gegen- und Kreuzstrom*, VDI-Forschungsheft 560, VDI-Verlag, Düsseldorf, 1973.
6. Spurlock, B. H., Performance of a forced draft cooling tower, *Heat. Pip. Air Cond.*, **25** (No. 6), 115–121 (1953).
7. Jackson, J., *Cooling Towers*, Butterworths Sci. Publ., London, 1951.
8. Kelly, N. W., and L. K., Swenson, Comparative performance of cooling tower packing arrangements, *Chem. Eng. Progr.*, **52**, 263–268 (1956).
9. Burger, R., Cooling towers can make money, *Chem. Eng. Progr.*, **78**, No. 2, 84–87 (1982).
10. Munters, C., and L. Lindqvist, A new cooling tower, *S. F. Review*, **5**, 2–15 (1958), *J. Refr.*, **4**, (No. 3), 59–63 (1961).
11. Berlin, P., *Kühltürme. Grundlagen der Berechnung und Konstruktion*, Springer-Verlag, Berlin, 1975.
12. Uchiyama, T., Cooling tower estimates made easy, *Hydrocarbon Processing*, **55**, Dec., 93–96 (1976).
13. Perry, R. H. and C. H., Chilton, *Chemical Engineers' Handbook*, 5th edn., pp. 12–17, McGraw-Hill, New York, 1973.

Appendix

Calculator program for calculators HP19C and HP29C for the integral in equation (5.32), temperatures θ_{w1} and θ_{w2} in °C without decimals.

Place the following 11 constants in storage:

θ_{w1}	STO 1
θ_{w2}	STO 2
h_{y1}	STO 3
$\dfrac{L}{G}c_w$	STO 4
237.7	STO 5
−4111	STO 6
23.7093	STO 7
18/29	STO 8
2501.6	STO 9
1.88	STO .0
P	STO .1

Intermittent storages used:

$\dfrac{1}{(h''_y - h_y)_{\theta_{w1}}}$	STO .2
$\dfrac{1}{(h''_y - h_y)_{\theta_{w1}+n}}$	STO .3
$\theta_{w1} + n$	STO .4

Step no.		Step no.		Step no.		Step no.		Step no.	
01	LBL 1	16	RCL .4	31	e^x	46	RCL .4	61	$x\rightleftarrows y$
02	0	**17**	$x=y$	32	STO 0	47	×	62	−
03	STO .2	18	GSB 5	33	RCL .1	48	RCL 0	63	1/x
04	STO .3	19	RCL 0	34	−	49	RCL 9	64	STO 0
05	RCL 1	20	STO +.3	35	CHS	50	×	65	RTN
06	STO .4	21	GSB 4	36	RCL 0	51	+	66	LBL 5
07	GSB 3	22	LBL 3	37	$x\rightleftarrows y$	52	STO 0	67	RCL .2
08	STO .2	23	RCL .4	38	÷	53	RCL .4	68	RCL 0
09	LBL 4	24	RCL 5	39	RCL 8	54	RCL 1	69	+
10	1	25	+	40	×	55	−	70	2
11	STO +.4	26	RCL 6	41	STO 0	56	RCL 4	71	÷
12	GSB 3	27	$x\rightleftarrows y$	42	RCL .0	57	×	72	RCL .3
13	$x<0$	28	÷	43	×	58	RCL 3	73	+
14	LOG	29	RCL 7	44	1	59	+	74	R/S
15	RCL 2	30	+	45	+	60	RCL 0		

CHAPTER 6
Drying of solids

Drying in this chapter refers to the removal of liquid from a solid by evaporation. In most cases the liquid is water.

Water removal by mechanical methods in equipment as filters, presses, and centrifuges is not considered as drying. It is, however, often used prior to drying, as it may be less expensive and easier to use than thermal methods.

Definitions

Some generally accepted definitions are given alphabetically below:[1]

Bound moisture in a solid is that liquid which exerts a vapour pressure less than that of the pure liquid at the given temperature. Liquid may become bound by retention in small capillaries, by solution in cell or fibre walls, or throughout the solid, and by chemical or physical adsorption on solid surfaces.

Capillary flow is the flow of liquid through the interstices and over the surface of a solid, caused by liquid–solid molecular attraction.

Constant rate period is that drying period during which the rate of vaporization per unit of drying surface area is constant.

Critical moisture content is the average moisture content when the constant-rate period ends.

Dry weight basis expresses the moisture content of wet solid as kg water per kg bone-dry solid.

Equilibrium moisture content is the limiting moisture to which a given material can be dried at a given air temperature and humidity.

Falling-rate period is a drying period during which the rate of vaporization continually decreases with time.

Free moisture content is that liquid which is removable at a given temperature and air humidity. It may include bound and unbound moisture.

Hygroscopic material is material which may contain bound moisture.

Moisture content of a solid is usually expressed as moisture quantity per unit weight of dry or wet solid. Use of the dry weight basis is recommended. Unless otherwise stated, the dry weight basis is used throughout this chapter.

Unbound moisture in a hygroscopic material is the moisture in excess of the equilibrium moisture content corresponding to saturation humidity. All water in a non-hygroscopic material is unbound water.

Heat and mass transfer

Drying of solids involves the simultaneous transfer of heat to evaporate the liquid

230

and transfer of moisture as liquid or vapour within the solid and vapour from the surface, usually into a hot carrier gas.

The heat is usually transferred to the surface of the solid. It may be by convection from hot air, by conduction through contact with heated surfaces, by radiation, or by a combination of these. Heat from the surface penetrates into the solid by conduction, and in some cases also by a sequence of vaporization and condensation.

The transfer of liquid inside the solid may occur by several mechanisms, such as diffusion in continuous homogeneous solids, capillary flow in granular and porous solids, flow caused by shrinkage and pressure gradients, and flow caused by a sequence of vaporization and condensation. Even if one mechanism predominates at a given time in a drying cycle, it is not uncommon for different mechanisms to predominate at other times.

The factors governing the rates of the two processes, heat transfer and mass transfer, determine the drying rate.

Periods of drying

Figure 6.1 shows the moisture content as a function of time in the general case when a wet solid is dried. Figure 6.2 is the derivative of the curve in Figure 6.1, the drying rate as a function of time. Section A–B represents the warming-up period, and B–C the constant rate period where moisture evaporates from a saturated surface of the solid. The curved portion C–D is the falling rate period which may consist of a period of evaporation from a saturated surface of gradually decreasing area, followed by a period when the water evaporates in the interior of the solid. The same points are shown in Figure 6.3 where the drying rate is plotted versus the moisture content.

The moisture content in point C is termed the 'critical moisture content.' Its value depends on the kind of solid material. It appears to increase with the

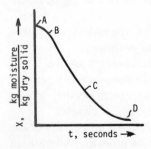

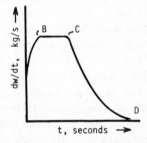

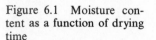

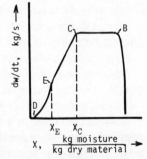

Figure 6.1 Moisture content as a function of drying time

Figure 6.2 Drying rate as a function of drying time

Figure 6.3 Drying rate as a function of moisture content

thickness of the solid and with the drying rate. Relatively coarse sand exhibits a critical moisture content in the range 3–10 per cent water (dry basis), fine precipitates and pigments in the range 40–60 per cent, and leather and wallboard in the range 100–160 per cent.[2]

Point E represents the point at which all the exposed surface is unsaturated; it is the start of the period where internal moisture movement controls the drying rate.

Constant rate period

In the constant rate period the moisture movement within the solid is sufficiently rapid to maintain a saturated condition at the surface. If the heat is supplied by convection from warmer air only, the surface temperature is the wet bulb temperature. (The process occurring at a wet bulb thermometer is in fact one of constant-rate drying.) This gives the rate of drying in kg/s as

$$\frac{dw}{dt} = \frac{-hA(T_a - T_w)}{\Delta h_v} \tag{6.1}$$

where h = heat transfer coefficient by convection, W/(°C m²), A = surface area for heat and mass transfer, m², $T_a - T_w$ = difference between the dry bulb (air) and the wet bulb temperature, °C or K, and Δh_v = enthalpy of vaporization at the wet bulb temperature, J/kg.

The rate of drying can also be expressed in terms of mass transfer,

$$\frac{dw}{dt} = -k_g A(p_s - p) \tag{6.2}$$

or

$$-\frac{dw}{dt} = \frac{hA}{\Delta h_v}(T_a - T_w) = k_g A(p_s - p) \tag{6.3}$$

where k_g = mass transfer coefficient, kg/(m²s Pa), p_s = vapour pressure of water at the temperature of the surface, N/m² = Pa, and p = partial pressure of water vapour in bulk of air, N/m² = Pa.

In practice, equation (6.1) provides a more reliable estimate of the drying rate than equation (6.2), because an error in estimating the surface temperature affects the temperature driving force, $T_a - T_w$, less than the partial pressure driving force, $p_s - p$.

If heat is supplied both by convection and by conduction and/or radiation, the surface temperature will be higher than the wet bulb temperature. In this case the surface temperature T_s can be determined by the equation

$$-\frac{dw}{dt} = \frac{\dot{Q}_r + \dot{Q}_{cd} + hA(T_a - T_s)}{\Delta h_v} = k_g A(p_s - p) \tag{6.4}$$

where $\dot{Q}_r$ = heat transferred by radiation, J/s = W, $\dot{Q}_{cd}$ = heat transferred by conduction, $J/s = W$, and $T_a = $ dry bulb (air) temperature, °C or K. The coefficient k_g in equation (6.4) may be estimated by the solution of equation (6.3) for convection only, and $T_a - T_w$ as the difference between the dry bulb and the wet bulb temperature. For calculation of Q_r and Q_{cd} see the literature[1.3]. Equation (6.4) is solved for the surface temperature T_s by a trial-and-error procedure with the saturation pressure given by the equation

$$p_s = \exp\left[23.7093 - 4111/(237.7 + \theta_s)\right] \tag{5.34}$$

for $0 < \theta_s < 57$ °C, and

$$p_s = \exp\left[23.1863 - 3809.4/(226.7 + \theta_s)\right] \tag{5.35}$$

for $57 < \theta_s < 135$°C,
where $\theta_s = $ surface temperature, °C.

If the air stream is constant, the drying time in the constant rate period is determined by integration of equation (6.2),

$$t_c = \frac{1}{k_g A(p_s - p)} \int_{w_2}^{w_1} dw \tag{6.5}$$

where $w_1 = $ initial moisture, kg, and $w_2 = $ moisture at the end of the constant rate period, kg.

Evaporation from liquid drops is important in spray drying. Below a Reynolds number of 20 for spherical particles, the Nusselt number is approximately 2.0,

$$Nu = hD_p/k_f \approx 2 \tag{6.6}$$

where $h = $ heat transfer coefficient referred to the surface area of the sphere, $W/(°C\, m^2)$, $D_p = $ diameter of sphere, m, and $k_f = $ thermal conductivity of the gas film, for dry air approximately $9.1 \times 10^{-7}\, T^{1.8}\, W/(°Cm)$ with T in K.

The rate of evaporation may be expressed in terms of heat or mass transfer. The heat transfer coefficient h from equation (6.6) and the surface area $A = \pi D_p^2$ inserted in equation (6.1) for heat transfer gives the rate of evaporation,

$$-\frac{dw}{dt} = \frac{2\pi k_f D_p(T_a - T_w)}{\Delta h_v} \tag{6.7}$$

The rate of evaporation may also be expressed in terms of change in moisture content,

$$-W_s \frac{dX}{dt} = \frac{2\pi k_f D_p(T_a - T_w)}{\Delta h_v} \tag{6.8}$$

where $W_s = $ mass of bone-dry particle, kg, and $X = $ water content, kg water/kg dry particle. This gives the drying time in the constant rate period,

$$t_c = \frac{W_s \Delta h_v}{2\pi k_f(T_a - T_w)} \int_{X_2}^{X_1} \frac{dX}{D_p} \tag{6.9}$$

where X_1 = initial water content, kg water/kg dry particle, X_2 = water content at the end of the constant rate period, kg water/kg dry particle, and Δh_v = enthalpy of vaporization, assumed constant, approximately 2.35×10^6 J/kg in the temperature range of interest. In equation (6.9) $T_a - T_w$ is assumed to be constant. This is not the case in many spray driers, and $T_a - T_w$ may be substituted by the logarithmic mean temperature difference ΔT_m.

In two cases the integral in equation (6.9) can be solved analytically. One case is when the solid in the liquid drop forms a rigid structure which gives a constant drop diameter during the drying.[†] The other case is when the volume of the drop decreases with the volume of the evaporated water, i.e.

$$\left(\frac{X}{\rho_w} + \frac{1}{\rho_s}\right) W_s = \frac{\pi}{6} D_p^3 \tag{6.10}$$

where ρ_w = density of water, ≈ 1000 kg/m^3, and ρ_s = density of solid, kg/m^3.

Solving for D_p, inserting in equation (6.9), and integrating, gives the drying time,

$$t_c = \frac{3 W_s^{2/3} \Delta h_v}{4 \times 6^{1/3} \pi^{2/3} k_f \Delta T_m} \left[\left(\frac{X_1}{\rho_w} + \frac{1}{\rho_s}\right)^{2/3} - \left(\frac{X_2}{\rho_w} + \frac{1}{\rho_s}\right)^{2/3}\right] \tag{6.11}$$

Equation (6.9) with constant D_p and equation (6.11) correspond to the two limiting cases.

In *through-circulation drying* the solids are supported on screens in shallow beds, and air is forced through the beds to pick up and carry away the moisture and to supply the heat of vaporization. In a deep bed of thoroughly wet particles of a size that is only a small fraction of the depth of the bed, a narrow zone of evaporation will develop at the bed surface and gradually move through the bed in the direction of the air flow. The air leaving the bed will be saturated, and at the adiabatic saturation temperature. This gives the rate of drying in the constant rate period,

$$-\frac{dw}{dt} = G(H'' - H) \tag{6.12}$$

or the drying time in the constant rate period,

$$t_c = \frac{w_1 - w_c}{G(H'' - H)} \tag{6.13}$$

where w_1 = initial water content, kg, w_c = critical water content, kg, G = rate of mass flow of dry air, kg/s, H'' = saturated humidity at the wet bulb temperature, kg water vapour/kg dry air, and H = humidity of entering air, kg water vapour/kg dry air.

Beds that are shallow relative to the particle size give less pressure drop. They

[†] Aqueous solutions of materials such as soap, glue, and water-solube polymers get a tough tenuous skin on spray drying, resulting in hollow spheres, sometimes with a hole caused by the internal vapour pressure.

are often made up of particles that are prefabricated from wet solids by extrusion or pelletizing. If the particles are initially saturated with water, the period of constant rate drying may be calculated by equation (6.1), giving the drying time,

$$t_c = \frac{(w_1 - w_c)\Delta h_v}{hA(T_a - T_w)} \tag{6.14}$$

Falling rate period

The falling rate period is indicated by curve C–D in Figures 6.1–6.3. In the most general case, this period can be divided into the zone of unsaturated surface drying, and the zone where the internal liquid flow is controlling.

The zone of unsaturated surface drying, shown as curve C–E in Figure 6.3, follows immediately after the critical point. The decrease in the rate of drying is caused by a decrease in the wetted surface area. Dry portions of the solid protrude into the air, reducing the rate of evaporation per unit surface area. The rate of drying in this zone is frequently a linear function of the moisture content X as shown by the straight line C–E in Figure 6.3,

$$-\frac{dw}{dt} = aX + b \tag{6.15}$$

where the constants a and b can be determined from two points on the curve C–E (Figure 6.3). This equation may be rewritten,

$$-\frac{dw}{dt} = a(X - X^*) \tag{6.16}$$

in which the constant b is replaced by $-aX^*$, and $X^* =$ moisture content of hygroscopic material in equilibrium with the air stream, kg water/kg dry material, and $X =$ moisture content in the zone of unsaturated surface drying, kg water/kg dry material.

The constant a may be determined from equation (6.1) applied to the point of critical moisture content X_C,

$$-\frac{dw}{dt} = \frac{hA(T_a - T_w)}{\Delta h_v} = a(X_C - X^*)$$

or

$$a = \frac{hA(T_a - T_w)}{\Delta h_v(X_C - X^*)} \tag{6.17}$$

where $T_a - T_w =$ the difference between the dry bulb and the wet bulb temperature. Equations (6.16) and (6.17) give

$$-\frac{dw}{dt} = \frac{hA(T_a - T_w)}{\Delta h_v} \frac{X - X^*}{X_C - X^*} \tag{6.18}$$

where $dw = W_s dX$ (W_s = mass of dry material, kg). Integrating equation (6.18) from moisture content X_E to moisture content X_C gives the drying time for the zone of unsaturated surface drying,

$$t = \frac{W_s \Delta h_v}{hA(T_a - T_w)}(X_C - X^*) \ln \frac{X_C - X^*}{X_E - X^*} \tag{6.19}$$

Care should be taken in using this equation. The integration is carried out for constant equilibrium moisture content X^*. This may be a rough approximation. The equilibrium moisture content is a function both of the humidity and the temperature of the air, and of the surface temperature of the solid. The latter varies from the wet bulb temperature at the parts of the surface with free water to higher values at islands with no water or only bound water. Hence, it is advisable to use values of X^* based on the fitting of equation (6.19) to experimental data obtained under conditions which simulate those in the drier in question. In doing this, equation (6.19) is sometimes used for the total falling rate period.

The zone where internal moisture movements control the drying rates is indicated by curve E–D in Figure 6.3. In drying to a low moisture content, this period may predominate in determination of the drying time.

The internal moisture movements are caused by mechanisms such as diffusion, capillary flow, and flow due to shrinkage and pressure gradients. In porous materials, drying may even take place by evaporation inside the solid instead of at the surface.

Materials such as soap and wood must be dried slowly in order to avoid cracks caused by too high moisture gradients. The water movements in these materials are due mainly to diffusion. Equation (1.32) gives the differential equation for one-dimensional, and equation (1.34) for three-dimensional diffusion. Equation (1.32) in terms of moisture content, is

$$\frac{\partial X}{\partial t} = D \frac{\partial^2 X}{\partial z^2} \tag{6.20}$$

where D = diffusion coefficient for moisture in solid, m^2/s, and z = distance in direction of diffusion, m. The solution of the corresponding equation for unsteady-state heat transfer is derived in reference 3 for a slab with a constant temperature gradient from the one surface to the other at time $t = 0$ and constant thermal diffusivity. The corresponding solution of equation (6.20) gives the average moisture content X_m after drying time t as

$$\frac{X_m - X^*}{X_E - X^*} = \frac{16}{\pi^3}\left\{ \exp[-(\pi/2L)^2 Dt] + \frac{1}{3^3}\exp[-(3\pi/2L)^2 Dt] \right.$$
$$\left. + \frac{1}{5^3}\exp[-(5\pi/2L)^2 Dt] + \ldots \right\} \tag{6.21}$$

where X_E = moisture content at the beginning of the drying period where internal moisture movement controls the drying rate, kg water/kg dry material, and L = thickness of slab when drying occurs from one side, m. When drying from both sides, the 2 in front of L is deleted.

Equation (6.21) is derived with the assumptions of the constant diffusion coefficient D and the constant equilibrium moisture content X^*. But even with materials such as soap and glue the diffusion coefficient does vary with temperature and moisture content.[4] Furthermore, X^* varies with the temperature. The saturation pressure is given by an Arrhenius-type equation[5],

$$p_s^* = p_s \exp\left[- (\Delta H_s / RT) \right] \tag{6.22}$$

where p_s = saturation pressure of pure water at temperature T, ΔH_s = sorption enthalpy, J/kmol, and R = gas constant, 8314.3 J/(K kmol).

For long drying times, only the first term in equation (6.21) is significant. With $Dt/L^2 > 0.02$ the error is less than 3 per cent, and equation (6.21) reduces to

$$\frac{X_m - X^*}{X_E - X^*} = \frac{16}{\pi^3} \exp\left[- (\pi/2L)^2 Dt \right] \tag{6.23}$$

or solved for the drying time,

$$t = \frac{4L^2}{\pi^2 D} \ln\left(\frac{16}{\pi^3} \frac{X_E - X^*}{X_m - X^*} \right) \tag{6.24}$$

For calculations that take into account variations in the diffusion coefficient, see reference 6.

Drying equipment

There are many types of drier on the market, each with its own speciality, and a few that are versatile enough to dry more than one type of wet material. This chapter will only give an overview of some of the most common types. Readers requiring more detailed information are referred to Perry[1] and the specialized literature.[7-13] Reference 9 also has a great deal of information on practical experience, mechanical performance, etc. and capital and operating costs.

The engineer concerned with the manufacture of a product from given raw materials or intermediates has an unenviable task in choosing the best drier to meet the often difficult specifications required by modern industry. His first

Table 6.1 Classification of driers by scale of production[9]

Small scale to 20–50 kg/h	Medium scale 50–1000 kg/h		Large scale > 1000 kg/h
Batch	Batch	Continuous	Continuous
Vacuum tray	Agitated	Fluidized bed	Indirect rotary
Agitated	Through-circulation	Vacuum bed	Spray
Convection tray	Fluidized bed	Indirect rotary	Pneumatic
Through-circulation		Spray	Direct rotary
Fluidized bed		Pneumatic	Fluidized bed
		Band conveyor	
		Tray	
		Through-circulation	

consideration may be to look into possibilities of reducing the water content by other means prior to the more energy-consuming drying. These can be steps such as dewatering in a press or in a centrifuge, or multistage vaporization. As an example, 85–90 per cent of the water in milk can be removed by heat pump or multistage vaporization before spray drying to give dry milk powder.

The next step is to prepare a list of driers that can handle the material to be dried, and that will discharge a product that is suitable for subsequent operations. An estimate of fixed and operating costs for each drier in the list will then make it possible to eliminate those with high total cost, i.e. two and a half times that of the lowest.

Nonhebel and Moss[9] give a classification in Table 6.1 that may serve as a guide for selection depending on the scale of operation, and Van't Land[14] gives criteria for preliminary selection of batch (Figure 6.4) and continuous (Figure 6.5) driers for particulate material. The diagrams can only be used as a rule-of-thumb. Thorough knowledge of the product is of major importance. The possibility of incrustations should also be kept in mind. Small-scale experiments can be carried out with a drier fed with a sidestream at the process plant, and pilot plant runs for a new product should include the drier.

In *direct heat* or *convection driers* the heat is transferred from the warm gas to the drying material by convection. This type of drier should be considered first because the initial cost for a given capacity is usually less than for most other

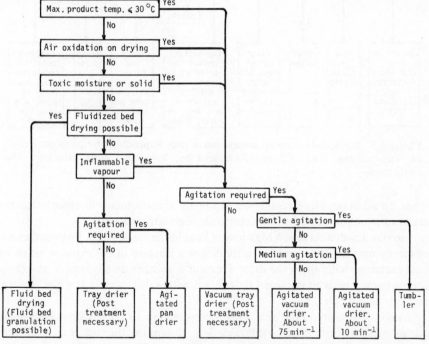

Figure 6.4 Particulate material batch drying. Reproduced by permission of C. M. Van't Land, Akzo Chemie Nederland bv, Research Centre Deventer, The Netherlands

238

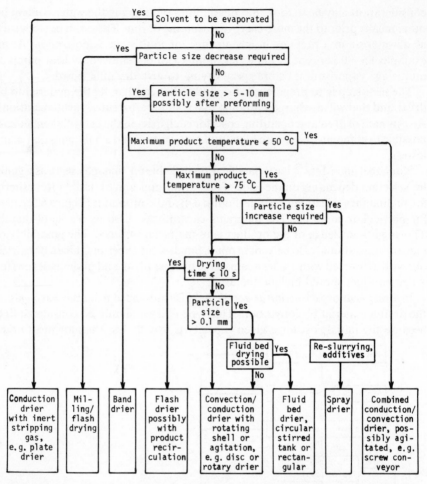

Figure 6.5 Particulate material continuous drying. Reproduced by permission of C. M. Van 't Land, Akzo Chemie Nederland bv, Research Centre Deventer, The Netherlands

types. In addition, close regulation of the air temperature will ensure that the material is not subjected to objectionable high temperatures.

A major disadvantage is a high loss of heat in the stack gases. But the thermal efficiency may be improved by recirculating a fraction of the exhaust air or by heat exchange with air to the drier. Dusting is another disadvantage, and some products may be contaminated by contact with substances in the gas. Solvent recovery is difficult.

In *indirect heat driers* the heat is transferred mainly by conduction. Their main advantages are minimum contamination, minimum dusting, easier solvent recovery, and usually a good thermal efficiency. They should be considered when any of these points are important and when a drying temperature equal to the boiling temperature of the liquid can be tolerated.

The main disadvantage is the price, and in some cases that additional boiler capacity has to be installed.

Radiation driers are used for drying of thin sheets and films such as textile fabrics, paints, and stove enamels, and for small-scale drying of industrial chemicals where other methods fail. A major disadvantage is high operating cost.

In *dielectric* or *microwave drying* high-frequency electricity generates heat internally. Until now it has had only occasional industrial applications, such as drying of beach shoe-lasts[11].

Freeze-drying is desiccation by sublimation of vapour from the solid state in a vessel evacuated to an absolute pressure usually in the range from 1 to 50 N/m^2. It is used for foodstuffs such as milk, onions, and coffee, in the pharmaceutical industry for virus solutions, serums, antibiotics, and bacterial cultures, and for animal tissues and extracts as skin, blood, tumours, arteries, and hormones.

Batch tray driers

The batch tray drier is mainly used for fragile or relatively expensive products such as dyestuffs and pharmaceuticals. Figure 6.6 is a schematic drawing of an *atmospheric tray drier*, also referred to as a shelf or cabinet drier, or a compartment oven. The drying air can pass across the trays as shown in Figures 6.6 and 6.7, or it can pass through the bed of material on perforated trays as shown in Figure 6.8. Standard trays are 800×400 mm and 30 mm deep.

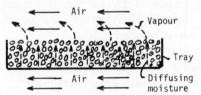

Figure 6.7 Solid-bottom tray where the moisture diffuses to the top

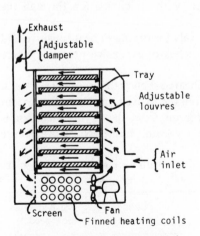

Figure 6.6 Batch atmospheric tray drier with adjustable louvres to promote uniform air flow and a damper to adjust recycling of drying air

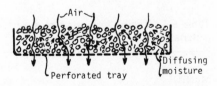

Figure 6.8 Perforated tray where the air passes through the bed. Used to a limited extent for granular or crystalline products

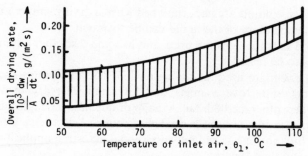

Figure 6.9 Typical overall drying rates on atmospheric tray driers for filter cakes of mud-like consistency. Initial moisture contents 50–500 per cent (dry basis), final moisture contents 0.5–3.0 per cent. Air recirculation in the range 80–95 per cent. Depth of mud on tray about 40 mm

Air-flow rate and temperature is usually limited by the fragile nature of the material to be dried. Thus, the air-flow rate is often kept below the free-falling velocity of the finest particles in the product. The common mean velocity over the trays is in the range from 1.0 to 1.5 m/s and air temperatures in the range from 40 to 100 °C with 70–95 per cent recycled air. Figure 6.9 gives typical overall drying rates (constant and falling rate periods) for drying of filter cakes of mudlike consistency on atmospheric tray driers[9].

Atmospheric tray driers are often equipped with adjustable louvres in order to avoid non-uniform air flow, which is one of the most serious problem in the operation of many direct heat driers.

Table 6.2 gives some data for tray driers with air circulation through the material (Figure 6.8).

Batch vacuum tray driers are used extensively for drying of heat-sensitive and easily oxidized materials, and for small batches of expensive products where material losses must be avoided.

The drier has an evacuated chamber with heated shelves that support the trays. The heat transfer is mainly by conduction from the shelves, and to a minor extent by radiation from the shelf above. The vapour from the drying chamber is withdrawn to a condenser connected with a vacuum pump. Normal absolute

Table 6.2 Typical data for atmospheric tray driers with air circulation through the material[9]

Material	Fine crystals	Polymer crumbs	Washed vege- table seed
Drying air temperature (°C)	100	88	36
Superficial air velocity (m/s)	0.25–0.5	0.9	1.0
Loading depth (mm)	50	65	38
Overall drying rate per unit tray area (kg/(h m^2))	1.07	0.88	1.12

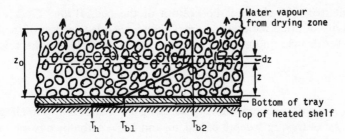

Figure 6.10 Drying bed with temperature T_{b1} in the bottom and T_{b2} in the drying zone. (T_{b2} = saturation temperature at the pressure in the vacuum drier)

pressure is in the range 6–30 kN/m² (0.06–0.27 bar). Normal bed depth is in the range 25–40 mm.

Some materials have a tendency to give case-hardening or skin formation.

Figure 6.10 is a simplified picture of drying in a vacuum tray drier where the saturation temperature of water is below the surface temperature of the shelf.

The following calculation of the unsteady-state drying is based on the simplifying assumptions that:

1. radiant heat is negligible;
2. sensible heat is negligible, i.e. only latent heat of vaporization is taken into account;
3. the thermal conductivity of the dried section (distance z in Figure 6.10) is constant;
4. the vapour can escape between particles, as indicated by dashed arrows in Figure 6.10.

The heat flux from the heated shelf to the drying zone is given by the equations[3]

$$\frac{dq}{dt} = U(T_h - T_{b1}), \qquad T_h - T_{b1} = \frac{1}{U}\frac{dq}{dt}$$

and

$$\frac{dq}{dt} = kA(T_{b1} - T_{b2})/z, \qquad \frac{T_{b1} - T_{b2} = \frac{z}{k}\frac{dq}{dt}}{\text{Sum: } T_h - T_{b2} = \left(\frac{1}{U} + \frac{z}{k}\right)\frac{dq}{dt}}$$

or

$$\frac{dq}{dt} = \frac{T_h - T_{b2}}{\dfrac{1}{U} + \dfrac{z}{k}} \qquad (6.25)$$

where T_h = temperature of the heated shelf, K or °C, T_{b2} = saturation temperature of water at the pressure in the drier, K or °C, U = overall heat transfer

242

coefficient from the shelf to the bottom of the bed, $W/(m^2\,°C)$, $k =$ thermal conductivity of dried material, $W/(m\,°C)$, and $z =$ dried section of the bed, m.

The layer dz dries in time dt. The latent heat of vaporization for this layer is

$$dq = \Delta h_v X_e g_s \frac{dz}{z_0} \qquad (6.26)$$

where $X_e =$ evaporated moisture, kg water/kg dry material, $g_s =$ loading, kg wet material/m^2 of tray, $z_0 =$ depth of bed, m, and $\Delta h_v =$ enthalpy of vaporization at temperature T_{b2}, J/kg water.

Equation (6.26) inserted in equation (6.25) gives

$$\int_0^t dt = \frac{\Delta h_v X_e g_s}{(T_h - T_{b2})z_0} \int_0^{z_0} \left(\frac{1}{U} + \frac{z}{k}\right) dz$$

or the drying time

$$t = \frac{\Delta h_v X_e g_s}{T_h - T_{b2}} \left(\frac{1}{U} + \frac{z_0}{2k}\right) z_0 \qquad (6.27)$$

The two constants U and k must be determined by measurements. By poor contact between shelves and trays, $1/U$ may be the dominating term in the parentheses, and equation (6.27) may be approximated without the term $z_0/2k$. With the material directly on the shelves, the term $1/U$ becomes zero.

Reference 9 gives some typical operating data for drying of organic chemicals in vacuum tray driers.

Batch agitated driers

The agitated batch drier consists of a heated, cylindrical vessel with rotating agitators. There are three main types on the market. One is a relatively short vertical cylinder with a flat bottom and a paddle agitator. The two others have a jacketed horizontal cylinder, the one with a scroll agitator and the other with radial paddle arms and rollers, as shown in Figure 6.11. Each set of radial arms

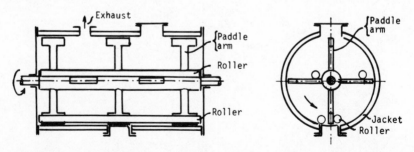

Figure 6.11 Horizontal agitated batch drier with radial paddles and rollers that have a grinding action[9]

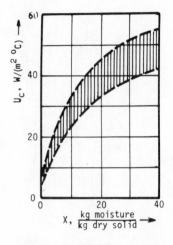

Figure 6.12 Mean overall heat transfer coefficient U_c as a function of initial moisture content of paste in agitated batch driers[9].

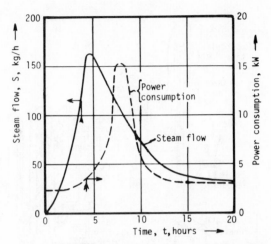

Figure 6.13 Typical steam and power consumption for the drier shown in Figure 6.11. Steam consumption for drier with diameter 1.2 m and length 5.5 m and power consumption for drier with diameter 1.35 m and length 3.8 m[9]

push a roller (tube or bar) that has a grinding effect. According to reference 9, for certain batch sizes, it is probably the cheapest form of non-continuous drying.

The dryer can be used under atmospheric pressure and under vacuum. Organic pastes, fine aromatic organic crystals, and anthracene are examples of material dried in batch agitated driers. Also, for design purposes, the rate of heat transfer is calculated by the approximate equation

$$\dot{Q} = U_c A \Delta T_m \tag{6.28}$$

where $\dot{Q}$ = the total heat transfer per second, W = J/s, U_c = mean overall heat transfer coefficient for the complete drying period, W/(m$^2\,°$C), A = area of drying surface, m^2, and ΔT_m = mean temperature difference between heating medium (steam or hot water) and batch, assuming that batch temperature equals saturation temperature at the pressure in the dryer, K or $°$C.

Drying under reduced pressure is normally carried out with absolute pressure in the range 13–30 kN/m^2 (0.13–0.3 bar) with drying temperatures between 50 $°$C and 110 $°$C and a temperature difference in the range 30–60 $°$C. Values for U_c for the types of material normally dried in agitated batch driers are given in Figure 6.12, and typical steam and power consumption curves in Figure 6.13.

Air-suspended systems

The most intimate contact between particles and drying air is obtained in driers where the particles are suspended in the air, as is the case in flash driers, spray driers, and fluidized-bed driers.

In the *flash drier* or pneumatic conveying drier, particulate solids are dried

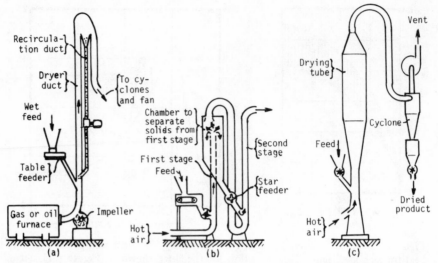

Figure 6.14   Three types of flash driers. Type a has means for recycling and an impeller or paddle-wheel slinger that distintegrates agglomerates. Type b has two stages to reduce height. Type c is operated with a gas velocity of about 30 m/s at the feed point, and about 6 m/s in the upper part of the drying tube[9]

while being transported in a stream of warm gas. Reference 9 gives diagrams and a brief description of six types of flash driers manufactured in the U.K. Figure 6.14 shows simplified diagrams of three of them.

In some driers the feed passes through a disintegrator, e.g. a cage mill, before it enters the drier. Typical applications of this combined milling and drying process are removal of the water of crystallization from copper sulphate monohydrate in order to produce a dry powder, and corn-mill air-lift drying systems grinding the coarse grain and dewatering it from about 65 per cent moisture to 8 per cent or less.[7]

In a simple drier without recirculation, the residence time in the drying tube is only 0.5–3 s. Hence, they can only handle easily dried, preferably non-porous materials with less than 70 per cent moisture in the feed. The other major limitations are particle size (usually less than 2 mm) and free-flowing properties (non-sticky material).

The drying time may be calculated by integration of equation (6.3) for the constant rate period after experimental determination of the products (hA) and (k_gA) per kg dry solid. Also, the enthalpy of vaporization Δh_v should be substituted by the sum of sensible heat and latent heat, calculated per kg evaporated water.

Table 6.3 gives an extract of some published performance data on flash driers.

Major advantages of the flash drier are the short residence time and concurrent flow that often permit satisfactory drying of heat-sensitive materials, and the high air inlet temperature that gives a high thermal efficiency. In addition, the capital cost can be low compared with other types of driers.

A *spray drier* may be considered a modification of a flash drier, in which the

Table 6.3   Performance data on flash driers.[9] The flow rates refer to dry solid and dry air

Material	Solid rate (kg/h)	Moisture% of dry solid		Solid temperature (°C)		Air temperature (°C)		Air/solid ratio (kg/kg)
		In	Out	In	Out	In	Out	
Ammonium sulphate	950	2.75	0.28	38.5	63	215	76	1.5
Sewage sludge filter cake	2270	80	10	15	71	700	121	7.2
Hexamethylene tetramine	2500	6–10	0.08–0.15	—	48	93	50	1.9
Coal 6 (mm)	50000	9	3	15	57	371	80	1.3

feed is introduced as a liquid that is atomized inside the drier, usually to give a product particle size in the range $1-300\,\mu$m. The drops fall down concurrent or countercurrent to the drying air, and are dried before they reach a wall. Spray drying as a special subject is described in reference 13 (which should be used together with reference 9 to supplement the information in this chapter).

The three types of atomizers used in spray driers are shown in reference 3 and listed in Table 6.4.

Figure 6.15 is a large spray drier for detergent with several spray nozzles and a particle size in the 1 mm range. The bulk density of the product is adjusted by adjustment of the air inlet louvres and the flow rate and temperature of the air. Figure 6.16 is a smaller drier with a spinning atomizer for products such as dried milk powder, dried egg powder, blood plasma, and pigments.

The spray-dried product is usually nearly spherical. Aqueous solutions of materials such as soap, gelatin, and water-soluble polymers usually acquire a

Table 6.4   Atomizers for spray driers

Type	Capacity	Drop size	Advantages and disadvantages
Spray nozzle (pressure to 700 bar)	5000 kg/h	Large	Cheap. Suitable for solid-free solutions of low or medium viscosity. Relatively inflexible, change in pressure changes both capacity and drop size
Rotating disc atomizer, (500–30000 r/min)	25 g/h to 7000 kg/h	Medium	Flexible, capacity and rotating speed can be varied independently. Best for low- and medium-viscosity liquids
Two-fluid (pneumatic) atomizer, air or steam under pressure	Small	Small	Best for particles less than $20\,\mu$m. Can handle highly viscous liquids and pastes (centrifuged pigments with viscosities up to $1000\,\text{N s/m}^2$)[15]

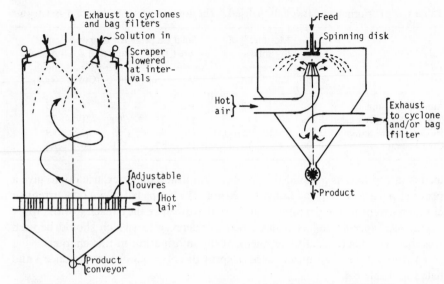

Figure 6.15 Large spray drier with spray nozzles and scraper for cleaning of walls

Figure 6.16 Small spray drier with rotating atomizer

tough tenuous outer skin on drying. This gives hollow spherical particles, often with a hole in the wall through which evaporated water has escaped.

Spray driers can handle feed with a moisture content higher than 80 per cent, liquid feeds, materials of which partially dry particles are sticky, and fragile particles, which flash driers cannot do.

The residence time is longer than in a flash drier, but still only a few seconds.

Spray drying gives a relatively low bulk density. This may be an advantage or a disadvantage, depending on the product. Restrictions on inlet and exit air temperatures may give a relatively low thermal efficiency. In addition, the capital cost is high per unit capacity, especially for small capacities.

Table 6.5 gives some typical operating characteristics of small and large spray driers.

Table 6.5 Operating characteristics for spray driers.[7] Reproduced by permission of McGraw-Hill, Inc.

	Small driers		Large driers	
	Typical	High	Typical	High
Operating temperature (°C)	260	540	260	540
Evaporative capacity (kg water/h)	180	450	3000	7000
Heat requirement (kW = kJ/s)	350	640	6000	9900
Power requirement (kW)	15	22	85	200
Relative equipment cost per unit evaporative capacity	3.08	1.30	1.00	0.45

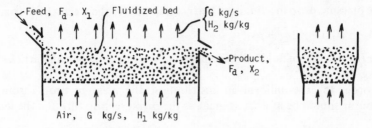

Figure 6.17 Fluidized-bed drier

In *fluidized-bed driers* the particulate solids are kept in suspension by a vertically rising stream of gas. Particles fed into a fluidized bed at one side of a vessel, as shown in Figure 6.17, 'float' in the air stream and 'flow' through the bed to the outlet.

The air velocity in a fluidized-bed drier is at or above the minimum fluidizing velocity and below the velocity of the particles in dilute suspension, i.e. it is a special case of hindered settling.[3] According to a widely used empirical correlation by Leva,[16] the mass flow rate required for incipient fluidization is estimated by the equation

$$G_M = 0.0093 \frac{d_m^{1.82}}{\mu^{0.88}} [\rho(\rho_p - \rho)]^{0.94} \tag{6.29}$$

where G_M = minimum mass flow rate of gas to give fluidization, kg/(s m²), d_m = mean particle diameter, m, μ = dynamic viscosity, Ns/m², ρ_p = density of particles, kg/m³, and ρ = density of gas, kg/m³.

The hindered settling velocity of particles in suspensions is estimated by the equation

$$V_h = \varepsilon^n V_t \tag{6.30}$$

where ε = void fraction = (volume of fluid)/(volume of suspension), n = empirical constant, according to Scholl[17] $n = 3.65$ for $\varepsilon > 0.6$, and V_t = terminal velocity in dilute suspension,[3] m/s. For

$$500 < Re_D < 200\,000, \qquad V_t = 1.74 \sqrt{\frac{D(\rho_p - \rho)g}{\rho}} \tag{6.31}$$

for

$$2 < Re_D < 500, \qquad V_t = \left[0.072 D^{1.6} \frac{\rho_p - \rho}{\mu^{0.6} \rho^{0.4}} g \right]^{1/1.4} \tag{6.32}$$

for

$$0.000\,01 < Re_D < 2, \qquad V_t = 0.0556 D^2 \frac{\rho_p - \rho}{\mu} g \tag{6.33}$$

where Re_D = Reynolds number for a sphere with diameter D_p giving the same volume as the volume of the particle, $Re_D = \rho V_t D_p / \mu$.

The pressure drop in a fluidized bed corresponds to the weight of the particles,

$$\Delta p = H_b(1 - \varepsilon)(\rho_p - \rho)g \tag{6.34}$$

where H_b = height of fluidized bed, m, and ρ_p = average density of particles in the bed, kg/m^3.

In order to give uniform air distribution, the pressure drop through the distributor should be at least as great as the pressure drop through the fluidized bed.

The hold-up in the bed is

$$W_b = Q_b(1 - \varepsilon)(\rho_p - \rho) \tag{6.35}$$

where Q_b = volume of the fluidized bed, m^3.

The movements of particles within the bed give a considerable variation in the residence times of solids. The average residence time is

$$t_r = (\rho_{pd}/\rho_p)W_b/F_d \tag{6.36}$$

where ρ_{pd} = density of bone-dry particles, kg/m^3, and F_d = feed rate, bone-dry material, kg/s.

A material balance for water (Figure 6.17) gives

$$F_d(X_1 - X_2) = G(H_2 - H_1) \tag{6.37}$$

The rate of mass transfer may be given by the equation

$$\frac{dw}{dt} = K_H W_{bd} \Delta H \tag{6.38}$$

where K_H = mass transfer coefficient per kg bone-dry solid with the humidity difference as driving force, kg water evaporated/(s kg dry solid); K_H must be determined experimentally, W_{bd} = hold-up in the bed, kg dry solid, and ΔH = mean humidity driving force, i.e. the difference in humidity of the drying air at the surface of the particles and in the bulk of the air stream, kg water vapour/kg dry air.

In a fluidized bed mixing and heat transfer is very rapid, and the mean humidity driving force may be approximated by

$$\Delta H_m \approx (H'' - H_2) \tag{6.39}$$

where H'' = saturation humidity at the surface temperature of the particles, kg/kg.

Fluidized-bed driers are suitable only for materials of size 0.05–15 mm, and the ratio of the largest to the smallest diameters of the particular product to be dried must be less than 8.[18] A large size ratio results in accumulation of the heavier particles at the base of the bed. Another drawback is the variations in moisture content of the dried particles caused by the considerable variations in the residence times of the solids. These variations are reduced by installation of baffles in troughs of the type shown in Figure 6.17 or by use of multistage (Figure 6.18) or cascade units (Figure 6.19).

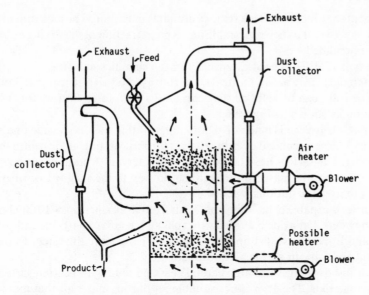

Figure 6.18 Two-stage fluidized-bed drier. The disengagement space over each bed is 1–2 m high

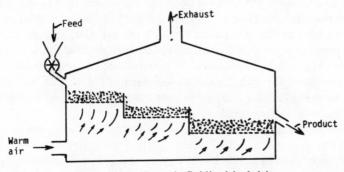

Figure 6.19 Cascade fluidized-bed drier

Fluidized-bed driers are used for drying of PVC and of polymer chips, and on a large tonnage scale for sand and coal. The fluidized-bed process is suitable also for cooling granular materials, such as NPK fertilizers leaving rotary driers. Batch-type fluidized driers are used extensively for heat-sensitive pharmaceutical products.

See reference 19 for further information.

Continuous tray and tunnel driers

The *tunnel truck drier* consists of a heating chamber where each truck with shelves enter through a door at one end of the chamber and advances slowly on rails to

250

the door at the other end. It is a semi-continuous operation. The moisture pick-up in each pass over the shelves is small, and a good fraction of the drying air or flue gas is recirculated.

This type of drier is used extensively in the ceramics industry.

In *continuous tray driers* the trays are conveyed either vertically or horizontally. The trays can be solid or they can have a perforated bottom for through circulation as shown in Figure 6.8.

The *turbo tray drier* is made up of annular, rotating shelves stacked one above the other.[7] Air is circulated over the shelves by fans rotating in the centre section of the drier. The shelves have radial slots with intervals. A stationary wiper over each tray pushes the material through the slots on to the shelf below where it is spread evenly by a stationary lever.

The drier is capable of handling materials from thick slurries ($> 10\,000\ \mathrm{N\,s/m^2}$) to fine powders. It may be operated with split air streams to both dry and cool the stock, and it may be sealed and operated as a closed-circuit system to recover vapours, or to dry in an inert atmosphere.

Major disadvantages are high capital cost and fouling of heating surfaces by dusting materials. The drier is not suitable for fibrous materials that mat, or for sticky or doughy substances.

Rotary driers The *direct-heat rotary drier* is the workhorse of continuous chemical driers. More chemicals are dried in rotary driers than in any other type.

The drier consists of a long, horizontal or slightly inclined revolving cylindrical shell. Heated air or flue gas passes through the cylinder, and the feed of wet material enters at one end and is discharged as dried product in the other end, as indicated in Figure 6.20. Flights welded to the inside of the shell lift the solids and shower them down through the gas, as indicated in Figure 6.21. Straight flights are used in the feed end for sticky or wet materials, and bent flights for free-flowing or almost dry material.

In driers with countercurrent flow the solids are moved forward by spiral flights and by gravity.

Rotary driers are applicable to the drying of relatively free-flowing granular material requiring drying times up to 1 h. Also, some wet and sticky pastes may be handled by special devices or precautions, such as premixing with a fraction of the

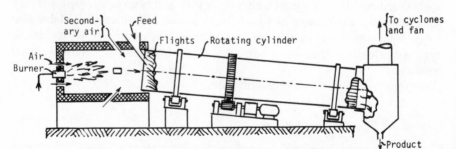

Figure 6.20 Direct-fired rotary drier with concurrent flow

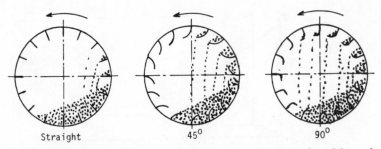

Figure 6.21 Flights with different geometry, straight flights for sticky and wet material, and bent flights for free-flowing or almost dry material

dried product. Within limits the tumbling action is beneficial, since any semi-permeable crust on the surface or case-hardening is disrupted, allowing easier escape of moisture. Heat-sensitive materials are often dried in concurrent flow where the warmest gases are in contact only with wet material. In addition, the product temperature is only 10–20 °C below the exit gas temperature, and it can be controlled by controlling this temperature. Countercurrent operation is used whenever it is necessary to heat the solid to a relatively high temperature to complete the drying process.

Examples of products dried in rotary driers are sand, certain ores, silica, sodium sulphate, ammonium sulphate, cellulose acetate, styrene, vinyl resins, oxalic acid, urea crystals, and fish meal. The thermal efficiency is in the range 60–70 per cent for products dried at high temperatures and 20–55 per cent for low-temperature drying.[7] As a special feature, drying and calcining may be carried out in the same unit.

Drier diameters are usually between 0.6 and 3.0 m, the length 4 to 15 times the diameter, speed of rotation 4–5 rpm, and air velocities 1.5–2.5 m/s. As a rule of thumb, the air velocity referred to the total cross-section can be 40–45 per cent of the terminal velocity of free-falling particles.

The residence time in seconds may be estimated by the formula published by Seaman and Mitchell,[20]

$$t_r = \frac{L}{anD(s + bV_m)} \tag{6.40}$$

where L = effective length of drier, m, D = diameter of drier, m, n = speed of rotation, s^{-1}, V_m = mean gas velocity, m/s (V_m is negative for countercurrent flow), s = slope of drier, m/m, a = constant depending on lifter design, 2.5 ± 0.5, and b = constant depending on particle size, 0.0009 for coarse inorganic materials (1.7–3.3 mm) and 0.0015 for fine materials (0.4–1.0 mm).

The hold-up in the drier in terms of kg bone-dry material is

$$W_{bd} = F_d t_r \tag{6.41}$$

where F_d = feed rate of bone-dry material, kg/s.

For calculation purposes a rotary drier can conveniently be divided in three

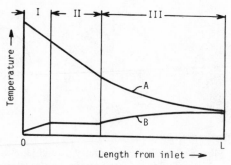

Figure 6.22 Temperatures in a concurrent rotary drier. A is air and B material temperature

zones (Figure 6.22). In zone I is preheating, evaporation in this zone being usually neglected. Zone II covers the period of constant rate drying, and zone III the falling rate period.

The design procedure outlined in reference 9 is as follows:

1. Assume exit temperature of product and air. In concurrent driers the exit air temperature may be 10–15 °C and in countercurrent driers 100 °C above the product temperature.
2. Find the equilibrium vapour pressure over the solid at the exit, and choose an exit air humidity.
3. Determine the dry air rate by an overall water mass balance.
4. Determine the inlet air temperature by a heat balance.
5. Calculate the drier diameter from the total air flow at exit conditions and air velocity 40–45 per cent of the terminal velocity of free-falling particles.
6. Calculate the heat transferred in the preheating zone by the following steps:
   Estimate the wet bulb temperature and calculate the heat required to heat the feed to the wet bulb temperature.
   Calculate the air temperature at the end of the preheat zone and check the wet bulb temperature. Recalculate if necessary.
7. Calculate the length of the preheat zone based on an experimental heat transfer coefficient per unit length of drier, h_L. Lacking experimental data the heat transfer coefficient per kg material, h_w, may be estimated as 2.3 W/(kg °C) for coarse material (1.7–3.3 mm), and 7.0 W/(kg °C) for fine material (0.4–1.0 mm). This gives

$$h_L = (W_{bd}/L)h_w \tag{6.42}$$

where W_{bd}/L is calculated by equations (6.40) and (6.41).
8. Calculate the length of the constant rate drying zone based on latent heat removed from the solid and the logarithmic mean temperature difference.
9. Divide the falling rate drying zone into sections having arbitrarily chosen the final moisture contents.
   Calculate the exit air humidity for one section at a time. Assume the solid

exit temperature and calculate the air exit temperature by a heat balance. Calculate the length of the section using the heat transfer equation with logarithmic mean temperature difference and heat transfer coefficient h_L.

With the same solid exit temperature and equation (6.2) written as

$$w_n = (k_g A_L) L_n \Delta p_m, \tag{6.43}$$

calculate the length L_n of section n.

Repeat the calculations with new solid exit temperatures until L_n calculated by equation (6.43) equals the length calculated by the heat transfer equation.

The symbols in equation (6.43) are w_n = moisture evaporated in section n, kg, k_g = mass transfer coefficient, $kg/(m^2\, s\, Pa)$, determined experimentally as the product $k_g A_L$ for the particles in question, as a function of temperature and moisture content, A_L = surface area of particles (hold-up) in 1 m drier length, m^2/m, and Δp_m = mean vapour pressure driving force, $N/m^2 = Pa$.

Major disadvantages of direct-heated rotary driers are that they are difficult to seal, that fine and dusty materials are blown out of the drier and require additional equipment for collection, and that the requirements for floor space and sturdy supports are fairly large.

Indirect heat rotary driers are used for materials with a long falling rate period and for materials with a dusting problem. In one type the flights or lifters of direct-heat rotary driers are replaced with one or two rows of tubes containing steam or hot water. Air is drawn through the drier at a rate sufficient to carry away the water vapour, and the heat is transferred by contact of the material with the tubes.

Other continuous driers

Examples of other continuous driers are screw conveyor, belt conveyor, vibrating conveyor, drum or film, and turbo tray or rotating-shelf driers.

The *screw-conveyor drier* is essentially a jacketed conveyor in which the material is heated and dried as it is conveyed through the unit. The screw can be hollow and heated with the same medium as the jacket, e.g. hot water, steam or Dowtherm. The drier can be air-swept to carry off the vapour, or completely closed for vacuum operation.

The drying may be carried out in several conveyors mounted horizontally one above the other with the material dropping from one unit to the next one below.

Screw-conveyor driers are used for granular, free-flowing substances where crystal structure and size specifications are unimportant. Pasty substances may be dried with recycling of part of the dry product in order to give a feed of proper consistency.

Heat transfer coefficients are reported in the range from 6 to 70 $W/(m^2\,°C)$ depending on the moisture content of the feed and the conveyor speed.

The *belt-conveyor drier* has a continuous belt where the material is spread out and passes through a drying chamber. The belt may have a solid surface, or it may

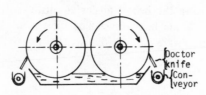

Figure 6.23   Twin-drum drier

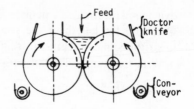

Figure 6.24   Double-drum drier

consist of woven wire or perforated metal plates to permit the drying air to pass through the bed of material.

The *vibrating-conveyor drier* has a perforated, vibrating conveyor screen through which hot air passes from a chamber below. Drying rates are high.

These driers are used primarily for free-flowing solids that contain mainly surface moisture. They are not suited for sticky substances or fibrous materials that mat.

A *drum drier* consists of one or more horizontal, rotating drums. The material to be dried is fed in in amounts to form a relatively thin layer or film that is scraped off with adjustable blades or a doctor knife pressed against the surface of the drum.

Figure 6.23 shows a twin-drum drier with dip feed. The steam-heated drums are immersed 10–25 mm in the solution or slurry to be dried. Figure 6.24 is a double-drum drier where the liquid feed is introduced into the V-shaped trough formed by the nip of the two drums and spring-loaded end plates. The clearance of the drums when hot is usually around 0.12 mm.

The drum speed is normally 5–15 rpm with drum diameters from 0.4 to 1.25 m and the length of the largest drums 3.0 m. In atmospheric driers the steam

Table 6.6   Performance of drum driers, based on data in reference 9

Type of material	Physical form of feed	% Water, wet basis Product	Feed	U in eq. (6.44) W/(m² °C)
Sodium acetate	Solution	0.4	80	300
	Solution	10	80	900
Organic salts	Solution	0.4–3	60–80	180–220
Organic salts	Solution	5	65	260
Organic salts	Solution	13	67	95
Organic salts	{ Thin slurry	1.7	80	140
	{ Thin slurry	3.1	80	280
Organic compounds	Viscous solution	10.5	72	160
	Thin slurry	1.2	70	200
Organic compounds, low surface tension	Solution	0.5	75	30
	Solution	0.5	80	35
	Thick slurry	2.5	70	170

pressure is usually in the range 4.5–6.5 bar absolute. The feed is normally solutions, slurries, or pastes of colloidal or poorly crystallizing materials with moisture content in the range 12–40 per cent (wet basis). Thick slurries may be splashed on to single- and twin-drum driers. But double-drum driers with nip feed can handle concentrated solutions and thick slurries and sludges as long as they do not have abrasive solids or lumps that may damage the drums.

In practice the thickness of the film is in the range 25–125 μm. The drying rate is calculated from the general heat transfer equation,

$$\dot{Q} = U_c A \Delta T_m \tag{6.44}$$

where A = surface area of the drum(s), m^2, ΔT_m = mean temperature difference between the condensing steam inside the drum and the vaporizing water, K or °C, and U_c = overall heat transfer coefficient for the entire drying period (see Table 6.6), $W/(m^2 \, °C)$.

Example 6.1 DRYING OF AMMONIUM NITRATE

Here, 1.5 kg/s wet crystalline ammonium nitrate is dried from 5 to 0.2 per cent by weight in countercurrent drying with air in a rotary drier. Pressure, temperature, and humidity of the inlet air are $P = 1.0$ bar (10^5 Pa), $T = 405$ K (132 °C), and $H = 0.007$ kg vapour/kg dry air. The temperature of the air from the drier is 355 K (82 °C), of ammonium nitrate to the drier 293 K (20 °C), and from the drier 340 K (67 °C). The specific heat capacity of ammonium nitrate is 1.66 kJ/(kg K) and of water 4.19 kJ/(kg K).

Calculate the mass flow rate of dry air:
(a) neglecting heat losses to the ambient air;
(b) assuming heat loss to the ambient air as approximately 8 per cent of the heat transferred to the ammonium nitrate.

Solution

(a) The feed contains

   $1.5 \times 0.95 = 1.425$ kg dry nitrate/s

and

   $1.5 \times 0.05 = 0.0750$ kg water/s

The product contains w kg water/s, i.e.

   $w/(1.425 + w) = 0.002$

   $w = 0.00286$ kg water/s

Evaporated,

   $0.075 - 0.00286 = 0.0721$ kg water/s

Humidity of exit air (G kg dry air/s),

   $H = 0.007 + 0.0721/G$ kg vapour/kg dry air $\tag{a}$

With 293 K (20 °C) as reference state, the enthalpy of the product per unit time is

   $(1.425 \times 1.66 + 0.00286 \times 4.19)(340 - 293) = 111.7$ kJ/s

Equation (5.9) gives the heat of vaporization at temperature 20 °C,

   $\Delta h_v = 2501.6 - 2.275 \times 20 - 0.0018 \times 20^2 = 2455.4$ kJ/kg

Equation (5.8) gives the enthalpy of the entering air,

$h_{y1} = (1.00 + 1.88 \times 0.007)(405 - 293) + 0.007 \times 2455.4 = 130.7 \, \text{kJ/kg dry air}$

Enthalpy of air from the drier (equations (5.8) and (a)) is

$h_{y2} = [1.00 + 1.88(0.007 + 0.0721/G)](355 - 293) + (0.007 + 0.0721/G)2455.4$

$= 80.0 + 185.4/G \, \text{kJ/kg dry air.}$

An enthalpy balance around the drier gives

$130.7G = (80.0 + 185.4/G)G + 111.7$

$G = 5.86 \, \text{kg dry air/s.}$

(b) Heat transferred to the ammonium nitrate,

$111.7 + 0.0721 \times 2455.4 = 288.7$

Heat losses, $0.08 \times 288.7 = 23.1 \, \text{kJ/s}$, giving the enthalpy balance

$130.7G = (80.0 + 185.4/G)G + 111.7 + 23.1$

$G = 6.32 \, \text{kg dry air/s.}$

Example 6.2 TRAY DRIER

Test drying of a food product was carried out in an insulated tray (Figure 6.2). The temperature of the drying air was 60 °C and the dew-point 20 °C. The dry weight of the material was 3.765 kg and the surface area 0.186 m². Test data are given in Table a.

Table a   Weight of humid material at different drying times

Time, t (h)	0	0.4	0.8	1.4	2.2	3.0	4.2	5.0	7.0	9.0	12.0	25.0
Weight, G (kg)	4.944	4.885	4.808	4.699	4.554	4.404	4.241	4.150	4.019	3.978	3.955	3.955

(a) Determine equilibrium moisture content.
(b) Determine the drying rate dw/dt in kg water per m² an hour as a function of the free moisture content X in the constant rate and in the falling rate period.
(c) Calculate the time needed for drying under the same conditions from total water content 0.25 to 0.10 kg/kg dry matter.
(d) Assuming total pressure $P = 1.0$ bar absolute (10^5 Pa), calculate the mass transfer coefficient k_g in kg water/(m² s Pa) in the constant rate period.

Solution

(a) Equilibrium moisture content seems to be obtained towards the end of the test (constant weight), giving

$X^* = (3.955 - 3.765)/3.765 = 0.0505 \, \text{kg water/kg dry material.}$

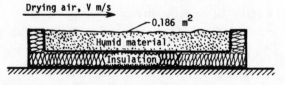

Figure 6.25   Test drying

(b) The data in Table a correspond to the free moisture content

$$X = \frac{G - 3.765}{3.765} - 0.0505 = \frac{G}{3.765} - 1.0505 \text{ kg water/kg dry material} \tag{a}$$

The decrease in moisture content per unit time between time t_n and time t_{n+1} is

$$-\frac{\Delta X}{\Delta t} = \frac{X_{n+1} - X_n}{t_{n+1} - t_n} \frac{\text{kg water/kg dry matter}}{h} \tag{b}$$

and the corresponding rate of evaporation in kg per m² and hour,

$$\frac{dw}{dt} = -\frac{3.765}{0.186} \frac{\Delta X}{\Delta t} = -20.24 \frac{\Delta X}{\Delta t} \frac{\text{kg}}{\text{m}^2 \text{h}} \tag{c}$$

Table b   Data calculated by equations (a), (b), and (c)

Time, t (h)	0	0.4	0.8	1.4	2.2	3.0	Notes
$X\left(\dfrac{\text{kg free water}}{\text{kg dry matter}}\right)$	0.263	0.247	0.227	0.198	0.160	0.120	eq. (a)
$-\dfrac{\Delta X}{\Delta t}\left(\dfrac{\text{kg/kg dry matter}}{h}\right)$		0.0400	0.0500	0.0483	0.0475	0.0500	eq. (b)
$dw/dt\,(\text{kg/(m}^2\text{h)})$		0.810	1.012	0.978	0.961	1.012	eq. (c)

Table b   (continued)

Time, t (h)	3.0	4.2	5.0	7.0	9.0	Notes
$X\left(\dfrac{\text{kg free water}}{\text{kg dry matter}}\right)$	0.120	0.0755	0.0518	0.0170	0.0061	eq. (a)
$-\dfrac{\Delta X}{\Delta t},\left(\dfrac{\text{kg/kg dry matter}}{h}\right)$		0.0371	0.0237	0.0174	0.0055	eq. (b)
$dw/dt\,(\text{kg/(m}^2\text{h)})$		0.751	0.481	0.352	0.110	eq. (c)

Here, dw/dt from Table b is plotted in Figure 6.26 together with two straight lines with approximately the same areas above and below the lines. The straight lines correspond to

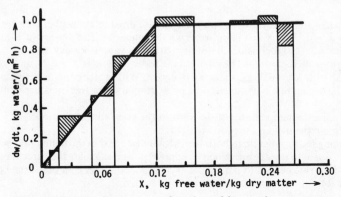

Figure 6.26   Drying rate as a function of free moisture content

the equation

$$dw/dt = 8X \text{ kg water}/(m^2 \, h) \tag{d}$$

in the falling rate period with $X < X_C$, where $X_C = 0.12 =$ the free moisture content, and

$$dw/dt = 0.96 \text{ kg water}/(m^2 \, h) \tag{e}$$

in the constant rate period with $X > X_C$.

(c) Surface area,

$$a = 0.196/3.765 = 0.0494 \, m^2/\text{kg dry matter}$$

Drying time in the constant rate period,

$$t_c = \frac{0.25 - (0.12 + 0.0505)}{0.96 \times 0.0494} = 1.68 \, h$$

In the falling rate period the free moisture content is reduced from $X_C = 0.12$ to $X = 0.10 - 0.0505 = 0.0495$ kg water/kg dry matter. With dw kg/m² substituted by $adX = 0.0494dX$ kg/kg dry matter, equation (d) gives the drying time,

$$t_f = \frac{1}{0.0494} \int\limits_{0.0495}^{0.12} \frac{dX}{8X} = 2.24 \, h$$

Total drying time,

$$t = t_c + t_f = 1.68 + 2.24 = 3.92 \, h$$

(d) With dew-point 20 °C and air temperature 60 °C, the humidity chart (Figure 5.4) gives the wet bulb temperature 30 °C. Equation (5.34) gives the vapour pressure,

$$p_s = \exp\left[23.7093 - 4111/(237.7 + 30)\right] = 4241 \, \text{Pa}$$

at 30 °C, and

$$p_s = \exp\left[23.7093 - 4111/(237.7 + 20)\right] = 2337 \, \text{Pa}$$

at 20 °C, corresponding to the driving force $4241 - 2337 = 1904$ Pa.

Equations (e) and (6.2) give the mass transfer coefficient,

$$k_g = 0.96/(3600 \times 1904) = 1.40 \times 10^{-7} \, \text{kg}/(m^2 \, s \, Pa)$$

Problems

6.1 Batches of 100 kg cotton yarn (dry weight) are dried in atmospheric air (pressure 1.0 bar) from $X = 0.53$ to $X = 0.11$ kg water/kg dry material. The drying air is at constant temperature 62 °C and constant humidity 0.0134 kg vapour/kg dry air.

The drying rate is constant 15 kg/h until the water content is $X = 0.23$, and equilibrium is obtained with $X^* = 0.05$ kg water/kg dry yarn. With water content between $X = 0.23$ and X^*, the drying rate is assumed to be proportional to the free moisture content in the yarn.

(a) Determine the temperature of the yarn in the constant rate period.
(b) Determine the drying time for each batch.
(c) Assuming the drying rate in both the constant rate and in the falling rate period to be proportional to the dry air velocity to the 0.8 power, calculate the percentage increase in air-flow rate needed to reduce the drying time to 75 per cent of the time calculated under (b).

6.2 2500 kg/h humid, hygroscopic material is to be dried from $X = 0.3$ to $X = 0.05$ kg water/kg dry material in a fluidized-bed drier. The air from the drier is assumed to be in equilibrium with the dried product, and the equilibrium moisture content is given by the equation

$$X^* = (0.35 - 0.002\theta)(H/H'') \text{ kg water/kg dry material}$$

where θ = temperature, $^\circ$C.

The feed to the drier is at temperature 20 $^\circ$C and the air to the drier at temperature 150 $^\circ$C, pressure 1.06 bar (106 kPa) and humidity $H = 0.009$ kg vapour/kg dry air. The specific heat capacity of the dry material is 2.1 kJ/(kg $^\circ$C).

(a) Neglecting heat losses, calculate the dry air-flow rate. (Hint: It is recommended to use a program.) *Ans.* 21 080 kg/h.

(b) It turns out that dry air-flow rates in excess of 16 000 kg/h give an unacceptable loss of fine particles. Calculate the inlet air temperature needed to obtain $X = 0.05$ kg water/kg dry material with $G = 15 000$ kg dry air/h.

6.3 1000 kg/h crystals of ammonium sulphate are dried from $X = 0.035$ to $X = 0.002$ kg water/kg dry material by direct contact with warm air in countercurrent flow in a 6.5 m long, uninsulated rotating drier with diameter 1.2 m. Air at pressure 1.02 bar (102 kPa), temperature 20 $^\circ$C and absolute humidity $H = 0.007$ kg vapour/kg dry air is heated to 90 $^\circ$C before it passes on to the drier. The air leaves the drier at 32 $^\circ$C. The temperature of the crystals to and from the drier is 20 $^\circ$C and 60 $^\circ$C respectively. The specific heat capacity of dry ammonium sulphate is 1.51 kJ/(kg $^\circ$C) and the overall heat transfer coefficient between drying air and ambient air at 20 $^\circ$C, is $U = 5.8$ W/(m^2 $^\circ$C).

(a) Calculate the mass flow rate of air, G kg dry air/h. *Ans.* 2650 kg/h.

(b) The drying air is heated from 20 $^\circ$C to 90 $^\circ$C in the tubes of a heat exchanger with condensing steam on the outside. Calculate the approximate amount of steam used per kg evaporated water, when the heat given off by the condensing steam is 2300 kJ/kg. *Ans.* Approx. 2.45 kg/kg.

Symbols

A	Area, m^2
A_L	Surface area of particles (hold-up) per unit length of drier, m^2/m
D	Diameter, m
D	Diffusion coefficient for moisture in solids, m^2/s
D_p	Diameter of spherical particle, m
d_m	Mean particle diameter, m
F_d	Feed rate of bone-dry material, kg/s
G	Mass flow of dry air, kg/s
G_M	Minimum mass flow rate of gas to give fluidization, kg/(m^2 s)
g	Acceleration of gravity, m/s^2
g_s	Loading of wet material, kg/m^2
H	Absolute humidity, kg water vapour/kg dry air
H''	Saturation humidity at the wet bulb temperature, kg water vapour/kg dry air
H_b	Height of fluidized bed, m
ΔH_s	Sorption enthalpy, J/kmol
h	Heat transfer coefficient, W/(m^2 K) = W/(m^2 $^\circ$C)
Δh_v	Enthalpy of vaporization at the wet bulb temperature, J/kg
h_w	Heat transfer coefficient per kg bone-dry material, W/(kg $^\circ$C)
k	Thermal conductivity of solid, W/(m K) = W/(m $^\circ$C)
k_f	Thermal conductivity of gas film, W/(m K) = W/(m $^\circ$C)
k_g	Mass transfer coefficient, kg/(m^2 s Pa)
L	Length, m
n	Speed of rotation, s^{-1}

p	Vapour pressure of water, $N/m^2 = Pa$
p_s	Saturation pressure of water, $N/m^2 = Pa$
$\dot{Q}$	Rate of heat transfer, $J/s = W$
Q_b	Volume of fluidized bed, m^3
q	Heat per unit area, J/m^2
R	Gas constant, $8314.3\,J/(kmol\,K)$
T	Temperature, K or $°C$
T_a	Dry bulb temperature (air), K or $°C$
ΔT_m	Mean temperature difference, K or $°C$
T_s	Surface temperature, K or $°C$
T_w	Wet bulb temperature, K or $°C$
t	Time, s
U	Overall heat transfer coefficient, $W/(m^2\,K) = W/(m^2\,°C)$
U_c	Mean overall heat transfer coefficient for the complete drying period, $W/(m^2\,K)$ $= W/(m^2\,°C)$
V_m	Mean gas velocity, m/s
V_t	Terminal sedimentation velocity of particle in dilute suspension, m/s
W_b	Hold-up, kg
W_{bd}	Hold-up of bone-dry material, kg
W_s	Mass of bone-dry material, kg
w	Water, kg or kg/m^2
w_c	Critical water content, kg
X	Water content, kg water/kg dry material
X^*	Equilibrium moisture content of hygroscopic material, kg water/kg dry material
X_C	Critical moisture content, kg water/kg dry material
z	Distance, m
z	Dried section of bed, m
ε	Void fraction $=$ (volume of fluid)/(volume of suspension)
ρ	Density of gas, kg/m^3
ρ_p	Density of particle, kg/m^3
ρ_{pd}	Density of bone-dry particle, kg/m^3
ρ_s	Density of solid, kg/m^3
ρ_w	Density of water, $\sim 1000\,kg/m^3$
μ	Dynamic viscosity, $N\,s/m^2 = kg/(m\,s)$
θ	Temperature, $°C$

Subscripts

cd	Conduction
r	Radiation

References

1. Perry, R. H., and C. H. Chilton, *Chemical Engineers' Handbooks*, 5th edn, McGraw-Hill, New York, 1973.
2. Tsao, G. T., and T. D. Wheelock, Drying theory and calculations, *Chem. Eng.*, **74** (19 June), 201–214 (1967).
3. Lydersen, A. L., *Fluid Flow and Heat Transfer*, J. Wiley & Sons, Chichester, 1979.
4. Hougen, O. A., H. J. Cauley, and W. A. Marshall, Jr., Limitations of diffusion equations in drying, *Trans. A.I.Ch.E.*, **34**, 183–209 (1940).
5. Ullmanns Enzyklopädie der technischen Chemie, *Verfahrenstechnik I*, 4th edn, Verlag Chemie, Weinheim, 1972.
6. Shoeber, W. J. A. H., and H. A. C. Thijssen, A short-cut method for the calculation of drying rates for slabs with concentration-dependent diffusion coefficient, *A.I.Ch.E. Symposium Series*, **73** (No. 163), 12–24 (1977).

7. Sloan, C. E., Drying systems and equipment, *Chem. Eng.*, **74** (19 June), 169–200 (1967).
8. Krischer, O., *Die wissenschaftliche Grundlagen der Trocknungstechnik*, 2nd edn, Springer-Verlag, Berlin, 1963.
9. Nonhebel, G., and A. A. H. Moss, *Drying of Solids in the Chemical Industry*, Butterworths, London, 1971.
10. Keey, R. B., *Drying: Principles and Practice*, Pergamon Press, Oxford, 1972.
11. Keey, R. B., *Introduction to Industrial Drying Operations*, Pergamon Press, Oxford, 1978.
12. Herron, D., and D. Hummel, How to select polymer drying equipment, *Chem. Eng. Progr.* **76** (No. 1), 44–52 (1980).
13. Marshall, W. R., Atomization and spray drying, *Chem. Eng. Progr. Monograph Series*, **50**, 2 (1954).
14. Van 't Land, Selection of industrial driers, paper to be published.
15. Cronan, C. S., New easy way changes paste to powder, *Chem. Eng.*, **67** (21 March), 83–84 (1960).
16. Leva, M., *Fluidization*, McGraw-Hill, New York, 1959.
17. Scholl, K. H.: Der Einfluss der Konzentration auf die Sinkgeschwindigkeit von Partikeln, *Chem.-Ing.-Techn.*, **48**, 149–150 (1976).
18. Van Heerden, C., *J. Appl. Chemistry, Suppl. Issue* 1, **2**, 57 (1952).
19. Vanecek, V., M. Markvart, and R. Drbohlav, *Fluidized Bed Drying*, Leonard Hill, London, 1966.
20. Seaman, W. C., and T. R. Mitchell, Jr., Analysis of rotary dryer and cooler performance, *Chem. Eng. Progr.*, **45**, 482–493 (1954).

CHAPTER 7
Adsorption and ion exchange

Adsorption

Adsorption is a separation process where a fluid phase (gas or liquid) is contacted with a porous particulate phase with the property of selectively taking up or storing one or more species originally contained in the fluid phase. It is frequently the most economical method of separation if the species to be removed are present at relatively low concentrations.

Adsorbents

Some common adsorbents are listed in Table 7.1, together with some characteristic data and examples of applications.

Activated carbons are sized granules, produced by pyrolysis of wood or coal, giving pores as indicated in Figure 7.1.[5] Activated carbon derived from bituminous coal is preferred for waste-water treatment. Its high hardness keeps down handling losses during reactivation.

Direct steaming is the most common method for regeneration of carbon beds.

Activated alumina and alumina gel are porous forms of aluminium oxide with high resistance to shock and abrasion. They can be reactivated an almost unlimited number of times by heating to temperatures above 177 °C.

Silica gel is an amorphous and extremely porous form of SiO_2, inert with most fluids, but not recommended for strong alkalis or hydrofluoric acid. It is normally regenerated by heating to 175–180 °C.

Molecular sieves are synthesized rod-like pellets or spherical beads with uniform pore size. Type 3A with pore size 0.3 nm is used for drying, type 5A with pore size 0.5 nm for separation of *n*-paraffins from mixed hydrocarbon streams, and type 13X with pore size 0.9–1 nm for removal of CO_2 and H_2O from air. Regeneration temperatures range from 200 °C to 315 °C for all types.

Figure 7.2 shows a section through a unit cell of type 4A molecular sieve cubic crystal with pore opening 0.42 nm.

The 1982 price of 4A molecular sieves is £0.75, of silica gel £0.35, and of F–1 activated alumina £0.12 per kg.[4]

Physical adsorption is due to intermolecular or van der Waals' forces. The process is analogous to condensation, and the heat of physical adsorption is approximately the same as the heat of liquefaction.

Chemisorption is adsorption accompanied by a chemical reaction or chemical bonding. It involves electron transfer or sharing, or the adsorbed species (the

Table 7.1 Typical data and examples of commercial applications of some common adsorbents[1-4]

Adsorbent	Size range (mm)	Surface area (m^2/g)	Bulk density (kg/m^3)	Adsorptive capacity (kg H_2O/kg)	Examples of applications
Activated carbon	0.6–5	600–1600	160–570	0.4–0.7[†]	Water treatment, gas purification, solvent recovery, alcohol purification, decolorizing
Aluminium oxide (activated alumina)	0.7–25	50–250	720–880	0.14–0.25	Drying gases and liquids, purifying, decolorizing and refining petroleum oils and waxes
Silica gel	0.1–3	320–900	400–830	0.3–0.6	Drying of gases, separation of hydrocarbons
Aluminosilicates (molecular sieves)	1.7–3	600–800	480–705	0.2–0.3	Selective adsorption based on molecular size and shape, as drying, separation of normal paraffins from branched and cyclic hydrocarbons, sweetening of natural gas (H_2S and mercaptan removal)
Fuller's earth and other clays	< 0.07	130–250	480–800		Purification of vegetable and animal oils

[†] Data for adsorption of CCl_4.

Figure 7.1 Cross-section of pore in activated carbon. Reproduced by permission of McGraw-Hill, Inc.

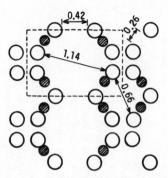

Figure 7.2 Cross-section of a type 4A molecular sieve cubic crystal. Small shaded molecules are aluminimum or silicon, large open atoms are oxygen. Distances are in nm. Reprinted with permission from ref. 6. Copyright 1956 American Chemical Society

adsorbate) breaks up into atoms or radicals which are bonded separately. Physical adsorption can involve the formation of multimolecular layers, while pure chemisorption is completed in a monolayer and gives a stronger bond.

Equilibrium and adsorption isotherms

Adsorption equilibrium is obtained when the number of molecules arriving on the surface is equal to the number of molecules leaving the surface to go into the fluid phase. In simple systems, the equilibrium concentration of the solute in the solid phase as a function of solute concentration or partial pressure is represented by a single curve for each temperature, known as an adsorption isotherm. Figure 7.3 shows the five types of isotherms obtained experimentally for physical adsorption. An example of type I is adsorption of oxygen on charcoal at $-183\,°C$, of type II nitrogen on iron catalysts at $-195\,°C$, of type III bromine on silica gel at $79\,°C$, of type IV benzene on ferric oxide gel at $50\,°C$, and type V water

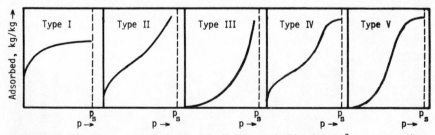

Figure 7.3 The five types of van der Waal's adsorption isotherms.[7] p_s = saturation pressure. Types I, II, and IV can be used to separate the adsorbate from the carrier fluid, while types III and V are of less interest

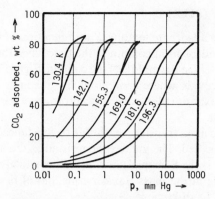

Figure 7.4 Isotherms for the adsorption of pure carbon dioxide on steam-activated coconut shell carbon with surface area $1800\,m^2/g$

vapour on charcoal at $100\,°C$. Adsorption isotherms with hysteresis are shown in Figure 7.4.[3]

An interpretation of the type I isotherm by adsorption from a gas was given by Langmuir,[8] and his isotherm equation for adsorption derived for a unimolecular layer can be adapted to type I and type III isotherms. Brunauer et al.[9] expanded the Langmuir isotherm to include multilayer adsorption,

$$v = \frac{v_m C p/p_s}{(1 - p/p_s)[1 + (C - 1)p/p_s]} \qquad (7.1)$$

where v = volume of gas $(0\,°C, 760\,mm\,Hg)$ adsorbed per unit mass of adsorbent, m^3/kg, v_m and C = empirical constants, p = partial pressure of adsorbate, Pa, and p_s = saturation pressure of adsorbate, Pa.

Isotherms for adsorption of organic material from aqueous solutions may be approximated by the Freundlich equation,[4]

$$X = kY^{1/n} \qquad (7.2)$$

where X = mass fraction of organic material on the adsorbent, kg/kg, Y = mass fraction of organic material in the solution, kg/kg, and k and n = constants (vary with temperature).

The constants k and n may be determined in a laboratory test. Bernardin[5] recommends that portions of the activated carbon ranging in weight from 0.1 to 20 g, for example, are placed in individual flasks. The same volume of liquid to be tested, e.g. $100\,cm^3$, is added to each of the flasks, and the flasks are placed in a shaker and left for about 2 h before the carbon is separated from the solutions on $0.45\,\mu m$ filters. The filtrate is analysed for a characteristic, such as total organic carbon or chemical oxygen demand, and a plot on log–log paper of the adsorbed amount per kg activated carbon versus concentration in the filtrate gives the empirical constants in equation (7.2).

266

Mass Transfer Zone (MTZ)

Figure 7.5 shows the adsorbate loading X as a function of the length (height) L from the inlet of an adsorption column at the time when the mass transfer zone (MTZ) is approximately in the middle of the column. Figure 7.6 is the same plot at the time of breakthrough, where *breakthrough* is taken as either the minimum detectable or the maximum allowable concentration of sorbable component in the adsorber effluent.

The S-shape of the mass transfer zone is due to resistance to mass transfer and to axial mixing in the bed. The greater the resistance, the axial mixing, and the flow rate, the longer is the S-shaped zone. A mathematical representation of the concentration profile in the mass transfer zone and of breakthrough curves is given by Linek and Dudukovic.[10]

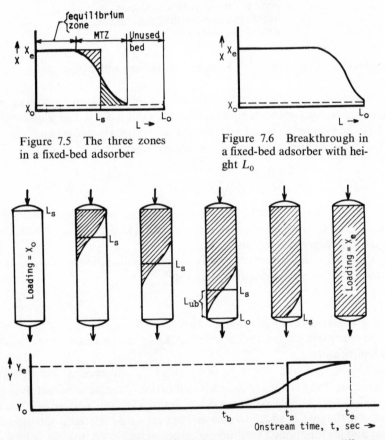

Figure 7.5   The three zones in a fixed-bed adsorber

Figure 7.6   Breakthrough in a fixed-bed adsorber with height L_0

Figure 7.7   Columns showing the loading of the adsorbent at different times. The horizontal lines L_s represent the stoichiometric front. The lower diagram gives the composition of the effluent versus time, where t_b is the breakthrough time and t_s stoichiometric time.[11] Reproduced by permission of the Union Carbide Corporation

The ideal case of no resistance to mass transfer corresponds to the vertical line in Figure 7.5 where the two hatched areas are equal. This vertical line with abscissa L_s is the *stoichiometric front*.

Figure 7.7 shows the mass transfer zone and the stoichiometric front at different times, and the corresponding adsorbate concentration in the effluent.

Practically all the adsorption takes place in the pores of the adsorbent, and the rate of adsorption is proportional to the rate of diffusion in the pores. Here, Stefan diffusion is dominating in pores larger than 100 times the mean free path of the molecules, and Knudsen diffusion in capillaries with diameter less than ten times the mean free path (see p. 8).

The time required to reach mass fraction X of the adsorbate on the adsorbent may be estimated by the equation

$$X = (1 - e^{-kt})X_e \tag{7.3}$$

where X_e = mass fraction absorbate on adsorbent in equilibrium with fluid with mass fraction Y_e, kg/kg, t = time, s, and k = constant, depending on pore depth and by Stefan diffusion proportional to the diffusion coefficient D_{AB} and by Knudsen diffusion D_{AK}, m^{2}s^{-1}.

The constant k must be determined experimentally for the fluid and the adsorbent in question, and equation (7.3) gives the time needed to obtain, for instance, 95 per cent of equilibrium. The length of the mass transfer zone is proportional to the product of this time and the mass flow rate.

Breakthrough

The amount of adsorbate adsorbed when the stoichiometric front advances dL, is

$$dw = (X_e - X_0)\rho_b A \, dL \tag{7.4}$$

and per unit time,

$$\frac{dw}{dt} = G(Y_e - Y_0)A \tag{7.5}$$

where dw/dt = mass adsorbed per unit time, kg/s, X_e = mass fraction of adsorbate on the adsorbent when in equilibrium with fluid with mass fraction Y_e, kg/kg, X_0 = original mass fraction of adsorbate on the adsorbent, kg/kg, Y_e = mass fraction of adsorbate in the entering fluid, kg/kg, Y_0 = mass fraction of adsorbate in fluid in equilibrium with adsorbent with mass fraction X_0, kg/kg, A = cross-sectional area of adsorbent bed, m^2, ρ_b = bulk density of adsorbent, kg/m^3, and G = mass velocity, kg/(m^2 s).

Combining the two equations gives

$$dL = \frac{G(Y_e - Y_0)}{\rho_b(X_e - X_0)} dt$$

Integration from time zero to the time t_s when the stoichiometric front reaches the exit (next to last column in Figure 7.7), gives

$$L_0 = \frac{G(Y_e - Y_0)}{\rho_b(X_e - X_0)} t_s \tag{7.6}$$

and integration to breakthrough gives

$$L_0 - L_{ub} = \frac{G(Y_e - Y_0)}{\rho_b(X_e - X_0)} t_b \tag{7.7}$$

The two last equations give the length of unused bed (Figure 7.7),

$$L_{ub} = L_0(1 - t_b/t_s) \tag{7.8}$$

This is the basic equation used for analysing breakthrough data. In applying this equation, however, caution should be exercised if the system involves chemisorption, large heat effects, high concentration of sorbable component, or the length of the bed is less than the MTZ.

When analysing the performance of a commercial unit, operation past breakthrough is not usually permissible, and equation (7.7) may be used alone, giving

$$L_{ub} = L_0 - \frac{G(Y_e - Y_0)}{\rho_b(X_e - X_0)} t_b \tag{7.9}$$

The drawback of this equation is the difficulty in determination of $(X_e - X_0)$ with

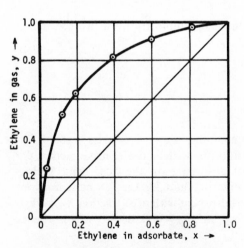

Figure 7.8 Ethylene–propylene mixtures. Mole fraction ethylene in gas versus mole fraction ethylene in adsorbate by adsorption on silica gel at 25°C and atmospheric pressure. Reprinted with permission from ref. 12. Copyright 1950 American Chemical Society

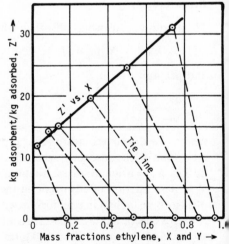

Figure 7.9 Ponchon–Savarit diagram for the same mixture and conditions as in Figure 7.8, calculated from data in reference 12

sufficient accuracy for commercial units that have been cycled for an extended period of time.

Calculation procedures for separation of binary mixtures by adsorption differ only in details from procedures in liquid–liquid extraction. Figure 7.8 is an example of an equilibrium diagram of the McCabe–Thiele type and Figure 7.9 the corresponding Ponchon–Savarit diagram.

Regeneration

Regeneration is carried out by one of the following four cycles:

In the *thermal-swing cycle*, desorption takes place at a temperature higher than that of the adsorption step. Cooling is usually required before the adsorbent is reused.

In the *pressure-swing cycle*, the desorption is carried out at a pressure lower than that of the adsorption step.

The *purge-gas stripping cycle* is based on the use of a purge gas that is not adsorbed, but lowers the partial pressure of the adsorbed component in the bulk of the fluid.

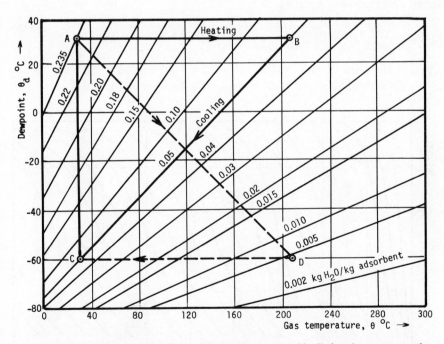

Figure 7.10   Regeneration of a molecular sieve, type 4A. Point A represents the equilibrium condition of the bed after the adsorption step, A–B heating in contact with air with constant dew-point, and B–C cooling to initial condition of the bed. A–D represents heating in contact with dried gas. Reproduced by permission of Union Carbide Corporation

The *displacement cycle* is one in which a second sorbable fluid displaces the initial adsorbate.

As an example, Figure 7.10 shows two ways of regenerating by heating to 209 °C of molecular sieves used for air at temperature 30 °C dried to a dew-point of − 60 °C.

During direct heating or cooling with a purge gas, a transient temperature front is formed within the packed bed. The overall particle heat transfer resistance is

$$1/U_v = 1/h_v + 1/h_s \qquad (7.10)$$

where h_v = volumetric fluid phase heat transfer coefficient, W/(m³ bed °C), and $1/h_s$ = resistance to heat conduction in the particles (m³ bed °C)/W. Experimental values of h_s for 3 mm diameter particles are from 45 000 to 150 000 W/(m³ bed °C).[11]

The heat transfer coefficient to spheres in a uniform bed may be estimated by the equation[13]

$$Nu = 1.03\, Re^{0.6}\exp(1.1 - 2.75\varepsilon) \pm 8 \text{ per cent} \qquad (7.11)$$

or with physical data for air at 30 °C passing through a bed of 3 mm diameter spheres and porosity $\varepsilon = 0.4$,

$$h \approx \frac{0.027}{0.003}1.03G^{0.6}\left(\frac{0.003}{0.000\,0187}\right)^{0.6} = 195G^{0.6} \text{ W/(m}^2\,°\text{C)}$$

The surface area in this bed is $6(1 - 0.40)/0.003 = 1200 \text{ m}^2/\text{m}^3$ bed, giving $h_v = 1200 \times 195G^{0.6} = 234\,000G^{0.6}$ W/(m² °C). Using a conservative value of h_s

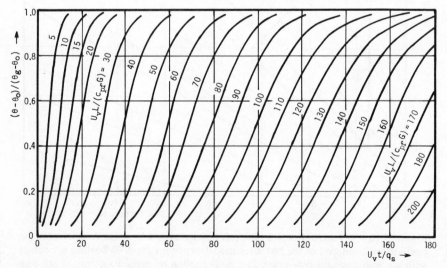

Figure 7.11 Temperature difference ratio $(\theta - \theta_0)/(\theta_g - \theta_0)$ where θ_g is the temperature of the purge gas and θ_0 the initial bed temperature, as a function of $U_vL/(c_{pf}G)$ and U_vt/q_s

in equation (7.10) gives

$$\frac{1}{U_v} = \frac{1}{234\,000\,G^{0.6}} + \frac{1}{45\,000} \quad \frac{m^2\,bed\,^\circ C}{W} \tag{7.12}$$

where G = mass velocity, kg/(m² s).

Figure 7.11 gives the temperatures in the bed during regeneration as a function of the two dimensionless parameters

$$U_v L/(c_{pf} G) \quad \text{and} \quad U_v t/q_s$$

where L = distance from entrance of bed, m, c_{pf} = specific heat capacity of fluid, J/(kg °C), t = time from start, s, and q_s = volumetric specific heat capacity of solid (includes particles, adsorber vessel, and latent heat converted to heat capacity), J/(m³ bed °C).

Large commercial adsorbers are usually designed for a temperature difference of 30 °C between inlet and outlet of the purge gas at the end of the regeneration cycle. Adsorbent loadings may be as indicated in Figure 7.12.

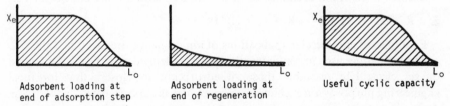

| Adsorbent loading at end of adsorption step | Adsorbent loading at end of regeneration | Useful cyclic capacity |

Figure 7.12 Capacity of bed regenerated by the thermal-swing cycle[11]. Reproduced by permission of Union Carbide Corporation

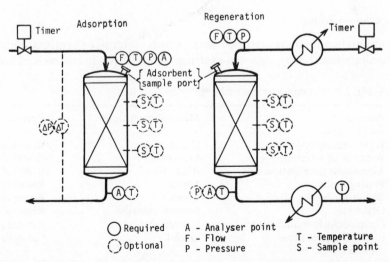

Figure 7.13 Timer-controlled adsorber system[11]. Reproduced by permission of Union Carbide Corporation

The cycle control may be semi-automatic or automatic as indicated in Figure 7.13.

Ion exchange

Ion exchange is a metathetical chemical reaction between an electrolyte in solution and an insolute electrolyte. The utility of ion exchange is due to the ability to use and reuse the ion-exchange material. For example, in water softening, the exchanger R in the sodium-ion form is able to exchange for calcium, thus removing calcium from hard water and replace it by sodium,

$$2\bar{R}^-\bar{N}a^+ + Ca^{2+} \rightleftharpoons \bar{R}_2^-\bar{C}a^{2+} + 2Na^+ \tag{7.13}$$

where the superscript bars stand for the solid phase.

Afterwards, the calcium-loaded exchanger is treated with a sodium chloride solution for regeneration to the sodium form for another cycle.

Complete deionization may be obtained using both an anion and a cation exchanger. This is also the case if a strong-acid resin is contacted with a base, or a hydroxide-form anion exchange resin is contacted with a mineral acid such as.

$$\bar{R}^+\overline{OH}^- + H^+Cl^- \rightarrow \bar{R}^+\overline{Cl}^- + H_2O \tag{7.14}$$

Examples of commercial applications of ion exchange are given in Table 7.2.

The techniques used in contacting the liquid electrolyte with the insoluble solid or gel electrolyte resemble those of adsorption so closely that for most engineering purposes ion exchange can be treated as a special case of adsorption. As in adsorption, diffusion in the pores of the solid is usually the rate-controlling step. In ion exchange, however, the relative magnitudes of the various ionic fluxes are determined by the constraint of electroneutrality.

Ion-exchange resins

The ion-exchange solids first used were porous, natural, or synthetic minerals containing silica. Today, most ion-exchange solids or gels in commercial use are

Table 7.2 Examples of ion exchange[14,15]

Process	Material exchanged	Purpose
Water treatment	Calcium ions	Removal
Water dealkylization	Dicarbonate	Removal
Aluminium anodization bath	Aluminium	Removal
Plating baths	Metal	Recovery
Rayon wastes	Copper	Recovery
Glycerine	Sodium chloride	Removal
Formaldehyde manufacture	Formic acid	Recovery
Ethylene glycol (from oxide)	Glycol	Catalysis
Fermentation broths	Streptomycin, neomycin	Recovery
Grapefruit processing	Pectin	Recovery
Citric acid	Inorganic salts	Removal

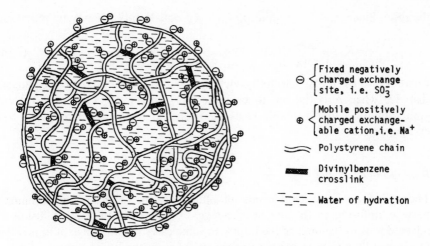

Figure 7.14 Schematic picture of a hydrated strong-acid cation exchange resin. Reproduced by permission of Academic Press Inc.

based on an organic polymer network. Styrene–divinylbenzene (DVB) is the most common base copolymer used as a cross-linked structure, with the functional ionic groups added to the benzene rings (Figure 7.14).[16] The resins are usually fine, granular solids or almost perfect spheres 0.044–1.0 mm diameter.

The number of ionic sites is usually expressed as hydrogen equivalents. Sulphonated styrene–DVB resins having one sulphonic acid group per benzene ring has 5.2 ± 0.1 milliequivalents per gram dry resin (meq/g). For chemical process work, wet-volume capacity of the swelled resin is conveniently expressed as equivalents per unit volume. The standard strong-acid exchange resins have a normality of about 2.0 eq/litre in sodium form, and strong-base resins 1.0 to 1.4 eq/litre in chloride form.[3]

With the present price of resins (£2500/m^3 for anion and £900/m^3 for cation exchangers), fixed bed ion-exchange columns are used only for an ionic concentration of the component to be removed of less than 0.3 eq/litre solution to be treated.

Ionic equilibria

Equilibrium for non-ionized products as shown by equation (7.14) corresponds to complete reaction until one of the reacting components is exhausted. Ionic equilibria as shown by equation (7.13) can be described by isotherms similar to those used for ordinary adsorption. Various empirical equations as well as the mass-action law have been proposed for description of the isotherms. With only two ions A and B of equal charge, the mass-action law gives

$$K_{AB} = \frac{\bar{x}_A x_B}{\bar{x}_B x_A} \tag{7.15}$$

where x_i = mole fraction of ions i in the liquid, and $\bar{x}_i$ = mole fraction of ions i in the solid. Since $x_B = 1 - x_A$ and $\bar{x}_B = 1 - \bar{x}_A$, equation (7.15) may be rewritten

$$\frac{\bar{x}_A}{1 - \bar{x}_A} = K_{AB}\frac{x_A}{1 - x_A} \tag{7.16}$$

where K_{AB} is the molar selectivity coefficient which should preferably be higher than 1.0 when A is the ion to be removed. Equations (7.15) and (7.16) are approximations, but valuable tools in the initial screening of proposed ion exchange processes.[3]

Rate of ion exchange

The rate of ion exchange depends on diffusion to the surface of the exchanger particle, diffusion to the site of exchange, exchange of ions, and diffusion of released ions to the bulk of the liquid. In some cases the exchange of ions is the rate-controlling step, in other cases the diffusion in the liquid phase.

An approximation by Gilliland and Baddour[17] is to some extent used for engineering calculations. It is based on a second-order chemical reaction, the film theory and experimental data with exchange of sodium and hydrogen ions, using Dowex 50 with particle diameters from 0.22 to 1.0 mm. Their correlation are based on the rate equation

$$\frac{d\bar{x}_B}{dt} = k\left[c_B(1 - \bar{x}_B) - \frac{\bar{x}_B}{K_{AB}}(c_{B0} - c_B) \right] \tag{7.17}$$

and the rate constant k given by

$$\frac{1}{kD^2} = \frac{10^7}{Re^{0.84} + 0.049} + 7.2 \times 10^6 \tag{7.18}$$

where t = time, s, k = rate constant, s^{-1}, c_B = concentration of B in bulk of solution, eq/m^3, c_{B0} = concentration of B in entering solution, eq/m^3, K_{AB} = molar selectivity coefficient, equation (7.15), (equation (7.17) is only valid for $K_{AB} > 1$), Re = Reynolds number, $\rho VD/\mu$, where V = superficial velocity, m/s, and D = particle diameter, m.

Scale-up

Most ion exchangers are vertical columns with liquid flow downwards through the bed, and scale-up can be made directly from laboratory data with for instance a 30 or 40 mm diameter column. For unchanged flow velocity and column height, the only condition for scale-up is that the feed is uniformly distributed over the resin surface and collected uniformly from across the bottom of the column. Optimum flow velocity in the laboratory column is optimum flow velocity in the full-sized column.

In ion exchange with a favourable selectivity coefficient ($K_{AB} > 1$), the exchange zone usually occupies only a small fraction of the column height.

Experience shows that scaling up the column height by a factor of two in such cases is more likely to improve than to reduce the preformance.[3]

Equipment for ion exchange

The majority of industrial ion exchangers consists of a vertical, cylindrical vessel with various distributors, as shown in Figure 7.15 for an anion or cation exchanger, and in Figure 7.16 for a mixed-bed exchanger. External piping and valves permit flushing and regeneration.

Applebaum[16] gives superficial flow velocities up to 0.005–0.007 m/s with all units on, 0.008–0.010 m/s with one unit out for regeneration, and 0.010–0.017 m/s for secondary, polishing units. Cation exchangers should have a bed depth of at least 0.6 m and a maximum working temperature of 120 °C. The corresponding data for anion exchangers are 0.75 m and 40 °C. Additional design data for water treatment are given in Table 7.3.

Figure 7.17 shows one of the arrangements used for mixed-bed ion exchangers for power stations.[19] The anion and cation exchange particles have different sizes and/or different densities. They are separated by fluidization with water before regeneration, and redistributed with air before a new service.

Continuous and semi-continuous ion-exchange contactors are in commercial use in water conditioning and chemical processing. The resin inventory is less than in fixed-bed exchangers, and continuous processes should be considered in cases where the component to be removed has a concentration much above 0.5 N or is more than 30 kg/h.[3]

In pulsed columns the resin is moved from plate to plate by periodic application of pressure or vacuum, while the solution passes through the resins between the pulses. Pulse intervals may be as short as 4–30 mins. In water

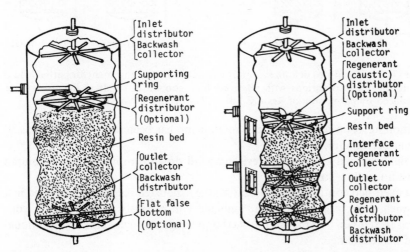

Figure 7.15 Anion or cation exchanger[18]

Figure 7.16 Mixed-bed deionizer[18]

Table 7.3   Design data for fixed-bed ion exchangers[15]

Type of resin	Flow velocity (m/s) Max.	Min.	Usable capacity $CaCO_3$ kg equivalents per m^3	Reagents
Weak-acid cation	0.0054	0.0007–0.0014	25–135	HCl or H_2SO_4 (110% of theoretical)
Strong-acid cation	0.005–0.008	0.0007–0.0014	40–75	175–530 kg $NaCl/m^3$
			25–45	70–420 kg H_2SO_4/m^3
			35–70	25–160 kg HCl/m^3
Weak-base cation	0.003–0.005	0.0007–0.0014	40–55	70–140 kg $NaOH/m^3$
Strong-base cations	0.003–0.005	0.0007–0.0014	18–37	140–280 kg $NaOH/m^3$
Mixed cation and strong-base anion				As for cation and anion individually

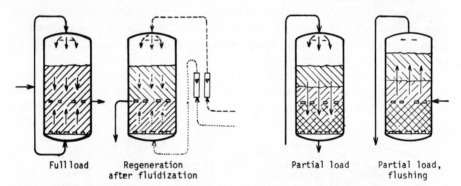

Figure 7.17   Mixed-bed deionizer for power station. The arrangement for partial load is used during start-up periods with relatively high content of solids in the condensate. Reproduced by permission of Bran & Lübbe, Norderstedt, West Germany

conditioning, resin replacement due to abrasion has been reported as high as 30 per cent per annum.

In fluidized-bed systems the resin may fall down through a baffled column against an upflow stream.

In mixed-bed continuous deionization the exhausted resins are separated in a backwash stage, regenerated, and remixed continuously. Water from a continuous mixed-bed unit may be polished by passage through a small conventional mixed-bed exchanger.

Calculations for continuous ion-exchange units may be carried out by the McCabe–Thiele method as shown in Example 7.3.

Example 7.1 BLEACHING

Table a gives the results of full-scale (50-ton charges) bleaching of neutralized and washed soya oil. The colours given are red units in the Lovibond colour system.

> Table a   Bleaching of neutralized soya oil with Fulmont C 725 bleaching clay. Temperature 94 °C, bleaching time 11–12 min, colour before bleaching 12.3 red units[20]

Clay (% by weight),	p	0.5	0.75	1.0	1.25	1.5
Colour (red) ± 0.2,	Y	5.0	4.25	3.7	3.4	3.3

Estimate per cent bleaching clay needed to reduce the colour to 3.5 red units:
(a) by a single-stage operation;
(b) in two stages with fresh clay added in each stage;
(c) in two countercurrent stages.

Solution

(a)

Table b   Recalculated data from Table a

Residual colour,	Y	5.0	4.25	3.7	3.4	3.3	Notes
Colour removed per % clay,	X	14.6	10.7	8.6	7.1	6.0	$X = (12.3 - Y)/p$

The data from Table b are plotted in the logarithmic diagram (Figure 7.18) where the length of the horizontal lines corresponds to the uncertainty of the reading (± 0.2 red units). The straight line corresponds to the equation

$$X = 0.706 Y^{1.88} \tag{a}$$

Equation (a) has the form of the Freundlich equation; $Y = 3.5$ inserted in equation (a) gives

$$X = 0.706 \times 3.5^{1.88} = 7.44 \text{ red units removed per per cent clay,}$$

giving

$(12.3 - 3.5)/7.44 = 1.18$ per cent bleaching clay.

(b) The minimum total amount of clay needed can be found by a trial-and-error procedure, selecting different values of the colour after the first stage as shown in Table c.

Table c   Two-stage cross-current operation

Colour after first stage, Y_1	5.75	6.0	6.25	6.5	Notes
Clay in first stage (%)	0.316	0.307	0.273	0.243	$(12.3 - Y_1)/(0.706 Y_1^{1.88})$
Clay in second stage (%)	0.302	0.336	0.370	0.403	$(Y_1 - 3.5)/(0.706 \times 3.5^{1.88})$
Total clay (%)	0.649	0.643	0.643	0.646	Sum both stages

Table c gives minimum consumption of clay with colour 6–6.25 after the first stage, 0.643 per cent by weight.

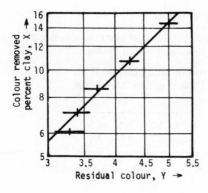

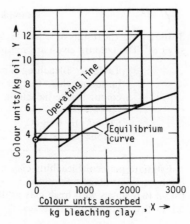

Figure 7.18 Adsorption isotherm for the soya oil

Figure 7.19 Equilibrium curve and operating line in Example 7.1

(c) With X given as colour units per kg clay used per kg oil, and Y per kg oil, equation (a) becomes

$$X = 70.6\, Y^{1.88} \tag{b}$$

The equilibrium curve given by this equation is plotted in Figure 7.19, together with the operating line. The latter is determined by trial, turning a ruler around the point $X = 0$, $Y = 3.5$, until it gives two theoretical stages as shown in Figure 7.19.

The slope of the operating line gives kg clay needed per kg oil.

$$\text{Slope} = \frac{12.3 - 3.5}{2280} = 0.0039$$

i.e. 0.39 per cent by weight bleaching clay

Note: The bleaching effect was influenced by factors such as the amount of residual soap after washing and the time between neutralization and bleaching. The data given in Table a were obtained shortly after neutralization and with residual soap less than 0.05 per cent.

Example 7.2 DRYING OF FLUE GAS

Flue gas from combustion is used for bright annealing of steel wire. The best results are obtained if the cleaned gas is dried to a dew-point of less than $-30\,°C$.

Measurements[21] with an activated alumina drier with 1.5–2.0 mm alumina particles with bulk density 765 kg/m³, bed depth 0.76 m, entering atmospheric air with moisture content 0.021 kg H_2O/m^3, temperature 24 °C, and flow rate 1.25 m³ air/(h kg alumina) indicate a short MTZ and dewpoint of exit air well below -30 °C after adsorption of 0.10 kg H_2O/kg alumina referred to the total charge.

(a) Based on these data, estimate the size of a drier to be operated for 8 h with 100 m³/h flue gas that enters the drier at atmospheric pressure and saturated with water vapour at 18 °C.

(b) The drier is reactivated by heating in 4 h with atmospheric air at 240 °C and cooled in another 4 h by internal finned cooling tubes. The heat capacity of alumina is 0.88, of steel 0.46 and of air 1.08 kJ/(kg °C). Assuming 65 kg steel (vessel and cooling tubes) to be heated together with the alumina, estimate the power input and the flow rate needed for regeneration in 4 h.

(c) By regeneration of the same kind of alumina with mass flow rate of air $G =$

0.144 kg/(m² s), bed temperature before regeneration 24 °C, and air temperature approximately 240 °C, reference 21 reports a temperature of 152 °C measured 280 mm from the entrance of the bed after 30 min. Compare this with what can be derived from Figure 7.11, assuming the highest value of h_s (150 000 W/(m³ bed °C)) for particles with diameter 1.75 mm.

Solution

(a) Equation (5.34) gives the vapour pressure

$$p_s = \exp\left[23.7093 - 4111/(237.7 + 18)\right] = 2063 \text{ Pa}$$

corresponding to

$$\rho = \frac{18 p_s}{RT} = \frac{18 \times 2063}{8314 \times 291} = 0.0153 \text{ kg H}_2\text{O/m}^3$$

The necessary amount of alumina,

$$0.0153 \times 100 \times 8/0.10 = 123 \text{ kg}$$

or

$$123/765 = 0.16 \text{ m}^3$$

$100/123 = 0.81$ m³/(kg alumina h) is well below the 1.25 given in reference 15 to give a short MTZ.
Keeping the bed depth the same as in the experimental work[21] gives column diameter

$$D = \left(\frac{4 \times 0.16}{0.76\pi}\right)^{1/2} = 0.52 \text{ m}$$

corresponding to superficial velocity

$$V = \frac{100}{3600\frac{\pi}{4}0.52^2} = 0.13 \text{ m/s}$$

which will give laminar flow and a very moderate pressure drop (equation (1.18) in reference 13).

(b) As an approximation the power input is estimated for heating from 18 to 240 °C, which also takes care of some loss to the surroundings. With heat capacity of water 4.19 kJ/(kg °C), of water vapour 1.88 kJ/(kg °C), and heat of vaporization at 0 °C 2501.6 kJ/kg (equation (5.37)), the energy consumption for vaporization and heating of H_2O is

$$0.10 \times 123[2501.6 + 1.88(240 - 18) - 4.19 \times 18] = 2844 \text{ kJ}$$

Heating alumina and steel,

$$(123 \times 0.88 + 65 \times 0.46)(240 - 18) = 30\,667 \text{ kJ}$$

Total energy required   33 511 kJ

$$\text{or over 4 h, } 33\,511/(4 \times 3600) = 2.33 \text{ kW}$$

Assuming average exit air temperature 30 °C gives

$$2.33/[1.08(240 - 30)] = 0.0103 \text{ kg air/s}$$

or

$$0.0103 \frac{8314 \times (273 + 240)}{29 \times 10^5} = 0.015 \text{ m}^3/\text{s}$$

(c) Equation (7.11) with $\varepsilon \approx 0.40$ and data for air at the average temperature $(240 + 18)/2 = 129\,°C$ from Appendix 5, gives,

$$\frac{0.00175h}{0.045} = 1.03\left(\frac{0.144 \times 0.00175}{2.3 \times 10^{-5}}\right)^{0.6} \exp(1.1-2.75 \times 0.40)$$

$$h = 11\,W/(m^2\,°C)$$

The surface area of the bed is

$$6(1 - \varepsilon)/D = 6 \times 0.60/0.001\,75 = 2060\ m^2/m^3\ bed,$$

giving

$$h_v = 111 \times 2060 = 229\,000\,W/(m^2\,°C)$$

Equation (7.10),

$$1/U_v = 1/229\,000 + 1/150\,000 = 1.1 \times 10^{-5}\ m^3\,°C/W$$

$$U_v = 90\,600\,W/(m^3\,°C)$$

$$\frac{U_v L}{c_{pf} G} = \frac{90\,600 \times 0.28}{1.08 \times 10^3 \times 0.144} = 163$$

The reported temperature corresponds to

$$\frac{\theta_g - \theta}{\theta_g - \theta_0} = \frac{240 - 152}{240 - 24} = 0.41$$

giving (Figure 7.11)

$$U_v t/q_s = 169$$

(or with $q_s \approx 33\,511/[0.16(240 - 18)] = 943\,kJ/(m^3\,°C)$

$$t = 169 \times 943\,000/90\,600 = 1760\,s\ or\ 29\ min$$

which happens to coincide completely with the experimental value.

Example 7.3   CONTINUOUS ION EXCHANGE COLUMN

A liquid with density $\rho = 1020\,kg/m^3$ and total ionic level 900 eq/m^3 passes continuously through the ion exchange column Figure 7.20 with volumetric flow rate 0.013 m^3/s, countercurrent to ion-exchange resin with total capacity 5.1 eq/kg dry resin. The ion exchange process is equimolal, and the selectivity coefficient in equation (7.16) is $K_{AB} = 2.5$. The regenerated resin fed to the column has ionic mole fraction $\bar{x}_{A0} = 0$ of ions A, and the liquid $x_{A1} = 1.0$. Assuming utilized 70 per cent of the ions A in the resin and ionic mole fraction $x_{A0} = 0.038$ in the liquid from the column, calculate:

(a) the flow rate of resin, kg dry resin/s;
(b) the number of theoretical stages.

Solution

(a) Exchanged ions from the liquid.

$$0.013 \times 900(1.0 - 0.038) = 11.3\ eq/s$$

which equals the ions taken up by $\bar{R}$ kg dry resin,

$$0.7 \times 5.1R = 11.3, \qquad R = 3.2\ kg\ dry\ resin/s.$$

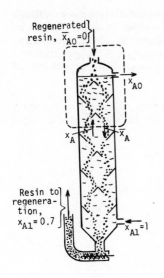

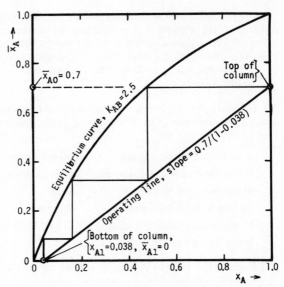

Figure 7.20 Baffled column for ion exchange

Figure 7.21 McCabe–Thiele diagram applied to ion exchange

(b) Equation (7.16) is rewritten as

$$\bar{x}_A = K_{AB}x_A/[1 + (K_{AB} - 1)x_A] \tag{a}$$

Different values of x_A and $K_{AB} = 2.5$ inserted in equation (a) give the equilibrium curve plotted in Figure 7.20.

The equation for the operating line is determined by a material balance as shown by the dotted cage in Figure 7.20,

$$5.1 \times 3.2\bar{x}_{A0} + 900 \times 0.013x_A = 5.1 \times 3.2\bar{x}_A + 900 \times 0.013x_{A0}$$

or, with $\bar{x}_{A0} = 0$ and $x_{A0} = 0.038$,

$$\bar{x}_A = 0.727x_A - 0.027 \tag{b}$$

This equation represents a straight line in Figure 7.21, with end points (0.038, 0) and (1.0, 0.70) at the bottom and at the top of the column.

Stepping off between the operating line and the equilibrium curve gives three theoretical stages.

Problems

7.1 Cartridges filled with 1.8 mm diameter spherical silica gel beads with bulk density 800 kg/m^3 have been used as desiccants in refrigeration systems. The inside diameter of the cartridges is 40 mm and the bed length 150 mm. Each cartridge with silica gel weighs approximately 0.55 kg. The heat capacity of the silica gel is 0.86, of the metal 0.46, and of air 1.08 kJ/(kg °C).

(a) The cartridges are heated from initial temperature 25 °C by air with temperature 182 °C and flow rate 2.0 kg air/h through each cartridge before it is sealed off and shipped. Estimate the heating time needed to give an exit air temperature of 175 °C, assuming void fraction $\varepsilon = 0.40$.

(b) Liquid refrigerant R 12 should contain less than 50 ppm water. Having one cartridge installed in the refrigerant loop, estimate the maximum amount of water that can be tolerated. The 'dry' silica gel contains 5 per cent by weight of water that must be retained to keep the adsorptive capacity. At 25 °C, equilibrium corresponds to approximately $X = 2.0Y$, where X is kg water/kg 'dry' silica gel and Y is kg water/m^3 R12. The density of R12 is 1400 kg/m^3.

7.2 Collins[22] gives the following laboratory test data for drying of nitrogen with type 4A molecular sieves at temperature 26 °C, operating pressure 5.93 bar absolute, and nitrogen rate 1.11 kg/(m^2 s):

Initial water content in gas,	1490 ppm by volume
Water content at breakthrough,	1 ppm by volume
Molecular sieve bed height,	0.44 m
Molecular sieve bulk density,	713 kg/m^3
Final water equilibrium loading,	0.215 kg H$_2$O/kg molecular sieve
Water breakthrough time,	15.0 h

Effluent water content in ppm by volume versus time in hours

Time	ppm	Time	ppm	Time	ppm	Time	ppm
13	<1	15.6	26	16.6	610	17.6	1355
14.5	<1	15.8	74	16.8	798	17.8	1432
15.0	1	16.0	145	17.0	978	18.0	1465
15.3	4	16.2	260	17.2	1125	18.3	1490
15.4	5	16.4	430	17.4	1245	18.5	1490

Estimate diameter and depth of bed required for drying of 200 kg/h nitrogen with the same molecular sieves operating under the same conditions, but with a moisture content of entering gas of 1300 ppm by volume and a breakthrough time of 24 h.

7.3 Using the data in Example 7.3, make up a program that gives the number of theoretical stages (including fraction of a stage) as a function of the mole fraction of ions x_{A0} in the liquid from the column. Insert $x_{A0} = 0.02$ and 0.05.

Symbols

A	Cross-sectional area, m^2
C	Constant
c	Concentration of ions, equivalents/m^3
c_{pf}	Specific heat capacity of fluid, J/(kg °C)
D	Particle diameter, m
G	Mass velocity, kg/(m^2 s)
h	Heat transfer coefficient, W/(m^2 °C)
h_s	Superficial heat transfer coefficient replacing heat conduction in solid particles, W/(m^2 °C)
h_v	Volumetric fluid phase heat transfer coefficient, W/(m^3 bed °C)
K_{AB}	Selectivity coefficient, equation (7.15)
k	Constant
k	Thermal conductivity, W/(m °C)
k	Rate constant, s^{-1}

L	Length (height) from inlet of bed, m
L_0	Length of bed, m
L_s	Length from inlet to stoichiometric front, m
L_{ub}	Length of unused bed, m
MTZ	Mass transfer zone, m
Nu	Nusselt number, hD/k
n	Constant
p	Partial pressure, $N/m^2 = Pa$
p_s	Saturation pressure, $N/m^2 = Pa$
q_s	Volumetric specific heat capacity of solids, $J/(m^3$ bed $°C)$
R	Mass of ion exchange resin, kg or kg/s
Re	Reynolds number, $\rho VD/\mu$
t	Time, s
t_b	Time to breakthrough, s
t_s	Time to stoichimetric front reaches end of bed, s
U_v	Volumetric overall heat transfer coefficient, $W/(m^3$ bed $°C)$
V	Velocity of fluid referred to total cross-section of bed, m/s
v	Volume of gas $(0\ °C, 760$ mm Hg) adsorbed per unit mass of adsorbent, m^3/kg
v_m	Constant
w	Mass adsorbate, kg
X	Mass fraction of adsorbate on adsorbent, kg/kg
X_e	Mass fraction adsorbate on adsorbent in equilibrium with fluid with mass fraction Y_e, kg/kg
X_0	Mass fraction adsorbate on adsorbent at time zero, kg/kg
x	Mole fraction of ions in fluid
$\bar{x}$	Mole fraction of ions in solid
Y	Mass fraction of adsorbate in solution, kg/kg
Y_e	Mass fraction adsorbate in entering fluid, kg/kg
Y_0	Mass fraction adsorbate in fluid in equilibrium with adsorbent with mass fraction X_0, kg/kg
ε	Void fraction
θ	Temperature, $°C$
θ_g	Temperature of entering gas, $°C$
θ_0	Initial bed temperature, $°C$
μ	Dynamic or absolute viscosity, Ns/m^2
ρ	Density of fluid, kg/m^3
ρ_b	Bulk density of adsorbent, kg/m^3

Subscripts

A	Component (ion) A
B	Component (ion) B
i	Component (ion) i

Superscript

    solid phase

References

1. Perry, R. H., and C. H. Chilton, *Chemical Engineers' Handbook*, 5th edn, McGraw-Hill, New York, 1973.

284

2. *Kirk-Othmer Encyclopedia of Chemical Technology*, Vol. 1, 3rd edn, John Wiley, New York, 1978.
3. Schweitzer, P. A. (ed.), *Handbook of Separation Techniques in Chemical Engineering*, McGraw-Hill, New York, 1979.
4. Moseman, M. H., and G. Bird, Desiccant dehydration of natural gasoline, *Chem. Eng. Progr.*, **78** (No. 2), 78–83 (1982).
5. Bernardin, F. E., Selecting and specifying activated-carbon-adsorption systems, *Chem. Eng.*, **83** (18 Oct.), 77–82 (1976).
6. Reed, T. B., and D. W. Breck, Crystalline zeolites. II. Crystal structure of synthetic zeolite, type A, *J. Am. Chem. Soc.*, **78**, 5972–5977 (1956).
7. Brunauer, S., L. S. Deming, W. E. Deming, and E. Teller, On a theory of the van der Waals adsorption of gases, *J. Am. Chem. Soc.*, **62**, 1723–1732 (1940).
8. Langmuir, I., The adsorption of gases on plane surfaces of glass, mica and platinum, *J. Am. Chem. Soc.*, **60**, 1361–1403 (1918).
9. Brunauer, S., P. H. Emmett, and E. Teller, Adsorption of gases in multimolecular layers, *J. Am. Chem. Soc.*, **60**, 309–319 (1938).
10. Linek, F., and M. P. Duduković, Representation of breakthrough curves for fixed-bed adsorbers and reactors using moments of impulse response, *Chem. Eng. J.*, **23**, 31–36 (1982).
11. Lukchis, G. M., Adsorption systems. Part I: Design by mass-transfer-zone concept, *Chem. Eng.*, **80** (11 June), 111–116; Part II: Equipment design, *Ibid.* (9 July), 83–87; Part III: Adsorbent regeneration, *Ibid.* (6 Aug.), 83–90 (1973)
12. Lewis, W. K., E. R. Gilliland, B. Chertow, and D. Bareis, Vapor-adsorbate equilibrium III, *J. Am. Chem. Soc.*, **72**, 1160–1163 (1950).
13. Lydersen, A. L., *Fluid Flow and Heat Transfer*, John Wiley & Sons, Chichester, 1979.
14. Henley, E. J., and J. D. Seader, *Equilibrium-Stage Separation Operations in Chemical Engineering*, John Wiley & Sons, New York, 1981.
15. Perry, R. H., and C. H. Chilton, *Chemical Engineers' Handbook*, 5th edn, McGraw-Hill, New York, 1973.
16. Applebaum, S. B., *Demineralization by Ion Exchange*, Academic Press, New York, 1968.
17. Gilliland, E. R., and R. F. Baddour, The rate of ion exchange, *Ind. Eng. Chem.*, **45**, 330–337 (1953).
18. Mead, W. J., *The Encyclopedia of Chemical Process Equipment*, Reinhold, New York, 1964.
19. Vollentsalzung und Kondensataufbereitung durch Ionenaustausch, Informationsheft W 2375, Bran & Lübbe, Norderstedt, West Germany, March 1975.
20. Fauli, R. O., M.Sc. Thesis, Norw. Inst. of Technology, Trondheim, Norway, 1963.
21. Carter, J. W., The industrial application of activated alumina to adsorption drying, *Proceedings Chem. Eng. Group, London*, **28**, 98–105 (1946).
22. Collins, J. J., The LUB/equilibrium section concept for fixed-bed adsorption, *Chem, Eng. Progr. Symp. Ser.*, **63** (74), 31–35 (1967).

CHAPTER 8

Crystallization

This chapter is only a brief introduction to the complicated process of formation of crystals of dissolved solids in solutions. Readers wanting introduction to the theory may start with Chapter 2.8 in Reference 1 and Section 2.4 in Reference 2. The books by Mullin,[3,4] de Jong,[5] and Smith[6] are recommended for further study, and Nývlt[7] gives an excellent presentation of the present state of the art.

Crystallization consists of crystal birth, or *nucleation*, and crystal growth. The driving force for both is *supersaturation*, obtained by vaporization, by cooling, or by adding a third component. The supersaturation may be expressed as the supersaturation ratio,

$$S = c/c_{eq} \geqslant 1.0, \tag{8.1}$$

relative supersaturation,

$$\sigma = S - 1 \geqslant 0, \tag{8.2}$$

or absolute supersaturation,

$$\Delta c = c - c_{eq} \tag{8.3}$$

where c = concentration, for instance kg solute/kg solvent

c_{eq} = concentration at equilibrium, in the same units as c.

Solutions vary in their ability to withstand supersaturation. Sugar solutions may have supersaturation ratios up to 2.0, while sodium chloride crystallizes with the same ratio close to 1.0.

Nucleation is promoted by the presence of small solid particles and by high shear rates in the liquid. The latter may result in crystal formation on propeller agitators. Under specific conditions it is possible to obtain supersolubility curves as shown in Figure 8.1.[8]

Homogeneous nucleation refers to nucleation in the bulk of the fluid phase without interference of solid–liquid interface.

Heterogeneous nucleation is caused by the presence of a foreign solid phase (e.g. dust or colloidal particles). An example is recovery of fluorine from an aluminium smelter by crystallization of aluminium fluoride on particles of aluminium oxide.[9]

Secondary nucleation is due to the presence of crystals of the crystallizing species itself. It is caused both by abrasion giving crystal fragments and also particles with size comparable with classical nuclei, and by nucleation in the boundary layer of the crystals.[7] Most industrial crystallizers are based on secondary nucleation.

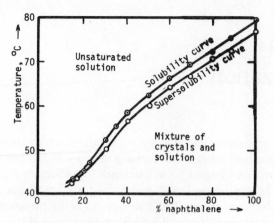

Figure 8.1   Miers' solubility–supersolubility
curves

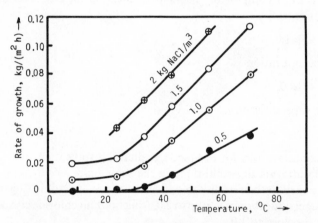

Figure 8.2   Rate of growth of sodium chloride crystals with
seed size 12–18 mesh (1.1 mm) placed in a tube with solution
velocity 0.03 m/s. Reproduced by permission of Institution of
Chemical Engineers

Crystal growth is a diffusional process. Solute molecules or ions diffuse to the
surface where they must be accepted by the crystal, and, as the next step, be
organized into the space lattice. The diffusional and the interfacial step are both
functions of the supersaturation and temperature. Figure 8.2 gives the growth
rate of sodium chloride crystals as a function of temperature with different
degrees of supersaturation.[10]

Crystal quality

Requirements to crystal quality, filterability, caking properties, etc. depend on
whether it is a final or intermediate product. Svanoe[11] defines crystal quality as:

Table 8.1 Classes of crystal sizes according to Nyvlt

Crystal class	Extremely large	Large	Fine	Very fine
Crystal size (mm)	$> 4 - 5$	$1 - 4$	$0.2 - 1$	< 0.2

1. Uniformity of the produced crystals.
2. Crystal purity.
3. Crystal hardness.

The desired size and size distribution of crystalline chemicals vary from $+ 10$ mesh (1.7 mm) and greater to $- 100$ mesh (0.15 mm). If the crystalline solid is a chemical intermediate, the important requirement will be economy in handling after the crystallization process, in steps such as filtering and washing.

There is a close interrelationship between supersaturation, crystal surface, amount of crystals in suspension, and size of crystals produced. For most inorganic products, the supersaturation has to be less than 1 per cent to give a reasonable crystal size. But even with close control, it is difficult to avoid variations. Product classification seems to be a major factor driving a stable system unstable.[12] Cycling in crystal size distribution (CSD) can be very regular, with peaks in crystal size hours apart.

Table 8.1 gives classes of crystal sizes.

Large crystals are produced in classifying crystallizers and crystallizers with low production rates. Fine crystals are produced at high production rates in most of agitated crystallizers. Batch crystallizers and batteries of crystallizers in series give larger crystals and more uniform crystal size than one single, continuous crystallizer. For large crystals, the change in crystallizer capacity with change in particle size may be estimated by the equation

$$\dot{m}_c = K/\bar{L} \tag{8.4}$$

where $\dot{m}_c =$ production rate, kg crystallized/(m^{3}s)

$L =$ mean crystal size, m, defined as the crystal size at which the oversize weight fraction is 65 per cent

$K =$ constant, often in the range $2.7 - 4.2 \times 10^{-5}$.

A well-formed crystal is nearly pure, but it retains mother liquor with its impurities. It is a question of liquor adhering to the surface, impurities attached to the crystal by adsorption, and mother liquor inclusions in the growing crystal. If the product is free of fines and the viscosity of the liquor not too high, the amount of liquor adhering to the surfaces can be reduced by centrifugation to 2 per cent, or in water solutions 1 per cent.[11]

Table 8.2 Example of impurities in crystals as a function of impurities in the mother liquor.[14]

	Impurity in crystal
Ratio impurity/solute in mother liquor 1/4	1/400
Ratio impurity/solute in mother liquor 1/14	1/3000

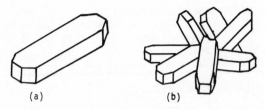

Figure 8.3 $(NH_4)_2SO_4$ crystals. Crystal (a) was grown at 0.5 per cent and (b) at 7 per cent supersaturation[5] (magnification 12 ×)

The amount of impurities due to adsorption can not be correlated on a theoretical basis,[13] but Table 8.2 may give an idea of orders of magnitude.

Even minor amount of impurities can have a marked effect in repressing growth on certain crystal faces. This is sometimes utilized to influence the crystal form, for instance by changing the crystal from long needles to shorter crystals resembling cubes. Reference 7 lists additives affecting the shape of 69 different crystals.

Higher ranges of supersaturation increase the tendency to obtain liquor inclusions in the growing crystals, in layers of crystallites, and between crystal aggregates, which usually leads to caking during storage.

The term 'crystal hardness' is used to indicate the resistance to abrasion or scratching. Crystals grown at very high supersaturation have a weak structure, often with spikes and aggregates which break up in centrifuging and drying operations, resulting in dust losses, caking problems, and change in bulk weight of the material. Figure 8.3 gives an approximate picture of ammonium sulphate crystals grown in solutions with 0.5 and 7 per cent supersaturation.

The acceptable degree of supersaturation must be determined experimentally. Additives are sometimes used for chemicals such as ammonium sulphate to stabilize or increase the allowable degree of supersaturation. In such cases, the crystal hardness will suffer slightly due to the increase in growth rate.

Crystal composition

Some solutes crystallize in a pure state, while others crystallize as hydrates. As an example, ammonium sulphate (Figure 8.4) has no water of crystallization, while magnesium sulphate has various hydrates depending on the concentration of the mother liquor, as shown in the equilibrium diagram Figure 8.5. In practical crystallization processes, however, the crystallized product often consists of a mixture of different hydrates.[17]

In solutions with two or more solutes, separation of the solutes may be obtained by *fractional crystallization*. The separation of KCl and NaCl from water solutions is an example of such a process used in large-scale operations. Figure 8.6 shows the solubility of the two salts at different temperatures. The heavy line A–B represents the mixed salt, and the crystallization of NaCl must be

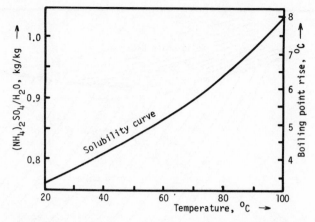

Figure 8.4 Solubility and boiling-point rise of saturated solutions of ammonium sulphate in water as a function of temperature[15]

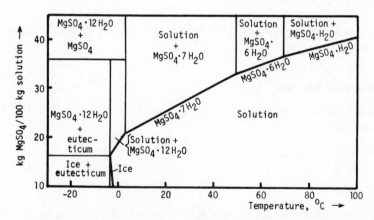

Figure 8.5 Phase diagram for magnesium sulphate in water[16]

above and of KCl below this line. Normally, it is possible to operate in a 95–98 per cent approach to the mixed salt line. Removing water by vaporization corresponds to a change in concentrations following a straight line through the origin. Dilution by adding water is represented by movement towards the origin.

An interesting separation process is shown with dotted lines in Figure 8.6, starting with saturated solution C at the normal boiling point 111.9°C. A small amount of water is added to dissolve traces of solid and give pure KCl by crystallization from point D. Evaporative cooling along the line D–E gives precipitation of KCl. During this cooling, some of the evaporated water has to be added in order to be well below the mixed salt line with coprecipitation of NaCl. The final solution at E may be concentrated by heating and removing water to get

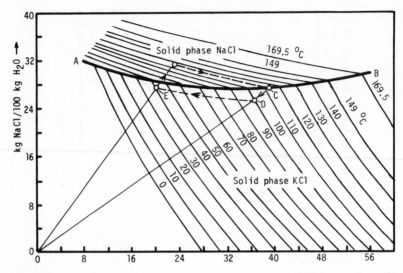

Figure 8.6   Phase diagram for KCl–NaCl–H$_2$O. Crystallization at the solid line A–B gives mixed salt[15]

well above the mixed salt line, followed by vaporization, giving precipitation of NaCl and ending up at point C.

Crystallization kinetics

Secondary nucleation and linear crystal growth rate may be described by the relationships

$$B^0 \sim \Delta c^g \tag{8.5}$$

and

$$G \sim \Delta c^u \tag{8.6}$$

where B^0 = nucleation rate, no./(m^{3}s), Δc = supersaturation, kg/m^3, g = kinetic order for nucleation, often between 3 and 6, $G = dL/dt$ = linear growth rate, m/s, in most cases between 2×10^{-8} and 2×10^{-7}, and u = kinetic order for linear growth, often between 1 and 2.

Combining the two relationships gives

$$B^0 \sim G^i \tag{8.7}$$

where $i = g/u$ = relative kinetic order, i.e. the ratio of the nucleation and growth orders.

This relationship is included in the commonly used semiempirical equation for continuous mixed suspension mixed product removal (MSMPR) crystallizer,[18]

$$B^0 = K_R M^j G^i \tag{8.8}$$

where M = magma density, kg crystals/m^3 magma, and K_R = rate constant,

Table 8.3 Published crystallization kinetics from MSMPR crystallizers[18,19].

Crystals in aqueous solution	Temp. (°C)	Time, (min.)	M, (kg/m³)	$G \times 10^8$ (m/s) $\times 10^8$	K_R	j	i
KCl	12–30	15–45	14–25	11.7–20.7	2.86×10^{21}	0	2.55
	32	—	60–250	2–12	7.12×10^{39}	0.14	4.99
KNO₃	21	15–45	25–87	4.8–13.3	1.66×10^{13}	1	1.38
	11–40	15–45	25–45	4.3–15.7	1.34×10^{14}	0	1.3
	20	14–29	10–40	7.7–13.2	3.85×10^{16}	0.5	2.06
	20	—	~10	3.7–12.6	1.18×10^{14}	1	1.76
KCr₂O₇	—	15–60	14–42	1.2–9.1	7.33×10^{4}	0.6	0.5
K₂SO₄	30	19–88	23–108	1.2–9.5	1.67×10^{3}	1	0
MgSO₄·7H₂O	25	—	—	2.6–7.1	9.65×10^{12}	0.67	1.24
NaCl	~50	10–54	25–200	3.5–13	$1.98 \times 10^{10} N^{2†}$	1	2
(NH₄)₂SO₄	18	8–20	38	15–23	2.94×10^{10}	0	1.03
	15	9–32	51–61	9.7–21	2.32×10^{13}	0	1.5
Citric acid ·1 H₂O	20	—	—	1.1–3.7	1.09×10^{10}	0.84	0.84

† N = revolutions of propeller, s⁻¹.

depending on temperature, degree of agitation, and the presence of impurities.

Published kinetic data have been collected and tabulated by Garside and Shah[18] and by Mersmann et al.,[19] and some of their data are given in Table 8.3.

Larson and Garside[19] derive equations for the design of both batch and continuous crystallizers, based on the gamma type distribution for crystal size distribution,[20] the same distribution in all spatial locations in the crystallizer, and classification only at the discharge. This gives the medium size, i.e. the size below which half of the mass exists, as

$$L_M = 3.67\, G\tau \tag{8.9}$$

where τ = mean retention time, s.

The following equation was derived for continuous, self-nucleating MSMPR crystallizers:[19]

$$P_c = 0.066\, 4k_v \rho_c B^0 L_M^4 / G \tag{8.10}$$

where P_c = production, mass of crystals per unit volume of crystallizer output magma, kg/m^3, k_v = volumetric shape factor, ratio of the volume of a crystal to the volume of a cube with sides L, $k_v = 1.0$ for cubes, and ρ_c = crystal density, kg/m^3.

Combining equations (8.8) and (8.10) for output magma identical to crystallizer magma, i.e. $P_c = M$, and $i \neq 1$, gives,

$$G = \left[\frac{15\, M^{1-j}}{K_R k_v \rho_c L_M^4} \right]^{1/(1-i)} \tag{8.11}$$

or

$$L_{,M} = \left[\frac{15}{K_R k_v \rho_c M^j G^{i-1}} \right]^{1/4} \tag{8.12}$$

An increase in crystal size is often obtained by removal or destruction of fines and by some classification at the outlet. Removal of fines may be taken care of by introduction of an effective nuclei population density,[19]

$$B_{eff}^0 = K_R M^j G^i \exp(- L_c / G\tau_f) \tag{8.13}$$

where L_c = 'cut' size, i.e. the largest crystal size destroyed, for instance 10^{-5} m, and τ_f = mean retention time of fines, s.

Combining equations (8.10) and (8.13) gives,

$$G = \left[\frac{15\, M^{1-j}}{K_R k_v \rho_c L_M^4 \exp(- L_c / G\tau_f)} \right]^{1/(i-1)} \tag{8.14}$$

or

$$L_M = \left[\frac{15}{K_R k_v \rho_c M^j G^{i-1} \exp(- L_c / G\tau_f)} \right]^{1/4} \tag{8.15}$$

The mass fraction with size $\leq L_c$ is normally negligible.

For a seeded batch crystallizer programmed to give constant linear growth rate, the time for crystallization of one batch is

$$\tau = (L_F - L_s)/G \tag{8.16}$$

where L_F = final crystal size, m, and L_s = seed size, m.

Enthalpy balances

Heat balances for crystallizers include the enthalpy of feed and products, where the latent heat of crystallization is an important part. It can be calculated from data on heats of solution and of dilution which are available,[16,21] together with data on the specific heats of the solutions and the crystals (see Example 8.1). For repeated calculations it is convenient to construct enthalpy–concentration diagrams based on the same data.[1]

Crystallization equipment

To obtain good quality crystals, the crystallizer should be based on low supersaturation, often less than 1 per cent and with crystal growth times in the range from 1 to 3 h, and with good control of nucleation. The type of crystallizer depends largely on the type of solubility, as indicated in Figure 8.7 and Table 8.4.

Batch and continuous crystallizers

The majority of crystallizers in industry is operated batch-wise.[22] In addition comes semi-continuous crystallizers with batch-wise or periodic product discharge.

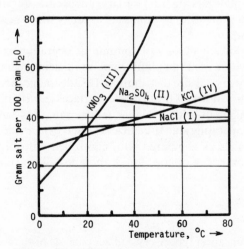

Figure 8.7 Effect of temperature on solubility in water[15]

Table 8.4 Guide for selection of type of crystallizer dependent upon the solubility characteristics of the solute[15]

Type solubility (Fig. 8.4)	Effect of temperature rise on solubility in water	Equipment to be used	Example
I	Small increase in solubility	Evaporator–crystallizer (salting-out evaporator)	Sodium chloride
II	Decrease in solubility (inverted solubility)	Evaporator-crystallizer	Anhydrous sodium sulphate, gypsum, iron sulphate monohydrate
III	Substantial increase in solubility	Vacuum crystallizer or water- or brine-cooled crystallizer	Potassium alum, glauber salt, copperas, potassium nitrate
IV	Moderate increase in solubility	Vacuum or evaporator–crystallizer	Potassium carbonate. sodium nitrate, potassium chloride

Advantages of batch crystallization are simple equipment, low requirements for maintenance and skill of operators, incrustations usually disappear when a fresh batch is started up, and automatic process control enables larger crystals to be produced. In addition, scale-up does not involve any substantial risks. Drawbacks are higher demands for operators' time, space requirements and intermediate storage. Also, the product quality may vary from batch to batch.

Continuous crystallizers are used for larger capacities. Nývlt[7] recommends 40 to 200 kg/h production rate to be minimum capacity where continuous operation may be considered.

Among the major advantages of continuous operation are lower operating costs, better utilization of mother liquors, more efficient separation and washing, and more constant crystal size distribution. Disadvantages are a tendency to the formation of incrustations on heat transfer surfaces as well as the solution level, sophisticated equipment with a higher failure rate, and substantially higher requirements for operating staff qualifications and experience, as well as smaller mean size of crystals as compared with controlled batch crystallization. The period of operation between stops for cleaning usually varies between 200 and 2000 hours.[7]

Tank crystallizers

The simplest and cheapest type of crystallizer consists of an open tank, which can be used either as an evaporative or as a cooling crystallizer. The stirred-tank crystallizer is mainly used for smaller capacities of products such as fine chemicals, pharmaceuticals, food products, and systems in which the liquor and not the crystals is the desired product. For agitated crystallizers Singh[2] gives the overall heat transfer coefficient typically in the range from 100 to 1000 W/(m² °C).

Scraped-surface crystallizers

High viscosity solutions can be handled in scraped-surface crystallizers where the supersaturation is generated by cooling. An example is the Swenson–Walker crystallizer. It consists of an open, round-bottomed trough 600 mm wide and 3–12 m long, jacketed for circulation of coolant, and with a ribbon agitator that scrapes the crystals from the cold walls and give them a forward motion. The heat transfer surface is 1.1 m² per m length and heat transfer coefficients are in the range from 55 to 140 W/(m² °C).

Continuous evaporator–crystallizers

Figures 8.8, 8.9, and 8.10 are examples of continuous evaporator–crystallizers. Figure 8.8 is a draft–tube–baffle (DTB) crystallizer where the supersaturation is obtained by vaporization from the liquor heated by the steam heater. A propeller in the internal draft tube gives good mixing of liquid and crystals, and the volume of settled crystals may be as high as 30–50 per cent of the total magma. Excess fines are withdrawn from the settling zone with the 'clear' recycle liquor that becomes unsaturated in the steam heater. The product slurry passes on to a filter or a centrifuge, and the mother liquor is recycled to the elutration leg. A fraction of this mother liquor may be bled off to keep the amount of impurities at an

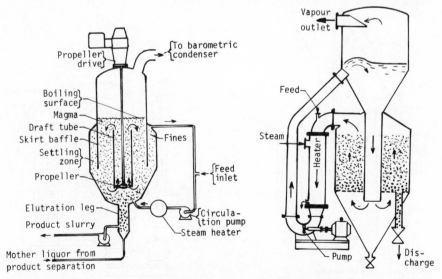

Figure 8.8 Draft–tube–baffle crystallizer with internal separation and removal of fines[23]

Figure 8.9 Krystal evaporator–crystallizer with growing crystals in fluidized bed[14]

acceptable level. In vacuum crystallization the vapour passes on to a barometric condenser, and in multistage crystallization to the next stage.

Figures 8.9 and 8.10 show two types of the Oslo or Krystal crystallizer, first developed by Jeremiassen in Norway during the First World War for the purpose of recovery of salt from sea-water. Today more than 100 materials are produced commercially in Krystal crystallizers, ranging from high-tonnage heavy chemicals such as ammonium sulphate, potassium chloride, and sodium chloride, to smaller quantities of speciality items such as silver nitrate and vanillin.

Feed liquor is mixed continuously with a large amount of circulating mother liquor supersaturated by vaporization. The supersaturated solution thus produced flows down the central pipe and upwards through the dense suspension of crystals in a suspension container with polished surfaces. The large fluidized surface keeps supersaturation in all parts of the crystallizer well below the limits required for spontaneous nucleus formation, resulting in favourable growth conditions and longer periods of operation between clean-outs.

The volume of the crystals in the fluidized bed may be in the range 15–35 per cent of the volume of the fluidized bed. A vapour velocity above the boiling surface giving moderate entrainment may be estimated by the equation

$$V = C \sqrt{\frac{\rho_1 - \rho}{\rho}} \tag{8.17}$$

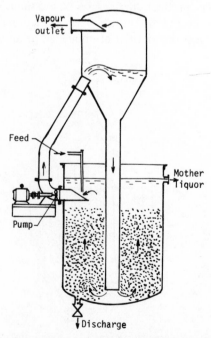

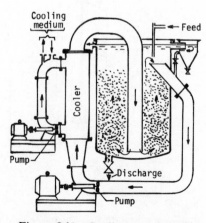

Figure 8.10   Vacuum crystallizer with open crystallizer tank[14]

Figure 8.11   Cooling crystallizer[14]

where $\rho_1 =$ liquid density, kg/m^3, $\rho =$ density of the vapour, kg/m^3, and $C =$ constant, $0.017 + 0.013\,p$ with p in bar for $p < 1.2$ bar, and 0.033 for $p > 1.2$ bar.

In the vacuum crystallizer (Figure 8.10) the crystallizer container is an open tank connected with the vaporizer through the barometric leg. The feed is mixed with the circulating mother liquor, and the sensible heat of the feed and heat of crystallization are utilized to vaporize water and cool the solution to the required temperature. If it is desired to vaporize more water, a tubular heater can be installed in the circulating line from the pump to the vacuum chamber.

Continuous cooling crystallizers

Figure 8.11 is an example of the general arrangement of a cooling crystallizer with a vertically arranged shell and tube heat exchanger. Water, brine, or some other cooling liquid can be used as cooling medium that is circulated by an extra pump to give good temperature control.

Example 8.1   FRACTIONAL CRYSTALLIZATION

Potassium chloride is to be produced by batch-wise vacuum crystallization from a brine with the composition and temperature given by point C in Figure 8.6 (temperature $113\,°C$, and 0.401 kg KCl and 0.277 kg NaCl per kg water). As a first step, water at a temperature of $30\,°C$ is added to give the composition in point D (0.369 kg KCl and 0.255 kg NaCl per kg water). Cooling is obtained by vaporization under vacuum and adding water (temperature $30\,°C$) to keep the concentration of NaCl at or below 97 per cent of the concentration that gives mixed salt. The process is continued until the boiling temperature is reduced to $40\,°C$.

Reference 21 gives the heat of a solution of KCl as 253 kJ/kg, and Reference 24 gives heat capacity data for the KCl-H_2O and NaCl-H_2O binary systems, indicating that the heat capacity of the brine in question is close to the heat capacity of the water in the brine.
(a) Calculate the temperature in point D in Figure 8.6 if no water is evaporated.
(b) Calculate the amount of water added and the amount evaporated to give $100\,°C$ in point D.
(c) Calculate the amount of crystals formed, amount of water evaporated, and amount of cold water added as the temperature is reduced from $100\,°C$ to $40\,°C$, following line D–E in Figure 8.6

Solution

(a) Basis: 1 kg KCl in the original brine, i.e. original brine with $1/0.401 = 2.494$ kg water, 1 kg KCl, and $0.277 \times 2.494 = 0.691$ kg NaCl.
Water added to give 0.369 kg KCl /kg water,

$$1/0.369 - 2.494 = 0.216 \text{ kg}$$

giving

$$0.691/(2.494 + 0.216) = 0.255 \text{ kg NaCl/kg } H_2O$$

and temperature $\theta_D\,°C$,

$$0.216 \times 4.19\,(\theta_D - 30) = 2.494 \times 4.19(113 - \theta_D)$$
$$\theta_D = 106.4°C$$

(b) Here, w_1 kg water is evaporated and w_2 kg added.

$$w_1 - w_2 = 0.216 \text{ kg}$$

An enthalpy balance gives

$$2.494 \times 4.19(113 - 100) = 4.19(100 - 30)(w_1 + 0.216) + (\Delta h_v)w_1$$

where $(\Delta h_v) =$ heat of vaporization, (equation (5.9)), $(\Delta h_v) = 2501.6 - 2.275 \times 100 - 0.0018 \times 100^2 = 2256$ kJ/kg, and $w_1 = 0.028$ kg water evaporated, and

$$w_2 = 0.216 + 0.028 = 0.244 \text{ kg water added}$$

(c) Table a gives compositions read at intersections between the dotted line D–E and the isotherms in Figure 8.6, together with calculations. Vaporized water w_1 is given by the enthalpy balance,

$$w_1(\Delta h_v)_\theta + (w_1 - w_3)4.19(\theta - 30) = Q_3 \tag{a}$$

where (Δh_v) is given by equation (5.9), w_3 is removed water, and Q_3 is removed heat.

Table a

Temperature, θ (°C)	100	90	80	70	60	50	40	Notes
$a \left(\dfrac{\text{kg NaCl}}{\text{kg H}_2\text{O}} \right)$	0.255	0.256	0.258	0.262	0.266	0.271	0.276	Figure 8.6
$b \left(\dfrac{\text{kg KCl}}{\text{kg H}_2\text{O}} \right)$	0.369	0.339	0.310	0.280	0.253	0.226	0.203	Figure 8.6
c (kg H$_2$O)	2.710	2.699	2.678	2.637	2.598	2.550	2.504	$c = 0.691/a$
d (kg KCl)	1.000	0.915	0.830	0.738	0.657	0.576	0.508	$d = bc$
Crystallized, e (kg KCl)	0	0.085	0.170	0.262	0.343	0.424	0.492	$e = 1 - d$
Removed water, w_3 (kg H$_2$O)	0	0.011	0.032	0.073	0.112	0.160	0.206	$w_3 = 2.710 - c$
Heat of crystallization, Q_1 (kJ)	0	21	43	66	87	107	124	$Q_1 = 253e$
Removed sensible heat, Q_2 (kJ)	0	113	224	331	435	534	630	$Q_2 = 4.19c(100 - \theta)$
Total heat, Q_3 (kJ)	0	125	267	397	522	641	754	$Q_3 = Q_1 + Q_2$
Vaporized, w_1 (kg H$_2$O)		0.050	0.109	0.164	0.216	0.265	0.311	eq.(a)
Added, w_2 (kg H$_2$O)		0.039	0.077	0.091	0.104	0.105	0.105	$w_2 = w_1 - w_3$

Data from the table are plotted in Figure 8.12, showing that most of the cold water is added in the beginning of the process, and nothing after the temperature has passed 60 °C.

Example 8.2 BATCH EVAPORATIVE CRYSTALLIZER

A stirred batch evaporative crystallizer for sodium chloride is to be operated at 50 °C under the following specifications:

Product crystal size, 10^{-3} m
Product per batch,   900 kg

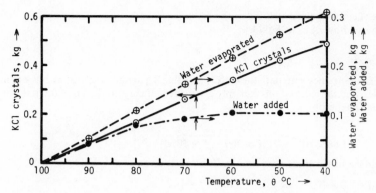

Figure 8.12   The crystallization process from point D to point E in Figure 8.6, as a function of the falling temperature. The numbers are all per kg KCl in the original brine

Crystals in discharged magma,	225 kg/m³ magma
Density of crystals	$\rho_c = 2165\,\text{kg/m}^3$
Solubility of NaCl at 50 °C,	37 kg/100 kg water
Density of saturated solution at 50 °C,	1200 kg/m³
Allowable growth rate,	$G = 8 \times 10^{-8}$ m/s
Seed size,	$L_s = 10^{-4}$ m

Determine
(a) crystallizer volume, and
(b) crystallization time per batch.

Solution

(a) Volume of final suspension, $900/225 = 4.0\,\text{m}^3$,
    containing $(4.0 - 900/2165)1200 = 4301$ kg liquid
    Evaporated water, $900/0.37 = 2432$ kg
Liquid volume when crystallization starts,
    $(900 + 4301 + 2432)/1200 = 6.36\,\text{m}^3$
(b) Equation (8.16) gives the time for crystallization,

$$\tau = (10^{-3} - 10^{-4})/(8 \times 10^{-8}) = 11\,250 \text{ seconds}$$
$$\text{or } 3.13 \text{ hours}$$

Example 8.3   CONTINUOUS COOLING CRYSTALLIZER

Potassium nitrate is to be produced continuously at a rate of 600 kg crystals/h from a brine containing 37 kg KNO_3 per 100 kg water in the cooling crystallizer Figure 8.11 operating at temperature 5 °C, equilibrium concentration 17 kg KNO_3 per 100 kg water and liquid density 1110 kg/m³. The following specifications are given:

Acceptable growth rate,	$G = 5 \times 10^{-8}$ m/s
Crystal density,	$\rho_c = 2110\,\text{kg/m}^3$
Magma density,	$M = 200$ kg crystals/m³ magma
Volume shape factor,	$k_v \approx 1$

In lack of other data, use the data in Table 8.3 to estimate the order of magnitude of
(a) medium crystal size L_M,
(b) mean retention time, τ,

(c) the effective magma volume in the crystallizer, and
(d) the overflow of clear liquid.

Solution

(a) The four lines with data for KNO_3 in Table 8.3 are all obtained with lower magma densities, and the calculations can only give orders of magnitude. The data in the first line inserted in equation (8.12) give

$$L_M = \left[\frac{15}{1.66 \times 10^{13} \times 2110 \times 200 \times (5 \times 10^{-8})^{0.38}} \right]^{1/4} = 1.9 \times 10^{-4}\,m$$

The data in the three next lines in Table (8.3) give

$$L_M = 3.0 \times 10^{-4},\ 9.2 \times 10^{-4},\ \text{and}\ 5.7 \times 10^{-4}\,m,$$

giving the average

$$L_M = \frac{(1.9 + 3.0 + 9.2 + 5.7)10^{-4}}{4} = 5 \times 10^{-4}\,m$$

(b) Equation (8.9) gives the mean retention time,

$$\tau = \frac{5 \times 10^{-4}}{3.67 \times 5 \times 10^{-8}} = 2725\ \text{s or } 0.76\,h.$$

(c) The discharge flow rate is

$$Q = 600/200 = 3\,m^3\,\text{magma/h}$$

and the effective magma volume in the crystallizer,

$$V_{eff} = 3 \times 0.76 = 2.3\,m^3$$

(d) The feed contains

$$600/(0.37 - 0.17) = 3000\,\text{kg water/h}$$

The discharge magma contains

$$(3 - 600/2110)1110 = 3014\,\text{kg liquid/h}$$

with

$$3014 \times 100/(100 + 17) = 2576\,\text{kg water/h}$$

Discharged in overflow,

$$3000 - 2576 = 424\,\text{kg water/h}$$

or

$$424(100 + 17)/100 = 496\,\text{kg liquid/h}$$
$$Q_v = 496/1110 = 0.45\,m^3/h$$

Problems

8.1 A crystallizer contains saturated brine of the composition given by point E Figure 8.6, i.e. 20.3 kg $KCl/100$ kg H_2O, 26.6 kg $NaCl/100$ kg H_2O, and temperature 40 °C. Based on data given in Example 8.1, calculate heat added or removed and water added or removed per 100 kg water in point E to change the composition to 22.9 kg $KCl/100$ kg H_2O and 31.2 kg $NaCl/100$ kg H_2O, and temperature 113 °C without any crystallization.

8.2. Calculate heat added or removed and water added or removed when the composition of the brine in Problem 8.1 (22.9 kg KCl and 31.2 kg NaCl per 100 kg water) is changed along the 113 °C isotherm until point C is reached with 40.1 kg KCl and 27.7 NaCl per 100 kg water. Heat of crystallization to be removed is 83.4 kJ/kg NaCl crystals.

Symbols

B^0	Nucleation rate, no. nuclei/(m^3 s)
c	Concentration, kg/kg or kg/m^3
c_{eq}	Concentration at equilibrium
G	Linear growth rate, dL/dt, m/s
g	Kinetic order for nucleation
i	Relative kinetic order, g/u
j	Exponent for magma density in nucleation rate equation
K	Constant
K_R	Rate constant in nucleation rate equation
k_v	Volumetric shape factor, ratio of volume of a crystal to the volume of a cube with sides L. $k_v = 1.0$ for cubes.
$\bar{L}$	Mean crystal size, m
L_c	Cut size, i.e. the largest crystal size destroyed, m
L_F	Final crystal size, m
L_M	Medium crystal size, i.e. the size below which half of the mass exists, m
L_s	Seed size, m
M	Magma density, kg crystals/kg magma
$\dot{m}_c$	Production rate, kg crystallized/(m^3 s)
P_c	Production, kg crystals per m^3 crystallizer output magma
S	Supersaturation ratio
u	Kinetic order of linear growth
ρ_c	Crystal density, kg/m^3
σ	Relative supersaturation
τ	Mean retention time, s
τ_f	Mean retention time of fines, s

References

1. McCabe, W. L., and J. C. Smith, *Unit Operations of Chemical Engineering*, 3rd edn, McGraw-Hill, New York, 1976.
2. Singh, G., Crystallization from solutions, Section 2.4 in P. A. Schweitzer (Ed.), *Handbook of Separation Techniques for Chemical Engineers*, McGraw-Hill, New York, 1979.
3. Mullin, J. W., *Crystallization*, 2nd edn, Butterworth, London, 1972.
4. Mullin, J. W. (Ed.), *Industrial Crystallization*, Plenum Press, New York and London, 1976.
5. de Jong, E. J., and S. J. Jančić (eds), *Industrial Crystallization*, North-Holland, Amsterdam, 1979.
6. Smith, A. L. (Ed.), *Particle Growth in Suspensions*, Academic Press, London, 1973.
7. Nývlt, J., *Industrial Crystallization*, Verlag Chemie, Weinheim and New York, 1978.
8. Miers, H. A., The growth of crystals in supersaturated liquids, *J. Inst. Metals*, **37**, 331–350 (1927).
9. Nordheim, E., Fluoride Recovery Plant at Mosjøen Aluminiumverk, Symp. on the Fluoride Problem in the Primary Aluminium Industry, Trondheim, Norway, 24–26 May 1972.
10. Rumford, F., and J. Bain, The controlled crystallizatiøn of sodium chloride, *Trans. Instn Chem. Engrs.*, **38**, 10–20 (1960).
11. Svanoe, H., Solids recovery by Crystallization, *Chem. Eng. Progr.*, **55** (May), 47–54 (1959).

302

12. Beckman, J. R., Experimental and theoretical study of KCl crystal-size distribution dynamics and control in a classified crystallizer with fines destruction, Ph.D. thesis, University of Arizona, 1976.
13. Buckley, H. E., *Crystal Growth*, John Wiley, New York, 1951.
14. Svanoe, H., 'Krystal' classifying crystallizer, *Ind. Eng. Chem.*, **32**, 636–639 (1940).
15. *Krystal Crystallizer*, Struthers Scientific & International Corp., Warren, Pa, Bulletin K-70.
16. Perry, R. H. and C. H. Chilton, *Chemical Engineers' Handbook*, 5th edn, McGraw-Hill, New York, 1973.
17. Cross, H. E., W. Krieger, A. Anschutz, L. Reh, and M. Hirsch, Process of producing magnesia with sulfuric acid recycle, U.S. Pat. 4,096,235 (1978).
18. Garside, J., and M. B. Shah, Crystallization kinetics from MSMPR crystallizers, *Ind. Eng. Chem. Progress Des. Dev.*, **19**, 509–514, (1980).
19. Larson, M. A., and J. Garside, Crystallizer design techniques using the population balance, *Chem. Eng.* (London), 318–328 (1973).
20. Randolph, A. D., and M. C. Larson, *Theory of Particulate Processes*, Academic Press, New York, 1971.
21. Bichowski, F. R. and F. D. Rossini, *Thermochemistry of Chemical Substances*, Reinhold, New York, 1936.
22. deJong, E. J., Entwicklung von Kristallisatoren, *Chem. Ing. Techn.*, **54**, 193–202 (1982).
23. *Swenson Crystallizing Equipment*, Bulletin 5 M779 APJ, Whiting Corp., Harvey, Illinois.
24. *Landolt-Börnstein Numerical Data and Functional Relationships in Science and Technology*, K.-H. Hellewege (Ed.), New Series, Group IV, Vol. 1, Part b, Springer-Verlag, Berlin, 1977.

APPENDIX 1

Experimental values of mutual diffusion coefficients for some binary gas mixtures at low pressures, PD_{AB} Pa m^2/s = (N/m^2)(m^2/s)

System	T (K)	PD_{AB} (Pa(m^2/s))	System	T (K)	PD_{AB} (Pa(m^2/s))
Acetone–hydrogen	296	4.30	Carbon dioxide–water	307.2	2.01
Air–carbon dioxide	276.2	1.44		328.6	2.60
	317.2	1.79		352.3	2.48
Air–ethanol	313	1.47	Carbon tetrachloride		
Air–helium	317.2	7.75	-oxygen	296	0.759
Air–n-hexane	294	0.81	Cylohexane–hydrogen	288.6	3.23
	328	0.94	Cyclohexane–nitrogen	288.6	0.741
Air–n-heptane	294	0.72	Cyclohexane–oxygen	288.6	0.756
Air–methanol	273	1.34	Ethanol–helium	423	8.32
Air–water	313	2.92	Helium–n-hexanol	423	4.752
Ammonia–diethylether	288.3	1.01	Helium–methane	298	6.84
	337.5	1.39	Helium–methanol	423	10.46
Ammonia–hydrogen	263	5.78	Helium–neon	242.2	8.02
	358	11.1		341.2	14.25
	473	18.8	Helium–oxygen	298	7.39
Ammonia–nitrogen	298	2.33	Helium–n-pentanol	423	5.14
Argon–carbon dioxide	276.2	1.35	Helium–isopropanol	423	6.86
Argon–helium	276.2	6.55	helium–n-propanol	423	6.85
Argon–hydrogen	295.4	8.4	Helium–water	307.1	9.14
Argon–krypton	273	1.21	Hydrogen–methane	288	7.03
Argon–methane	298	2.05	Hydrogen–nitrogen	298	7.94
Argon–neon	273	2.80		358	10.66

Appendix 1 (*Contd.*)

System	T (K)	PD_{AB} (Pa(m²/s))	System	T (K)	PD_{AB} (Pa(m²/s))
Argon–xenon	329.9	1.39	Hydrogen–pyridine	318	4.43
Benzene–helium	423	6.18	Hydrogen–sulphur dioxide	285.5	5.32
Benzene–hydrogen	311.3	4.09	Hydrogen–thiophene	302	4.05
Benzene–nitrogen	311.3	1.03	Hydrogen–water	307.1	9.27
Benzene–oxygen	311.3	1.02	Hydrogen–xenon	341.2	7.61
n-Butanol–helium	423	5.95	Krypton–neon	273	2.26
Carbon dioxide–helium	298	6.20	Methyl chloride –sulphur dioxide	323	0.78
Carbon dioxide–nitrogen	298	1.69	Methane–water	352.3	3.61
Carbon dioxide–oxygen	293.2	1.55	Nitrogen–sulphur dioxide	263	1.05
Carbon dioxide –sulphur dioxide	263	0.65	Nitrogen–water	352.1	3.64
	473	1.98	Oxygen–water	352.3	3.57

APPENDIX 2

Experimental values of diffusion coefficients for gases and non-electrolytes at low concentrations in water, $D_{AB}^0 (m^2/s)$

Solute	Temp. (°C)	$D_{AB}^0 \times 10^9$	Solute	Temp. (°C)	$D_{AB}^0 \times 10^9$
Acetic acid	20	1.19	Furfural	20	1.04
Acetone	20	1.16	Glycerol	15	0.72
Acetonitrile	15	1.26	Glycine	25	1.06
Air	20	2.5	Helium	25	6.28
Allyl alcohol	15	0.90	Hydrogen	25	4.50
Ammonia	12	1.64	Krypton	20	1.68
i-Amyl alcohol	15	0.69	Methane	20	1.49
Aniline	20	0.92		60	3.55
Argon	25	2.00	Methanol	15	1.26
Benzene	2	0.58	Neon	20	3.00
	20	1.02	Nitric oxide	20	2.07
	60	2.55	Nitrogen	20	2.6
Benzoic acid	25	1.21	Nitrous oxide	25	2.67
Benzyl alcohol	20	0.82	Oxalic acid	20	1.53
n-Butane	20	0.89	Oxygen	25	2.10
	60	2.51	i-Pentanol	15	0.69
i-Butanol	15	0.77	Propane	20	0.97
n-Butanol	15	0.77	i-Propanol	15	0.87
Carbon dioxide	25	1.92	n-Propanol	15	0.87
Chlorine	25	1.25	Propylene	25	1.44
Diethyl amine	20	0.97	1, 2-Propylene glycol	20	0.88
Ethane	20	1.20	Pyridine	15	0.58
Ethanol	20	0.84	Salicylic acid	25	1.06
Ethyl acetate	20	1.00	Urea	20	1.20
Ethylene	25	1.87	Urethane	15	0.80
Ethylene glycol	20	1.04	Vinyl chloride	25	1.34
	55	2.26		75	3.67
			Xenon	25	0.60

APPENDIX 3

Experimental values of diffusion coefficients, $D_{AB}^0 \, m^2/s$, in organic solvents at infinite dilution

Solute A	Solvent B	T(K)	$D_{AB}^0 \times 10^9$	Solute A	Solvent B	T(K)	$D_{AB}^0 \times 10^9$
Acetic acid	Acetone	298	3.31	Carbon dioxide	Ethanol	290	3.20
Benzoic acid	Acetone	298	2.62	Carbon dioxide	Ethanol	298	3.42
Cyclohexane	Acetone	298	2.22				
Carbon dioxide	Amyl alcohol	298	1.91	Glycerol	Ethanol	293	0.51
Water	Aniline	293	0.70	Pyridine	Ethanol	293	1.10
Acetic acid	Benzene	298	2.09	Urea	Ethanol	285	0.54
Carbon tetra-chloride	Benzene	298	1.92	Water	Ethanol	298	1.132
Cinnamic acid	Benzene	298	1.12	Water	Ethylene glycol	293	0.18
Ethylene chloride	Benzene	281	1.77	Water	Glycerol	293	0.0083
Ethanol	Benzene	288	2.25	Carbon dioxide	Heptane	298	6.03
Methanol	Benzene	298	3.82	Carbon tetra-chloride	n-Hexane	298	3.70
Naphthalene	Benzene	281	1.19	Toluene	n-Hexane	298	4.21
Carbon dioxide	i-Butanol	298	2.20	Carbon dioxide	Kerosene	298	2.50
				Water	n-Propanol	288	0.87
Acetone	Carbon tetra-chloride	293	1.86	Water	1, 2-Propylene glycol	293	0.075
Benzene	Chlorobenzene	293	1.25				

Acetone	Chloroform	288	2.36	Acetic acid	Toluene	298	2.26
Benzene	Chloroform	288	2.51	Acetone	Toluene	293	2.93
Ethanol	Chloroform	288	2.20	Benzoic acid	Toluene	293	1.74
Acetone	Cyclohexane	298	4.06				
Carbon-tetra chloride	Cyclohexane	298	1.49	Chlorobenzene	Toluene	293	2.06
				Ethanol	Toluene	288	3.00
Azobenzene	Ethanol	293	0.74	Carbon dioxide	White spirit	298	2.11
Camphor	Ethanol	293	0.70				

APPENDIX 4

Units, prefixes, and conversion factors

The calculations in this book are carried out in the SI system (*Système International d'Unités*) as given in the International Standard ISO 1000–1973(E) and with symbols as given in ISO Recommendations 31.

The base and supplementary units of this system are listed in Table A4.1.

Table A4.1   Base and supplementary units in the SI system

Quantity	Symbol	SI unit Name	SI unit Symbol
Length	l	metre	m
Mass	m	kilogram	kg
Time	t	second	s
Electric current	I	ampere	A
Thermodynamic temperature	T	kelvin	K
Amount of substance	n	mole	mol
Luminous intensity	I_v	candela	cd
Plane angle	α	radian	rad
Solid angle	ω	steradian	sr

Units

The metre is the length equal to 1 650 763.73 wavelengths in vacuum of the radiation corresponding to the transition between the levels $2p_{10}$ and $5d_5$ of the krypton-86 atom.

The kilogram is the mass of the international prototype.

The second is the duration of 9 192 631 770 periods of the radiation corresponding to the transition between the two hyperfine levels of the ground state of the caesium-133 atom.

The ampere is that constant electric current which, if maintained in two straight parallel conductors of infinite length, of negligible circular cross-section, and placed 1 metre apart in vacuum, would produce between these conductors a force equal to 2×10^{-7} newton per metre of length.

The kelvin, unit of thermodynamic temperature, is the fraction 1/273.16 of the thermodynamic temperature of the triple point of water.

The mole is that amount of substance of a system which contains as many elementary entities as there are atoms in 0.012 kilogram of carbon 12.

The candela is the luminous intensity perpendicular to a surface of $1/600\,000$ square metre of a black body at the temperature of freezing platinum under a pressure of $101\,325$ newtons per square metre.

The radian is the plane angle between two radii of a circle which cut off an arc on the circumference which equal in length to the radius.

The steradian is the solid angle which, having its vertex in the centre of a sphere, cuts of an area of the surface of the sphere equal to that of a square with sides of length equal to the radius of the sphere.

Prefixes

The multiples of SI units, symbols, and prefixes, are given in Table A4.2. Values in parentheses should not be introduced except where they are already in common use.

Note: Multiplying prefixes are printed adjacent to the SI unit symbol with which they are associated. The multiplication of units is usually indicated by leaving a gap between them, i.e. mN means millinewton, m N means metre times newton.

To avoid errors in calculations, it is recommended that coherent SI units rather than their multiples and submultiples be used, i.e. $10^6\,\mathrm{N/m^2}$ instead of $\mathrm{MN/m^2}$ or $\mathrm{N/mm^2}$.

A prefix applied to a unit is part of that unit when applied to a power, i.e. $\mathrm{mm^2} = (\mathrm{mm})^2 = 10^{-6}\,\mathrm{m^2}$.

Table A4.2   Prefixes and symbols for multiples of SI-units.

Prefix	Symbol	Value	Prefix	Symbol	Value
exa	E	10^{18}	(deci	d	10^{-1})
peta	P	10^{15}	(centi	c	10^{-2})
tera	T	10^{12}	milli	m	10^{-3}
giga	G	10^9	micro	μ	10^{-6}
mega	M	10^6	nano	n	10^{-9}
kilo	k	10^3	pico	p	10^{-12}
(hecto	h	10^2)	femto	f	10^{-15}
(deca	da	10^1)	atto	a	10^{-18}

Conversion factors

There is often a need for conversion between the different systems, SI, cgs, British, and technical metric system. Table A4.3 gives a selection of conversion factors for quantities used in this text. The kilocalories and the calories used in the table are the IT calories (International Table calories). The two other calories not used in this text are the 15°C calorie equal to 4.1855 Joule and the thermochemical calorie exactly equal to 4.184 Joule. Joule in this text is N m and not the international Joule of 1948 which is $1.000\,17\,\mathrm{N\,m}$.

Table A4.3 Conversion factors. Figures with an asterisk are exact. kp = kg force

Quantity	given in	multiplied by	gives
Absorption, coefficient of (see mass transfer)			
Acceleration	ft/s^2	0.304 8*	m/s^2
Area	in^2	0.000 645 16*	m^2
	ft^2	0.092 903 04*	m^2
	yd^2	0.836 127 36*	m^2
acre	4 046.9		m^2
	da	1 000*	m^2
Density	lb/ft^3	16.018 4	kg/m^3
Diffusivity [see viscosity (kinematic)]			
Elasticity, modulus of (see pressure)			
Energy	erg	10^{-7}*	N m = J = W s
	kp m	9.806 65	N m = J = W s
	kcal	4 186.8*	N m = J = W s
	cal	4.186 8*	N m = J = W s
	ft lb	1.355 8	N m = J = W s
	ft pdl	0.042 139	N m = J = W s
	Btu	1 055.056	N m = J = W s
	therm	1.055×10^8	N m = J = W s
Enthalpy	kcal/kg	4 186.8*	$\dfrac{\text{N m}}{\text{kg}} = \dfrac{\text{J}}{\text{kg}} = \dfrac{\text{W s}}{\text{kg}}$
	cal/g	4 186.8*	$\dfrac{\text{N m}}{\text{kg}} = \dfrac{\text{J}}{\text{kg}} = \dfrac{\text{W s}}{\text{kg}}$
	Btu/lb	2 326*	$\dfrac{\text{N m}}{\text{kg}} = \dfrac{\text{J}}{\text{kg}} = \dfrac{\text{W s}}{\text{kg}}$
Entropy (see heat capacity)			
Force	dyn	0.000 01*	N
	kp = kg$_f$	9.806 65*	N
	Ib$_f$	4.448 22	N
	pdl	0.138 255	N
Heat capacity	kcal/(kg °C)	4 186.8*	J/(K kg)
	cal/(g °C)	4 186.8*	J/(K kg)
	Btu/(Ib°F)	4 186.8*	J/(K kg)
Heat flux	kcal/(m^2 h)	1.163*	W/m^2 = J/(m^2 s)
	cal/(cm^2 s)	41 868*	W/m^2 = J/(m^2 s)
	Btu/(ft^2 hr)	3.154 6	W/m^2 = J/(m^2 s)
Heat transfer coefficient	kcal/(m^2 h °C)	1.163*	W/(m^2 K) = J/(s m^2 K)
	cal/(cm^2 s °C)	41 868*	W/(m^2 K) = J/(s m^2 K)
	Btu/(ft^2 hr °F)	5.678 4	W/(m^2 K) = J/(s m^2 K)

Table A4.3 (*Contd.*)

Quantity	given in	multiplied by	gives
Henry's law constant	$atm/(g/cm^3)$	101.325*	N m/kg
	$bar\,(kg/m^3)$	100 000*	N m/kg
	$atm/(kg/m^3)$	101.325*	N m/kg
	$atm/(lb/ft^3)$	6 325.8	N m/kg
	$psi/(lb/ft^3)$	430.43	N m/kg
Length	Å	0.1*	nm
	mil	25.4*	μm
	in	0.025 4*	m
	ft	0.304 8*	m
	yd	0.914 4*	m
	mile	1 609.3	m
Mass	grain	0.064 798 9	g
	oz	28.349 5	g
	troy oz	31.103 486	g
	lb	0.453 592 37*	kg
	slug	14.594	kg
	cwt	50.802	kg
	tonne	1 000	kg
	long ton	1 016.05	kg
	short ton	907.18	kg
Mass flow rate	kg/h	0.000 277 78	kg/s
	lb/hr	0.000 126 0	kg/s
	lb/s	0.453 592 37*	kg/s
Mass flux	$kg/(m^2\,h)$	0.000 277 78	$kg/(m^2\,s)$
	$lb/(ft^2\,hr)$	0.001 356 2	$kg/(m^2\,s)$
	$lb/(ft^2\,s)$	4.882 4	$kg/(m^2\,s)$
Mass transfer coefficient (a) dimensionless driving force	$kg/(m^2\,h)$	0.000 277 8	$kg/(m^2\,s)$
	$g/(cm^2\,s)$	10*	$kg/(m^2\,s)$
	$lb/(ft^2\,hr)$	0.001 356 2	$kg/(m^2\,s)$
(b) concentration as driving force	$kg/[m^2\,h(kg/m^3)]$	0.000 277 8	m/s
	$g/[cm^2\,s(g/cm^3)]$	0.01*	m/s
	$lb/[ft^2\,hr(lb/ft^3)]$	84.667×10^{-6}	m/s
(c) Pressure as driving force	$kg/(m^2\,h\,atm)$	$2.741\,3 \times 10^{-9}$	kg/(s Pa)
	$kg/(m^2\,h\,bar)$	$2.777\,8 \times 10^{-9}$	kg/(s Pa)
	$g/(cm^2\,s\,atm)$	$9.868\,7 \times 10^{-5}$	kg/(s Pa)
	$lb/(ft^2\,hr\,atm)$	$1.338\,4 \times 10^{-8}$	kg/(s Pa)
	$lb/[ft^2\,hr\,(lb_f/in^2)]$	$1.967\,05 \times 10^{-7}$	kg/(s Pa)

Table A4.3   (*Contd.*)

Quantity	given in	multiplied by	gives
Power	erg/s	10^{-7}*	N m/s = J/s = W
	kp m/s	9.806 5	N m/s = J/s = W
	Hp (metric)	735.5	N m/s = J/s = W
	Hp (British)	745.7	N m/s = J/s = W
	cal/s	4.186 8*	N m/s = J/s = W
	kcal/h	1.163 *	N m/s = J/s = W
	ft lb/min	0.022 597	N m/s = J/s = W
	ft lb/s	1.355 8	N m/s = J/s = W
	Btu/hr	0.293 1	N m/s = J/s = W
	Btu/s	1 055.1	N m/s = J/s = W
	tons refr.	3 516.85	N m/s = J/s = W
Pressure	dyn/cm^2	0.1*	N/m^2 = Pa
	Pa	1.0*	N/m^2 = Pa
	kp/m^2	9.806 7	N/m^2 = Pa
	mm water	9.806 7	N/m^2 = Pa
	mmHg = torr	133.322 4	N/m^2 = Pa
	bar	0.1*	M N/m^2 = MPa
	kp/cm^2	0.098 066 5	MN/m^2 = MPa
	at(techn.)[†]	0.098 066 5	MN/m^2 = MPa
	atm	0.101 325	MN/m^2 = MPa
	pdl/ft^2	1.488 1	N/m^2 = Pa
	lb$_f$/ft^2	47.88	N/m^2 = Pa
	inches water	249.09	N/m^2 = Pa
	in Hg	3 386.6	N/m^2 = Pa
	psi	6 894.8	N/m^2 = Pa
Shear stress (see pressure) Stress (see pressure)			
Surface tension	dyn/cm	0.001*	N/m
	kp/m	9.806 65	N/m
	lb$_f$/ft	14.593 9	N/m
	lb$_f$/in	175.127	N/m
Temperature (see after the table)			
Thermal conductivity	cal/(cm s °C)	418.68	W/(m °C) = W/(m K)
	kcal/(m h °C)	1.163*	W/(m °C) = W/(m K)
	Btu/(ft hr °F)	1.730 8	W/(m °C) = W/(m K)
	Btu/[ft^2 hr (°F/in)]	0.144 2	W/(m °C) = W/(m K)
	Btu/(ft s °F)	6 232	W/(m °C) = W/(m K)
Thermal diffusivity	m^2/h	0.000 277 8	m^2/s
	ft^2/sec	0.092 903	m^2/s
	ft^2/hr	0.000 025 81	m^2/s

Table A4.3 (*Contd.*)

Quantity	given in	multiplied by	gives
Velocity	km/h	0.2778	m/s
	ft/hr	84.67×10^{-6}	m/s
	ft/min	0.00508*	m/s
	knots	0.51444*	m/s
Viscosity			
(absolute =	cP	0.001*	$N s/m^2 = kg/(m s)$
dynamic)	P	0.1*	$N s/m^2 = kg/(m s)$
	$kp s/m^2$	9.80665	$N s/m^2 = kg/(m s)$
	lb/(ft hr)	0.00041338	$N s/m^2 = kg/(m s)$
	lb/(ft sec)	1.4882	$N s/m^2 = kg/(m s)$
(kinematic)	cSt	$10^{-6}*$	m^2/s
	$St = cm^2/s$	0.0001*	m^2/s
	ft^2/hr	0.00002581	m^2/s
	ft^2/sec	0.092903	m^2/s
Volume	ounces (Brit. fluid)	28.414×10^{-6}	m^3
	ounces (U.S. fluid)	29.576×10^{-6}	m^3
	in^3	16.387×10^{-6}	m^3
	gallons (Brit.)	0.004546	m^3
	gallons (U.S.)	0.0037853	m^3
	ft^3	0.028317	m^3
	bushels (Brit.)	0.036369	m^3
	bushels (U.S.)	0.035239	m^3
	barrels (Brit.)	0.163659	m^3
	barrels (U.S. petr.)	0.15898	m^3
	barrels (U.S. dry)	0.11563	m^3
	barrels (U.S. liq.)	0.11924	m^3
	yd^3	0.76455	m^3

[†] German and Russian publications use 'at' for technical atmospheres. In German 'ata' is technical atmospheres absolute and 'atü' technical atmospheres gauge pressure, while Atm with capital A means atmospheres equal to 760 mm mercury.

Temperature:

$$1\,°R = \tfrac{5}{9}\,K$$
$$\theta\,°F = (459.67 + \theta)°R = [(5/9)(\theta - 32)]\,°C$$
$$\theta\,°C = (273.15 + \theta)\,K$$

The gas constant, $R = 8314.3\,N m/(K\,kmol)$.
Volume of an ideal gas:
    at 1 atm and 0 °C, $22.414\,m^3/kmol$;
    at 30 inches mercury and 60 °F, saturated with water vapour (Brit. gas industry), $386.23\,ft^3/lb$-mole;
    at 1 atm and 60 °F, dry gas (U.S.), $379.34\,ft^3/lb$-mole.
Acceleration due to gravity, 'normal', $g = 9.80665\,m/s^2$.

APPENDIX 5
Physical properties of gases and liquids

Main sources: R. C. Reid, J. M. Prausnitz, and T. K. Sherwood: *Properties of Cases and Liquids*, McGraw-Hill, New York, 3rd edn, 1977, and *VDI-Wärmeatlas*, VDI-Verlag, Düsseldorf, 3rd edn, 1977.

The data for gases in Tables A5.1 and A5.2 are for low and moderate pressure. Constant α W/(m K^2) for thermal conductivity of liquids is for a limited temperature range. T is the absolute temperature in K.

Table A5.1   Dynamic viscosity and liquid heat capacity

			Dynamic viscosity in $N s/m^2 = cP \times 10^{-3}$				
			Liquid $\mu = Ae^{B/T}$		Gas $\mu \approx \dfrac{bT^{1.5}}{S+T}$		Liquid heat capacity at 20°C
Compound	Temperature range liquid (°C)		$A \times 10^6$	B	$b \times 10^6$	S	c_{20}(kJ/(kg K))
Acetic acid	20 to	100	13	1328			2.00
Acetone	− 75 to	50	17.1	864	1.17	527	2.16
Air					1.46	110	
Ammonia	− 50 to	50	8.6	813	1.56	492	4.61
Benzene	20 to	170	8.7	1264	0.99	366	1.73
n-Butanol	− 50 to	100	1.06	2324			2.35
Carbon dioxide	− 25 to	20	0.71	1345	1.50	220	2.28 (0°C)
Carbon tetrachloride	0 to	175	11.3	1302	1.17	300	0.85
Chlorine	− 100 to	0	68	470	1.66	336	0.94
Diethyl ether	− 100 to	70	14.5	826	0.94	343	2.34
Ethane	− 150 to	− 50	22.8	358	0.97	229	4.31
Ethanol	− 50 to	100	4.0	1660	1.32	514	2.41
Ethylene					1.04	225	2.41 (− 100°C)
n-Heptane	− 20 to	100	13	1010	0.88	513	2.36
Hydrogen					0.65	67	
Methane					1.00	168	4.82 (− 100°C)
Methanol	− 75 to	50	7.6	1275	1.44	477	2.50
Nitrogen					1.40	108	2.15 (− 180°C)
Oxygen					1.69	126	1.70 (− 180°C)
Propane	− 40 to	40	7.8	746	0.86	241	
n-Propanol	− 75 to	100	1.4	2154	1.13	479	2.35
Toluene	− 25 to	200	13.3	1110	0.97	420	1.72
Water	10 to	115	2.6	1750	1.75	626	4.19

Table A5.2   Gas heat capacity and liquid and gas thermal conductivity

Compound	Gas heat capacity $c_p = A + BT + CT^2 + DT^3$ (kJ/(kg K))				Thermal conductivity (W/(m K))		
					Liquid $k_2 = k_{25} - \alpha(T_2 - 298)$		Gas $k_g \approx k_{g25} \left(\dfrac{T_2}{298}\right)^{1,8}$
	A	$B \times 10^3$	$C \times 10^6$	$D \times 10^{10}$	k_{25}	$\alpha \times 10^4$	k_{g25}
Acetic acid	0.081	4.24	-2.92	8.24	0.159	2.0	
Acetone	0.108	4.49	-2.16	3.51	0.160	5.0	0.013
Air	1.057	-0.37	0.86	2.42			0.026
Ammonia	1.604	1.40	1.00	-6.96	0.516	(10)	0.024
Benzene	-0.434	6.07	-3.86	9.13	0.145	3.3	0.013
n-Butanol	0.044	5.64	-3.02	6.32	0.153	2.4	0.013
Carbon dioxide	0.450	1.67	-1.27	3.90	0.08	(10)	0.016
Carbon tetrachloride	0.265	1.33	-1.48	5.75	0.104	1.8	0.0067
Chlorine	0.380	0.48	-5.46	2.18			0.0093
Diethyl ether	0.289	4.53	-1.40	-1.26	0.133	0.7	0.015
Ethane	0.180	5.92	-2.31	2.90			0.021
Ethanol	0.196	4.65	-1.82	0.30	0.167	5.6	0.015
Ethylene	0.136	5.58	-2.98	6.26	0.119	3.5	0.021
n-Heptane	-0.051	6.75	-3.64	7.64	0.123	3.3	0.012
Hydrogen	13.5	4.60	-6.85	37.9	0.118	($-253\,°C$)	0.181
Methane	1.200	3.25	0.75	-7.05			0.034
Methanol	0.660	2.21	0.81	-8.90	0.201	3.0	0.016
Nitrogen	1.112	-0.48	0.96	4.17			0.026
Oxygen	0.878	0	0.55	-3.33			0.026
Propane	-0.096	6.95	-3.60	7.29	0.08		0.018
n-Propanol	0.041	5.53	-3.09	7.15	0.156	1.7	0.014
Toluene	-0.264	5.56	-3.00	5.33	0.140	1.6	0.015
Water	1.790	0.11	0.59	-2.00	0.605	1.5	0.019

Table A5.3   Saturation pressure p_s and specific volume v'' of saturated steam

θ (°C)	p_s (kN/m^2)	v'' (m^3/kg)	θ (°C)	p_s (MN/m^2)	v'' (m^3/kg)	θ (°C)	p_s (MN/m^2)	v'' (m^3/kg)
10	1.228	106.4	90	0.0701	2.361	240	3.348	0.0597
15	1.705	77.96	100	0.1013	1.673	250	3.978	0.0500
20	2.337	57.84	110	0.1433	1.210	260	3.694	0.0421
25	3.166	43.41	120	0.1985	0.891	270	5.505	0.0355
30	4.241	32.94	130	0.2701	0.668	280	6.419	0.0301
35	5.621	25.26	140	0.3614	0.508	290	7.445	0.0255
40	7.374	19.56	150	0.4760	0.392	300	8.592	0.0216
45	9.581	15.28	160	0.6180	0.307	310	9.870	0.0183
50	12.33	12.05	170	0.7920	0.243	320	11.29	0.0155
55	15.74	9.583	180	1.003	0.194	330	12.87	0.0130
60	19.92	7.682	190	1.255	0.1564	340	14.61	0.0108
65	25.01	6.205	200	1.555	0.1273	350	16.54	0.0088
70	31.16	5.048	210	1.908	0.1043	360	18.67	0.0070
75	38.55	4.135	220	2.320	0.0861	370	21.05	0.0050
80	47.36	3.410	230	2.798	0.0715	374	22.09	0.0036

Index

318